A SHORT COURSE IN

SOIL–STRUCTURE ENGINEERING

OF DEEP FOUNDATIONS, EXCAVATIONS AND TUNNELS

CHARLES W. W. NG, NOEL SIMONS and BRUCE MENZIES

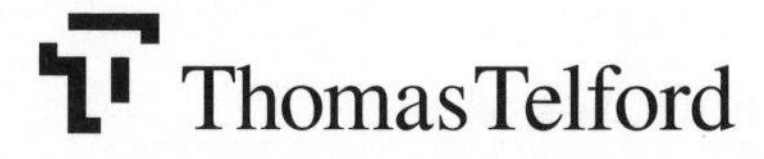

Published by Thomas Telford Publishing, Thomas Telford Ltd, 1 Heron Quay, London E14 4JD.
www.thomastelford.com

Distributors for Thomas Telford books are
USA: ASCE Press, 1801 Alexander Bell Drive, Reston, VA 20191-4400, USA
Japan: Maruzen Co. Ltd, Book Department, 3–10 Nihonbashi 2-chome, Chuo-ku, Tokyo 103
Australia: DA Books and Journals, 648 Whitehorse Road, Mitcham 3132, Victoria

First published 2004

A catalogue record for this book is available from the British Library

ISBN: 0 7277 3263 3

Typeset by Academic + Technical, Bristol
Printed and bound in Great Britain by MPG Books, Bodmin

Cover photograph
Lion Yard excavation showing a vertical concrete bored pile and horizontal steel props

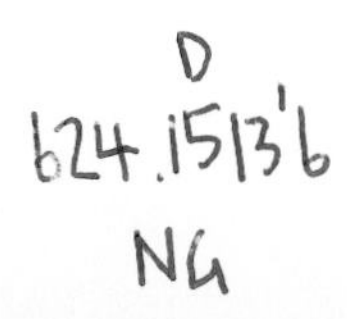

Preface

This book is divided into three parts – effectively three short courses – focusing on the following three major geotechnical challenges of static soil–structure interaction problems:

- Part 1 Deep foundations – piles and barrettes
- Part 2 Multi-propped deep excavations
- Part 3 Bored and open-face tunnels below cities.

Part 1 *Deep foundations – piles and barrettes* concentrates on axially loaded long piles and barrettes. We were unable to include laterally loaded piles because of space limitations. The introductory sections of Part 1 draw mainly on the state-of-the-art papers by Poulos (1989) and Randolph and Wroth (1979). We then focus on the long piles that are typical in the Far East. Long piles have a particular character because their elastic shortening under load makes the pile head load–displacement test more difficult to interpret. Pile failure load criteria, therefore, need re-examining and we provide new displacement failure criteria based on many pile tests. For 'floating' barrettes and long piles, where their load capacity is derived from side resistance only, the degree of mobilisation of side resistance is clearly critical and so Mobilisation Rating is a key concept that is explained.

Part 2 *Multi-propped deep excavations* uses an extended case study as the vehicle for explaining the design concepts – this is Lion Yard, Cambridge. Preliminary design methods are presented as 'handwritten' Short Course Notes. Plastic mechanisms of failure are explained drawing on the work of Bolton (1993). Numerical modelling is used to obtain vertical and horizontal ground movements associated with the separate effects of panel trenching and installation (including bentonite support and tremie concreting), and of propping and excavating deep excavations retained by diaphragm walls. In particular, the realistic Wall-Installation-Modelled (WIM) method is introduced and compared with the more common and unrealistic Wished-In-Place (WIP) method of analysis.

Part 3 *Bored and open-face tunnelling below cities* begins with an introduction from Bolton (1979) showing how easily soil arching supports tunnels – making sprayed concrete linings quite acceptable as used by the New Austrian Tunnelling Method (NATM). The main body of Part 3 draws extensively on the state-of-the-art paper by Mair and Taylor (1997) and concentrates on the evaluation of the surface settlement trough above a tunnel heading. The settlement trough shape follows the classic Gaussian probability curve – all that is required to define the trough are the parameters of either maximum settlement or ground loss, and the point of inflexion of the Gaussian curve. Numerical modelling is used to model the interaction between two parallel tunnels being driven with one lagging behind the heading of the other. The left and right tunnels are even driven in left and right drifts. This demonstrates the quite detailed and numerous installation steps that can be numerically modelled. We also show the effect of driving a tunnel near a pile tip. The pile capacity is first assessed by modelling an axial loading test and so providing a load–displacement curve from which a working load is found. A tunnel is driven near the tip of the pile where plastic yielding then occurs. The loss of working load is found to be an alarming two-thirds!

Several detailed case studies are considered. These include a well-instrumented test barrette in Hong Kong together with 15 other test barrettes, plus an extensive review of 28 large-diameter floating bored pile tests. These studies help to establish the Mobilisation Ratings that are critical in design. We also include the extended case studies of the Heathrow Express Trial Tunnel (Deane and Bassett, 1995) and the Lion Yard deep excavation (Ng, 1992, 1998). Finally, we briefly review the precautions taken to protect the Big Ben Clock Tower from ground movements caused by the Jubilee Line Extension tunnels (Burland *et al.*, 2001). These measures included monitoring of the Tower tilt and compensation grouting by *tubes à manchette* (TAMs).

This book will give both student and practising civil engineers a useful review of the state-of-the-art of designing deep foundations, excavations and tunnels. In addition, the case studies and numerical modelling presented will give valuable insights into the challenges of soil–structure engineering.

Charles W. W. Ng, Noel Simons, Bruce Menzies

Hong Kong University of Science and Technology
Hong Kong 2003

Acknowledgements

We acknowledge permissions from authors and publishers to make verbatim extracts from their published works as follows:

Professor Braja Das and Brookes/Cole, Publishers, from:

- Das, B. M. (2000). *Fundamentals of Geotechnical Engineering*. Brookes/Cole, Thomson Learning.

Professor Harry Poulos and Thomas Telford, Publishers, from:

- Poulos, H. G. (1989). Twenty-ninth Rankine Lecture: Pile behaviour – theory and application. *Géotechnique*, **39**(3), 363–415.

Professor Harry Poulos and John Wiley, Publishers, from:

- Poulos, H. G. and Davis, E. H. (1980): *Pile Foundation Analysis and Design*. John Wiley and Sons, New York.

Professor Mark Randolph and Thomas Telford, Publishers, from:

- Randolph, M. F. and Wroth, C. P. (1979). An analysis of the vertical deformation of pile groups. *Géotechnique*, **29**(4), 423–439.

Mr Tommy Tomlinson and Prentice Hall, Publishers, from:

- Tomlinson, M. J. (2001). *Foundation Design and Construction*. 7th edn. Prentice Hall, London.

Professor Malcolm Bolton and M. D. and K. Bolton, Publishers, from:

- Bolton, M. D. (1979). *A Guide to Soil Mechanics*. M. D. and K. Bolton, Cambridge, 439 pp.

Professor Malcolm Bolton and the Japanese Society of Soil Mechanics and Foundation Engineering, Publishers, from:

- Bolton, M. D. (1993). Mechanisms of ground deformation due to excavation in clay. *Excavation in Urban Areas, KIGForum '93*, Toshihisha Adachi (ed.), Japanese Soc. of Soil Mech. and Fndn Engng, 1–33.

Professors Robert Mair and Neil Taylor and © Swets & Zeitlinger, Publishers, from:

- Mair, R. J. and Taylor, R. N. (1997). Theme lecture: Bored tunnelling in the urban environment. In *Proc. 14th Int. Conf. Soil Mech. & Fdn Engng*, **4**, 2353–2385, Balkema, Rotterdam.

Mr Tony Deane, Dr Dickie Bassett and Thomas Telford, Publishers, from:

- Deane, A. P. and Bassett, R. H. (1995). The Heathrow Express Trial Tunnel. *Proc. Inst. Civ. Engrs*, **113**, July, 144–156.

Professor John Burland, Dr Jamie Standing, Mr Fin Jardine and Thomas Telford, Publishers, from:

- Burland, J. B., Standing, J. R. and Jardine, F. M., eds (2001). *Building Response to Tunnelling*. Thomas Telford, London.

Professors C. W. W. Ng and G. T. K. Lee and Elsevier, Publishers, from:

- Ng, C. W. W. and Lee, G. T. K. (2002). A three-dimensional parametric study of the use of soil nails for stabilising tunnel faces. *Computers and Geotechnics*, **29**(8), 673–697.

Mr David Nash and Mr Martin Lings, University of Bristol and JT Design and Build, for:

- Photographs of Lion Yard, Cambridge.

The American Society of Civil Engineers, from:

- Ng, C. W. W., Yau, T. L. Y., Li, J. H. M. and Tang, W. H. (2001c). New failure load criterion for large diameter bored piles in weathered geomaterials. *J. Geotech. & Geoenv. Engng, ASCE,* **127**(6), 488–498, Tables 1–2, Figs 3, 4, 6.
- Ng, C. W. W. (1998). Observed performance of multi-propped excavation in stiff clay. *J. Geotech. & Geoenv. Engng, ASCE,* **124**(9), 889–905, Figs 1–23, Table 1.
- Ng, C. W. W. (1999). Stress paths in relation to deep excavations. *J. Geotech. & Geoenv. Engng, ASCE,* **125**(5), 357–363, Fig. 1.
- Ng, C. W. W. and Lei, G. H. (2003a). Performance of long rectangular barrettes in granitic saprolites. *J. Geotech. & Geoenv. Engng, ASCE,* **129**(8), 685–696, Figs 2–9, Tables 1–3.
- Ng, C. W. W. and Yan, W. M. (1998c). Stress transfer and deformation mechanisms around a diaphragm wall panel. *J. Geotech. & Geoenv. Engng, ASCE,* **124**(7), 638–648, Figs 1–15.
- Ng, C. W. W., Rigby, D., Ng, S. W. L and Lei, G. (2000b). Field studies of a well-instrumented barrette in Hong Kong. *J. Geotech. & Geoenv. Engng, ASCE,* **126**(1), 60–73, Figs 2–5, 7–9, 11–12.

- Ng, C. W. W., Yau, T. L. Y., Li, J. H. M. and Tang, W. H. (2001a). Side resistance of large diameter bored piles socketed into decomposed rocks. *J. Geotech. & Geoenv. Engng, ASCE,* **127**(8), 642–657, Tables 2–4, Figs 2, 4, 7.
- Ng, C. W. W. and Lings, M. L. (1995). Effects of modelling soil nonlinearity and wall installation on back-analysis of deep excavation in stiff clay. *J. Geotech. & Geoenv. Engng, ASCE,* **121**(10), 687–695, Figs 3–14, Tables 1–2.

Contents

PART I

Deep foundations – piles and barrettes

Kentledge reaction stack for testing piles in Hong Kong

Overview

The major challenges facing designers of deep foundations, particularly in crowded urban areas, are:

1. How are piles analysed for their load-carrying capacity and how is a safe working load established?
2. How are loading tests on barrettes and long piles interpreted and of the many failure load criteria which one should be used?
3. Are dynamic pile-driving formulae reliable?
4. For 'floating' barrettes (i.e. that rely on side resistance for their load capacity), how are lateral stresses and pore water pressures distributed and how do they change during a load test?
5. What is the importance of degree of mobilisation of side resistance in floating piles and barrettes?
6. How are rock-socketed or rock end-bearing piles assessed for their load capacity in weathered rocks in particular (as distinct from sedimentary rocks)?
7. What are realistic design values of side resistance and settlement?
8. What is the influence of construction factors such as duration, use of bentonite and grouting?

This Part 1 – effectively a short course on deep foundations – seeks to answer these questions by correlating theory with field measurement. Numerous case studies of both barrettes and long bored piles are considered and their implication for design practice is made clear. The French term 'barrette' refers to a concrete replacement pile formed in a short and deep trench excavated under bentonite or polymer slurry by diaphragm walling equipment.

Sources

In addition to the authors' papers, this Part 1 makes verbatim extracts from the following:

- Poulos, H. G. and Davis, E. H. (1980). *Pile Foundation Analysis and Design*. John Wiley and Sons, New York.
- Poulos, H. G. (1989). Twenty-ninth Rankine Lecture: Pile behaviour – theory and application. *Géotechnique*, **39**(3), 363–415.
- Randolph, M. F. and Wroth, C. P. (1979). An analysis of the vertical deformation of pile groups. *Géotechnique*, **29**(4), 423–439.
- Tomlinson, M. J. (2001). *Foundation Design and Construction*, 7th edn. Prentice Hall, London.

We gratefully acknowledge permissions to make verbatim extracts from these sources from Professor Harry Poulos, Professor Mark Randolph and Mr 'Tommy' Tomlinson.

CHAPTER ONE

Uses and types of piles and pile groups

Deep foundations are almost invariably piles, caissons or barrettes. Piles that are cast-in-place concrete are generally circular in cross-section. Driven piles may be circular or square in section. Caissons may be any section and are often installed in stages. Barrettes are generally rectangular or cruciform in section.

Pile foundations are needed in many circumstances. Das (2000) defines the following cases in which piles may be considered for the construction of a foundation.

- When the upper soil layer or layers are highly compressible and too weak to support the load transmitted by the superstructure, piles are used to transmit the load to underlying bedrock or a stronger soil layer, as shown in Fig. 1.1(a). When bedrock is not encountered at a reasonable depth below the ground surface, piles are used to transmit the structural load to the soil gradually. The resistance to the applied structural load is derived mainly from the frictional resistance developed at the soil–pile interface (Fig. 1.1(b)).
- When subjected to horizontal forces (see Fig. 1.1(c)), pile foundations resist by bending while still supporting the vertical load transmitted by the superstructure. This situation is generally encountered in the design and construction of earth-retaining structures and foundations of tall structures that are subjected to strong wind and/or earthquake forces.
- In many cases, the soils at the site of a proposed structure may be expansive or collapsible. These soils may extend to a great depth below the ground surface. Expansive soils swell and shrink as the moisture content increases and decreases, and the swelling pressure of such soils can be considerable. If shallow foundations are used, the structure may suffer considerable damage. However, pile foundations may be considered as an alternative when piles are extended beyond the active zone, which swells and shrinks (Fig. 1.1(d)). Soils such as loess may be collapsible. When the moisture contents of these soils increase, their structures may break down. A sudden

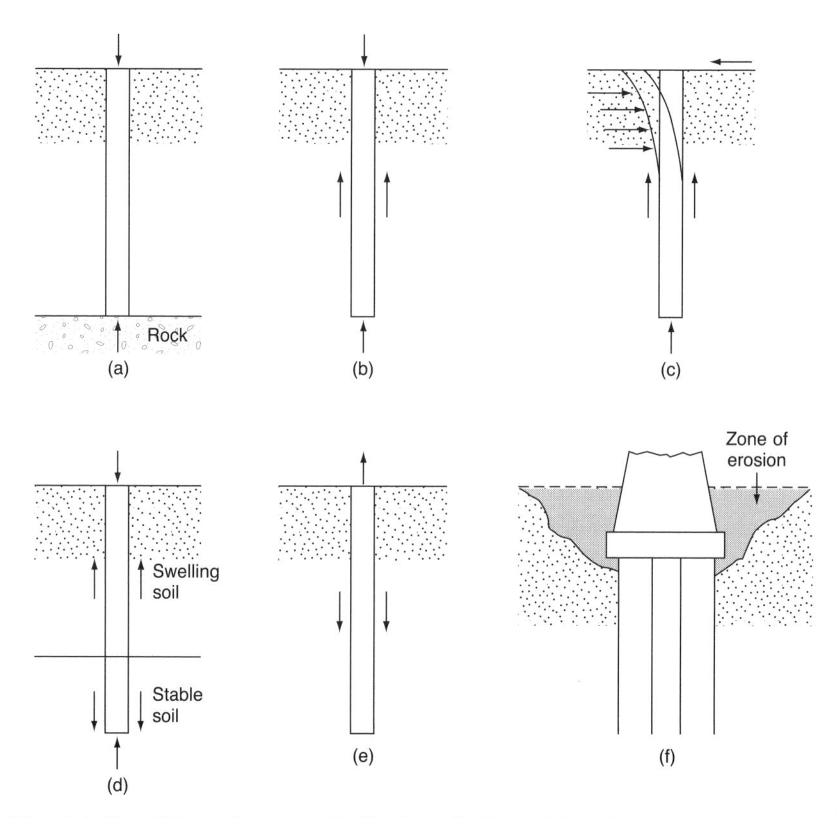

Fig. 1.1 Conditions for use of pile foundations (after Das, 2000)

decrease in the void ratio of soil induces large settlements of structures supported by shallow foundations. In such cases pile foundations may be used, in which piles are extended into stable soil layers beyond the zone of possible moisture change.

- The foundations of some structures, such as transmission towers, offshore platforms, and basement mats below the water table, are subjected to uplifting forces. Piles are sometimes used for these foundations to resist the uplifting force (Fig. 1.1(e)).
- Bridge abutments and piers are usually constructed over pile foundations to avoid the possible loss of bearing capacity that a shallow foundation might suffer because of soil erosion at the ground surface (Fig. 1.1(f)).

Although numerous investigations, both theoretical and experimental, have been conducted to predict the load–settlement characteristics and the load-bearing capacity of piles in soils and rocks, the mechanisms are not yet entirely understood and may never be clear due to the difficulty of estimating the effects of construction and workmanship. The design of pile foundations may therefore be considered somewhat of an 'art' as a result of the uncertainties involved in working with some subsoil conditions and construction methods.

Piles can be classified according to the type of material forming the piles, the mode of load transfer, the degree of ground displacement during pile installation or the method of installation. Pile classification in accordance with material type (e.g. steel and concrete) has drawbacks because composite piles are available. A classification system based on the mode of load transfer will be difficult to set up because the proportions of side resistance and end resistance that occur in practice usually cannot be reliably predicted.

In the installation of piles, either displacement or replacement of the ground will predominate. A classification system based on the degree of ground displacement during pile installation, such as that recommended in BS 8004 (BSI, 1986), encompasses all types of piles and reflects the fundamental effect of pile construction on the ground which in turn will have a pronounced influence on pile performance. Such a classification system is therefore considered to be the most appropriate.

In this section, piles are classified into the following three types:

- Large-displacement piles, which include all solid driven piles, including precast concrete piles, and steel or concrete tubes closed at the lower end by a driving shoe or a plug, that is, driven cast-in-place piles.
- Small-displacement piles, which include rolled steel sections such as H-piles and open-ended tubular piles. However, these piles will effectively become large-displacement piles if a soil plug forms.
- Replacement piles, which are formed by machine boring, grabbing or hand-digging. The excavation may need to be supported by bentonite slurry or lined with a casing that is either left in place or extracted during concreting for reuse.

Special piles, which are particular pile types or variants of existing pile types, are introduced from time to time to improve efficiency or overcome problems related to special ground conditions.

Large-displacement piles

Advantages and disadvantages

The advantages and disadvantages of large displacement piles are summarised in Table 1.1 (GEO, 1996).

Precast reinforced concrete piles

These piles are commonly in square sections ranging from about 250 mm to about 450 mm with a maximum section length of up to about 20 m.

Table 1.1 Advantages and disadvantages of displacement piles (after GEO, 1996)

Types of pile	Advantages	Disadvantages
Large-displacement piles	(a) Material of preformed section can be inspected before driving (b) Steel piles and driven cast-in-place concrete piles are adaptable to variable driving lengths (c) Installation is generally unaffected by groundwater condition (d) Soil disposal is not necessary (e) Driving records may be correlated with in situ tests or borehole data (f) Displacement piles tend to compact granular soils thereby improving bearing capacity and stiffness (g) Pile projection above ground level and the water level is useful for marine structures and obviates the need to cast in situ columns above the piles (h) Driven cast-in-place piles are associated with low material cost	(a) Pile section may be damaged during driving (b) Founding soil cannot be inspected to confirm the ground conditions as interpreted from the ground investigation data (c) Ground displacement may cause movement of, or damage to, adjacent piles, structures, slopes or utility installations (d) Noise may prove unacceptable in a built-up environment (e) Vibration may prove unacceptable due to presence of sensitive structures, utility installations or machinery nearby (f) Pile cannot be easily driven in sites with restricted headroom (g) Excess pore water pressure may develop during driving, resulting in negative skin friction on piles upon dissipation (h) Length of precast concrete piles may be constrained by transportation or size of casting yard (i) Heavy piling plant may require extensive site preparation to construct a suitable piling platform in sites with poor ground conditions (j) Underground obstructions cannot be coped with easily (k) For driven cast-in-place piles, the fresh concrete is exposed to various types of potential damage, such as necking, ground intrusions due to displaced soil and possible damage due to driving of adjacent piles
Small-displacement piles	(a) As (a), (b), (c), (d), (e) and (g) for large-displacement piles (b) Cause less ground disturbance and less vibration	(a) As (a), (b), (d), (f) and (i) for large-displacement piles

Other pile sections may include hexagonal, circular, triangular and H shapes. Maximum allowable axial loads can be up to about 1000 kN. The lengths of pile sections are often dictated by practical considerations including transportability, handling problems in sites of restricted area, and facilities of the casting yard. Reinforcement may be determined by handling stresses.

These piles can be lengthened by coupling together on site. Splicing methods include welding of steel end plates or the use of epoxy mortar with dowels. Specially fabricated joints have been successfully used in other geographical regions such as Hong Kong and Scandinavia.

This type of pile is not suitable for driving into ground which contains a significant amount of boulders or stones.

Prestressed concrete tubular piles

Precast prestressed concrete piles are typically tubular sections of 400 mm to 600 mm diameter with maximum allowable axial loads up to about 3000 kN. Pile sections are normally 12 m long and are usually welded together using steel end plates. Pile sections up to 20 m can also be specially made.

Prestressed concrete piles require high-strength concrete and careful control during manufacture. Casting is usually carried out in a factory where the curing conditions can be strictly regulated. Special manufacturing processes such as compaction by spinning or autoclave curing can be adopted to produce high-strength concrete up to about 75 MPa.

Closed-ended steel tubular piles

Steel tubular piles have high bending and buckling resistance, and have favourable energy-absorbing characteristics for impact loading. Steel piles are generally not susceptible to damage caused by tensile stresses during driving and can withstand hard driving. Driving shoes can be provided to aid penetration.

The tubular piles may be infilled with concrete after driving, as appropriate.

Driven cast-in-place concrete piles

Driven cast-in-place concrete piles are formed by driving a steel tube into the ground to the required set or depth and withdrawing the tube after concrete placement. The tube may be driven either at the top or at the bottom with a hammer acting on an internal concrete or compacted gravel plug. A range of pile sizes is available, up to 600 mm in diameter. The maximum allowable axial load is about 1400 kN.

Small-displacement piles

Overview, advantages and disadvantages

Small-displacement piles are either solid (e.g. steel H-piles) or hollow (open-ended tubular piles) with a relatively low cross-sectional area. This type of pile is usually installed by percussion methods; however, a soil plug may be formed during driving, particularly with tubular piles, and periodic drilling out may be necessary to reduce the driving resistance. A soil plug can create a greater driving resistance than a closed end, because of damping on the inner-side of the pile.

The advantages and disadvantages of small-displacement piles are also summarised in Table 1.1 (GEO, 1996).

Steel H-piles

Steel H-piles are widely used worldwide because of their ease of handling and driving. Compared with concrete piles, they generally have better driveability characteristics and can generally be driven to greater depths. H-piles can be susceptible to deflection upon striking boulders, obstructions or an inclined rock surface. In areas underlain by marble, heavy-section H-piles with appropriate tip strengthening are commonly used to penetrate the karst surface and to withstand hard driving.

A range of pile sizes is available, with different grades of steel. The maximum allowable axial load is typically about 3000 kN. Very large H sections (283 kg/m) with a working load of about 3600 kN have been used in some projects.

Open-ended steel tubular piles

Driven open-ended tubular piles have been used in marine structures and in buildings on reclaimed land. This type of pile has been driven to over 50 m. A plug will form when the internal side resistance exceeds the end-bearing resistance of the entire cross-sectional area of the pile. Driving resistance can be reduced by pre-boring or by reaming out the plug formed within the pile. Typical diameters range from 275 mm to about 2 m with a maximum allowable axial load of about 7000 kN. Maximum pile diameter is often governed by the capacity of the driving machine available.

Replacement piles

Overview, advantages and disadvantages

Replacement, or bored, piles are formed by excavation or boring techniques. When constructed in water-bearing soils which are not self-supporting, the pile bore will need to be supported using steel casing, concrete rings or drilling fluids such as bentonite slurry, polymer mud, etc. Excavation of the pile bore may also be carried out by hand-digging in the dry; and the technique involving manual excavation is known as hand-dug caissons.

The advantages and disadvantages of replacement piles are summarised in Table 1.2 (GEO, 1996).

Machine-dug piles

Machine-dug piles are formed by rotary boring, or percussive methods of boring a hole, and subsequently filling the hole with concrete. Piles 600 mm or less in diameter are commonly known as small-diameter piles. Piles greater than 600 mm diameter are referred to as large-diameter piles.

Small-diameter bored piles have sizes ranging from about 300 mm to 600 mm with working loads up to about 1500 kN.

One proprietary form of small-diameter bored pile involves the use of drop tools for excavation and compressed air to compact the concrete in the pile shaft. The common sizes of this type of pile range from 325 mm to 508 mm, with working loads up to about 1000 kN. These piles can be installed in sites where the headroom is limited. They are sometimes constructed without reinforcement and the integrity of such unreinforced piles when subject to ground movements arising from adjacent construction activities should be considered.

Another proprietary piling system is the continuous-flight auger (CFA) type pile. In this system, the bore is formed using a flight auger and concrete or grout is pumped in through the hollow stem. Sizes of PIP piles range from 300 mm to 700 mm diameter and lengths are generally less than 30 m. The CFA piles have considerable advantages over conventional bored piles in water-bearing and unstable soils by eliminating the need for casing and the problems of concreting under water. The piles can be installed with little noise and vibration and are therefore suited for sites in urban areas. However, this type of pile cannot cope with boulders. The lack of penetration under continuous rotation due to a hard layer or an obstruction can lead to soil flighting up the auger, causing ground loss and settlement.

Large-diameter machine-dug piles and pile groups are used to support heavy columns of tall buildings (e.g. Holt *et al.*, 1982; Fraser, 1985; Ng *et al.*, 2001a,b,c; Zhang *et al.*, 2001b) and highways structures such as viaducts (Fraser and Kwok, 1986). Typical sizes of these piles range from 1 m to 3.5 m, with lengths up to about 100 m and working loads up to about 40 000 kN. Special mechanical tools are readily available for belling out the base.

Barrettes

The French term 'barrette' refers to a concrete replacement pile formed in a short and deep trench excavated under bentonite or polymer slurry by

Table 1.2 Advantages and disadvantages of replacement piles (after GEO, 1996)

Type of pile	Advantages	Disadvantages
Machine-dug piles	(a) No risk of ground heave induced by pile driving (b) Length can be readily varied (c) Spoil can be inspected and compared with site investigation data (d) Structural capacity is not dependent on handling or driving conditions (e) Can be installed with less noise and vibration compared to displacement piles (f) Can be installed to great depths	(a) Risk of loosening of sandy or gravelly soils during pile excavation, reducing bearing capacity and causing ground loss and hence settlement (b) Susceptible to waisting or necking during concreting in unstable ground (c) Quality of concrete cannot be inspected after completion except by coring (d) Unset concrete may be damaged by significant water flow (e) Excavated material requires disposal, the cost of which will be high if it is contaminated
Hand-dug caissons	(a) As (a) to (e) for machine-dug piles (b) Use of comparatively economical labour force (c) Versatile construction method requiring minimal site preparation and access (d) Removal of obstructions or boulders is relatively easy through the use of pneumatic drills or, in some cases, explosives (e) Generally conducive to simultaneous excavation by different gangs of workers (f) Not susceptible to programme delay arising from machine downtime (g) Can be constructed to large diameters	(a) As (a), (c) and (e) for machine-dug piles (b) Hazardous working conditions for workers and the construction method has a poor safety record (c) Liable to base heave or piping during excavation, particularly where the groundwater table is high (d) Possible adverse effects of de-watering on adjoining land and structures (e) Health hazards to workers, as reflected by a high incidence rate of pneumoconiosis and damage to hearing of caisson workers

diaphragm walling equipment. It is actually identical in form to a short discrete length or a panel of a diaphragm wall (Ng *et al.*, 1999; Ng *et al.*, 2000b). Functionally, it is used as a pile foundation that transfers the load from the superstructure to the ground rather than as a soil-retaining structure supporting a mass excavation.

Conventional circular piles sometimes become impractical and uneconomical when very heavy column loads are to be transmitted to a very deep bearing stratum. Barrettes provide an alternative to large-diameter bored and cast-in-place piles or drilled shafts. This is because a rectangular barrette has a higher specific surface than has a circular pile for resisting large vertical loads by side resistance. In addition to this, the construction method used to construct barrettes satisfies reasonable noise and other environmental stipulations (Ramaswamy and Pertusier, 1986). In particular, barrettes are often selected on projects that require diaphragm wall construction for supporting the excavation of a foundation pit, because they are installed using the same equipment, thus avoiding mobilising other equipment for the foundation installation. Substantial savings in both mobilisation costs and time can be achieved. Coordination and scheduling of operations may also be improved if the same subcontractor installs the diaphragm wall and the foundation units (Johnson *et al.*, 1992).

Over the past three decades, barrette foundations have been widely used all over the world (Ng *et al.*, 2000b; Ng and Lei, 2003a). The fact that a single barrette can replace a group of conventional circular piles in many applications results in a more compact, economical and reliable foundation system. A good example is the use of shaft-grouted floating barrettes (varying from 40 m to 105 m in depth) in the foundation of the world's tallest building prior to the 21st century, the Petronas Twin Towers in Kuala Lumpur, Malaysia (Baker *et al.*, 1994).

Hand-dug caissons

Hand-dug caissons have been used as foundations or earth-retaining structures with a diameter typically ranging from 1.5 m to 2.5 m, and an allowable load of up to about 25 000 kN. Hand-dug caissons of a much larger size, of between 7 m and 10 m in diameter, have also been constructed successfully (e.g. Humpheson *et al.*, 1987; Barcham and Gillespie, 1988). The advantages and disadvantages of hand-dug caissons are summarised in Table 1.2.

Hand-dug caisson shafts are excavated using hand tools in stages with depths of up to about 1 m, depending on the competence of the ground. De-watering is facilitated by pumping from sumps on the excavation floor or from deep wells. Advance grouting may be carried out to provide support in potentially unstable ground. Each stage of excavation is lined with in situ concrete rings (minimum 75 mm thick) using tapered steel forms which provide a key to the previously constructed rings. When the diameter is large, the rings may be suitably reinforced against stresses arising from eccentricity and non-uniformity in hoop compression. Near the bottom of the pile, the shaft may be belled out to enhance the load-carrying capacity.

The isolation of the upper part of hand-dug caissons by sleeving is sometimes provided for structures built on sloping ground to prevent the transmission of lateral loads to the slope or conversely the build-up of lateral loads on caissons by slope movement (GCO, 1984). The influence of sleeved piles and pile groups on slope stability has been reported by Ng *et al.* (2001d) and the performance of laterally loaded sleeved piles in sloping ground has been described by Ng and Zhang (2001). However, there is a lack of instrumented data on the long-term performance of the sleeving.

Given the disturbingly high accident rate and the health hazard in installation and construction (Ng *et al.*, 1987) of caissons, their use should be discouraged.

CHAPTER TWO

Vertical pile load transfer mechanisms

Load–settlement relationships

Load–settlement curve

As described by Tomlinson (2001), the load–settlement relationship typical for a single pile driven into a sand when subjected to vertical loading to the point of failure is shown in Fig. 1.2(a). At the early stages of loading, the settlement is very small and is due almost wholly to elastic movement in the pile and the surrounding soil. When the load is removed at a point such as A in Fig. 1.2(a) the head of the pile will rebound almost to its original level. If strain gauges are embedded along the length of the pile shaft they will show that nearly the whole of the load is carried by side resistance on the upper part of the shaft (Fig. 1.2(b)). As the load is increased, the load–settlement curve steepens, and release of load from a point B will again show some elastic rebound, but the head of the pile will not return to its original level, indicating that some 'permanent set' has taken place. The strain gauge readings will show that the shaft has taken up an increased amount of side resistance but the load carried by the shaft will not equal the total load on the pile, indicating that some proportion of the load is now being carried in the end-bearing. When the load approaches failure point, C, the settlement increases rapidly with little further increase of load. A large proportion of the ultimate load is now carried by end-bearing.

The relative proportions of load carried in side resistance and end-bearing depend on the shear strength and stiffness of the soil. Generally, the vertical movement of the pile which is required to mobilize full end resistance is much greater than that required to mobilize full side resistance.

Ultimate bearing capacity

At the limit state corresponding to point C in Fig. 1.2(a) the ultimate bearing capacity of a pile is given by the following equation and is illustrated in Fig. 1.3.

$$Q_u = Q_s + Q_b - W_p \tag{1.1}$$

where Q_s = ultimate side resistance, Q_b = ultimate end-bearing resistance, W_p = weight of pile. Usually W_p is small in relation to Q_u and is often

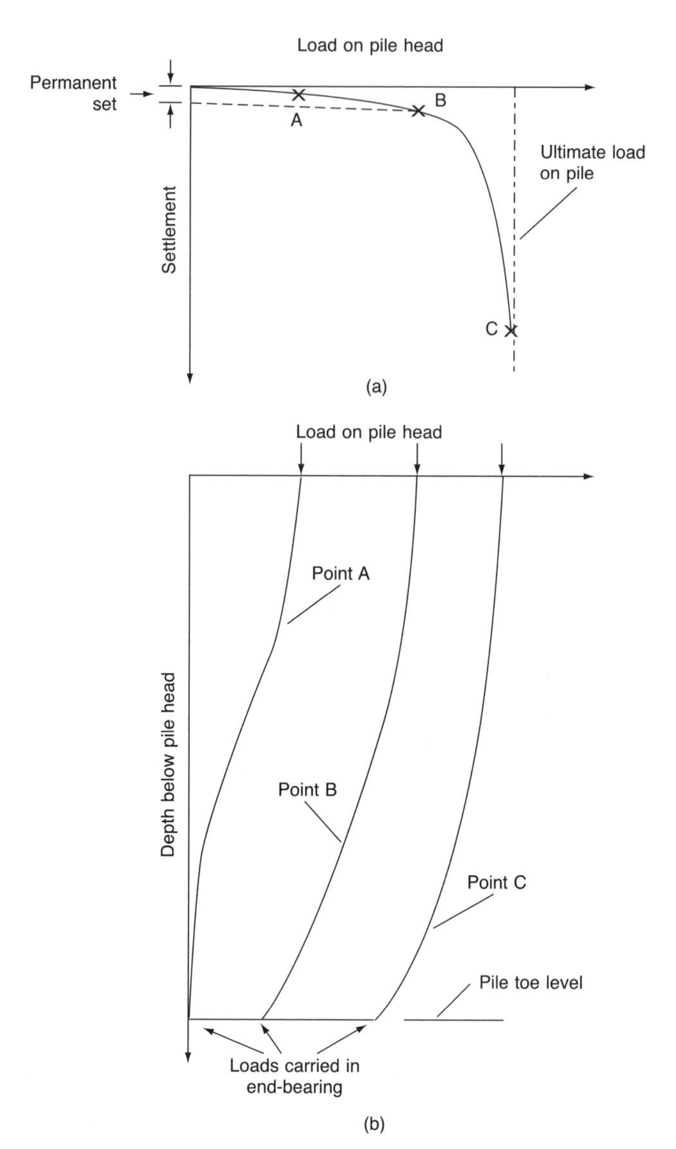

Fig. 1.2 Effects of loading a pile. (a) Load–settlement curve. (b) Distribution of load over depth of pile shaft for various stages of loading (after Tomlinson, 2001)

neglected because it is not much greater than the weight of the displaced soil. However, it must be taken into account for marine piling where a considerable proportion of the pile length extends above the sea-bed.

Eurocode 7 (BSI, 1997) refers to the 'design bearing resistance', R_{cd}, which is derived from:

$$R_{cd} = R_{sk}/\gamma_s + R_{bk}/\gamma_b \tag{1.2}$$

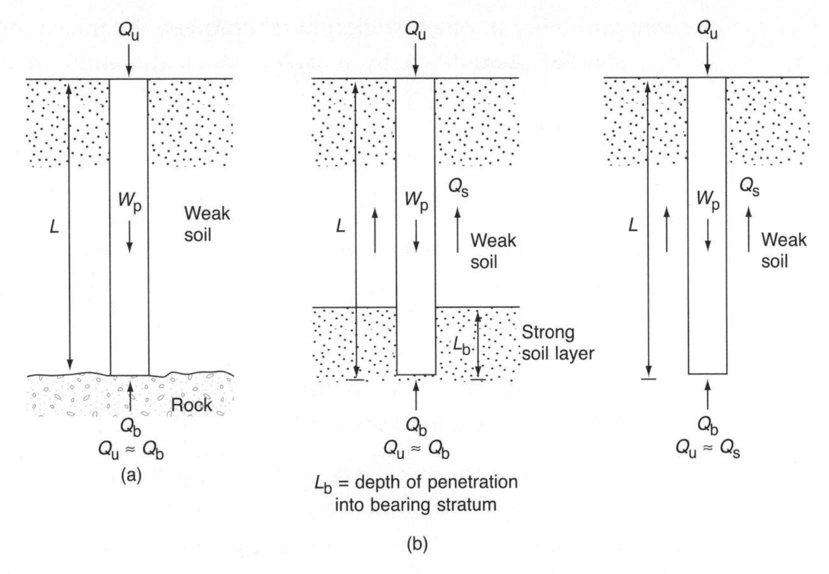

Fig. 1.3 (a) and (b) End-bearing piles. (c) Floating piles (after Das, 2000)

where γ_s and γ_b are the partial safety factors for side resistance and base resistance respectively, and

$$R_{sk} = \sum_{i=1}^{\pi} q_{sik} A_{si} \tag{1.3}$$

and the base resistance is given by

$$R_{bk} = q_{bk} A_b \tag{1.4}$$

where A_{si} is the nominal surface area of the pile in soil layer i, A_b is the nominal plan area of the pile base, q_{sik} is the characteristic value of the resistance per unit of the shaft in layer i, and q_{bk} is the characteristic value per unit area of base.

Eurocode 7 (BSI, 1997) requires that the characteristic values q_{sk} and q_{bk} do not exceed the measured bearing capacities used to establish the correlation divided by 1.5 on average. For example if q_b is shown by correlation with loading tests to be equal to nine times the average undrained shear strength of the soil below the pile base the characteristic value should be obtained from $9 \times s_u$ (average) divided by 1.5. This additional factor is used essentially to allow for uncertainties in the calculation method or scatter in the values on which the calculation was based.

Where the ultimate limit state pile load is determined from loading tests, Equation (1.2) can be used to obtain R_{cd} provided that the components of base resistance and side resistance are obtained either by instrumentation or by a graphical interpretation. To allow for the variability of the ground and variability in pile installation techniques, the measured or estimated R_{cm} should be reduced by a factor which depends on the number of loading tests performed.

The Institution of Civil Engineers' specification for piling (1996a) defines the *allowable pile capacity* as

> *...a capacity which takes into account the pile's bearing capacity, the materials from which the pile is made, the required load factor, settlement, pile spacing, down-drag, the overall bearing capacity of the ground beneath the piles and other relevant factors. The allowable pile capacity indicates the ability of the pile to meet the specified loading requirements and is therefore required to be not less than the specified working load.*

Hence this definition is broader than the Eurocode (BSI, 1997) term 'design bearing resistance', which refers only to the bearing capacity of an individual pile and takes no account of the serviceability limit state of the structure supported by the piles. The Institution of Civil Engineers' definition corresponds to that of 'allowable load' in BS 8004: Foundations (BSI, 1986).

Pile groups

When piles are arranged in close-spaced groups (Fig. 1.4) the mechanism of failure is different from that of a single pile. The piles and the soil contained within the group act together as a single unit. A slip plane occurs along the perimeter of the group and 'block failure' takes place when the group sinks and tilts as a unit. The failure load of a group is not necessarily that of a single pile multiplied by the number of piles in the group. In sand it may be more than this; in clays it is likely to be less. The 'efficiency' of a pile group is taken as the ratio of the average load per pile when failure of the group occurs to the load at failure of a comparable single pile.

It is evident that there must be some particular spacing at which the mode of failure changes from that of a single pile to block failure. The

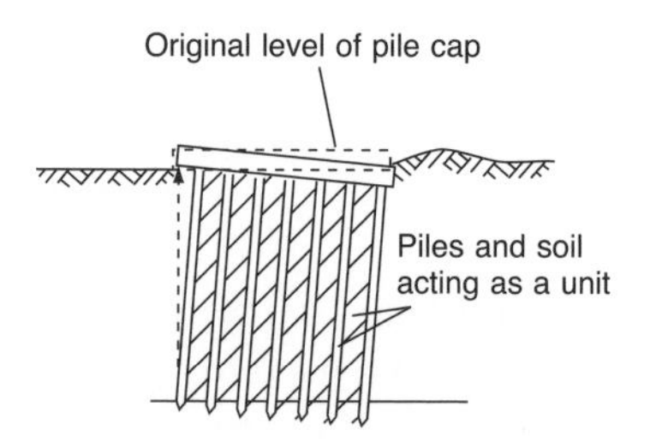

Fig. 1.4 Failure of piles by group action (after Tomlinson, 2001)

change is not dependent only on the spacing but also on the size and shape of the group and the length of the piles.

Group effect is also important from the aspect of consolidation settlement, because in all types of soil the settlement of the pile group is greater than that of the single pile carrying the same working load as each pile in the group. The ratio of the settlement of the group to that of the single pile is proportional to the number of piles in the group – that is, to the overall width of the group. The group action of piles in relation to carrying capacity and settlement will be discussed in greater detail later in this Part 1.

CHAPTER THREE

Failure load on piles: definitions, interpretation and criteria

Definitions

In the foregoing discussion, the failure load is taken as the load causing ultimate failure of the pile material, or the load at which the bearing resistance of the soil is fully mobilised. However, in the engineering sense, failure may have occurred long before the ultimate load is reached when the settlement of the structure has exceeded tolerable limits.

Terzaghi's suggestion that, for practical purposes, the ultimate load can be defined as that which causes a settlement of one-tenth of the pile diameter or width, is widely accepted by engineers. However, if this criterion is applied to piles of large diameter and a nominal safety factor of 2 is used to obtain the working load, then the settlement at the working load may be excessive.

In almost all cases where piles are acting as structural foundations, the allowable load is governed solely by considerations of tolerable settlement at the working load. An ideal method of calculating allowable loads on piles would be one which enabled the engineer to predict the load–settlement relationship up to the point of failure, for any given type and size of pile in any soil or rock conditions.

In most cases, a simple procedure is to calculate the ultimate bearing capacity of the isolated pile and to divide this value by a safety factor that experience has shown will limit the settlement at the working load to a value which is tolerable to the structural designer. But where settlements are critical it is necessary to evaluate separately the proportions of the applied load carried in side resistance and end-bearing and then to calculate the settlement of the pile head from the interaction of the elastic compression of the pile shaft with the elasto-plastic deformation of the soil around the shaft and the compression of the soil beneath the pile base.

In all cases where piles are supported wholly by soil and are arranged in groups, the steps in calculating allowable pile loads are as follows:

- Determine the base level of the piles that is required to avoid excessive settlement of the pile group. This level is obtained by the methods described in Tomlinson (2001). The practicability of attaining this level with the available methods of installing the piles must be kept in mind.

- Calculate the required diameter or width of the piles such that settlement of the individual pile at the predetermined working load will not result in excessive settlement of the pile group.
- Examine the economics of varying the numbers and diameters of the piles in the group to support the total load on the group.

The general aim should be to keep the numbers of piles in each group as small as possible – that is, to adopt the highest possible working load on the individual pile. This will reduce the size and cost of pile caps, and will keep the settlement of the group to a minimum. However, if the safety factor on the individual pile is too low, excessive settlement, leading to intolerable differential settlements between adjacent piles or pile groups, may result.

In the case of isolated piles, or piles arranged in very small groups, the diameter and length of the piles will be governed solely from consideration of the settlement of the isolated pile at the working load.

Interpretation and criteria

Overview

'Failure criteria' or 'ultimate load criteria' are methods for interpreting the 'failure load' of a pile loading test, which usually refers to the maximum test load that can be used for design purposes. Here, as suggested by Hirany and Kulhawy (1989), the term 'interpreted failure load' will be used to emphasise that the load is an interpreted value. In a load test, the load–movement curve may show that the load is still increasing at the termination of the test and at large movements. The recorded load–movement behaviour may be subject to errors due to creep characteristics of the soil and fluctuations in the applied load (Hirany and Kulhawy, 1989). Therefore, failure criteria are often formed with two considerations in mind, namely practicality and consistency. Practicality is a measure of whether a failure criterion can be achieved in practice. Consistency refers to the similarity of the results from a criterion for any number of loading tests, with regard to the position of the failure load on the load–movement relationship – that is, the degree of mobilisation and the level of yield. A high level of consistency may also mean a criterion is more effective in providing predictable and safe movements at working load after the interpreted failure load is combined with a factor of safety.

For different pile types and sizes, construction methods, testing methods and soil types, variations in the characteristics of the load–movement relationships can be expected. This may help to explain the considerable number of different failure criteria which have been proposed in the literature and which are in use in specifications and codes of practice around the world. Criteria are sometimes selected on a project-by-project basis and are specified by government departments responsible for overseeing a particular type of work. In the United States for example, a limiting plastic settlement of 0.25 inches (about 6 mm) has been adopted by the American Association of State Highway and Transportation Officials (AASHTO), New York State, Louisiana (ASCE, 1993) and Butler and Hoy's method has been adopted by the Federal Highway Administration (Butler and Hoy, 1977). In Hong Kong, separate general failure criteria are set out by the Buildings Department (BD) for private developments (BD, 1997a, b), by the Housing Authority (HA) for public housing works (HA, 1998) and by the Civil Engineering Department (CED) for civil engineering works (HKG, 1992).

Methods and criteria for interpreting failure load

Brinch Hansen (1963) described two methods for interpreting failure. The '90% criterion' defines failure as the stress at which the strain is twice the strain at a 10% smaller stress. The '80% criterion' defines failure as the stress at which the strain is equal to four times the strain at 20% smaller stress. The criteria assume hyperbolic relationships of stress and strain. They were originally not specifically intended for use with piles, but were proposed for stress–strain relationships of soils in general. However, the 90% criterion has gained widespread use in Scandinavian countries as a pile load test failure criterion (Fellenius, 1990). Brinch Hansen (1963) suggested that the definition of failure using the 80% criterion is useful for practical cases in which the stress–strain curve does not have any vertical tangent and where a point of failure must be defined more or less arbitrarily. The criteria can be used to extrapolate the load–movement relationship if so desired. However, Fellenius (1990) recommended that an extrapolated failure load should not be used for design purposes due to the uncertainty of the pile behaviour after reaching the maximum test load.

De Beer (1967) described a method for slow[1] tests and interpreted failure load as the load at the intersection of two straight lines, which are plotted on the load–displacement relationship in double logarithmic scale. The criterion was applied to tests on Franki piles with over-expanded bases.

[1] Slow tests refer to maintained load tests in which load is applied in increments. Each applied load is held until the rate of movement has reduced to an acceptable low value before the next load increment is applied. It is usual practice to include a number of loading and unloading cycles in a load test programme. Different loading procedures have been proposed in the literature (GEO, 1996). The entire load test programme may last a few days.

Chin (1970) proposed a method of finding failure load which involved full mobilization of end-bearing capacity using extrapolation for tests not carried to failure. Similar to the Brinch Hansen criterion, the method assumes that when the failure load is approached, the load–settlement curve is hyperbolic in shape. The suitability of the method can be measured by how close the load–settlement relationship of a pile is to a hyperbolic curve. Chin (1970) conducted a series of laboratory tests on a 1.94 in (about 49 mm) diameter model steel piles and three field tests on a 14 in (about 356 mm) reinforced concrete driven piles to verify his method; all the load increments were carried out in a constant time interval (48 h). This method is applicable to both quick[2] and slow tests, provided constant time increments are used (Fellenius, 1990).

Fuller and Hoy (1970) proposed a criterion for use with the quick load test method, which was applied to driven piles installed for the Texas Highway Department. The failure load is interpreted to be the load at the maximum slope of the load–movement curve or the load at which the slope of the load–displacement curve is greater than 0.05 in/t (about 1.2 mm/1000 kg). The purpose of the criterion was to produce an interpreted failure load, which could correlate well with the interpreted failure load obtained by using slow maintained load testing methods. The interpreted failure load is found graphically by extending a line back from the point at which 'plunging failure' occurs. Plunging failure is defined as the load on the pile that requires constant pumping of the hydraulic jacks on the pile head to be maintained, and the interpreted failure load will usually be slightly less than this load. Butler and Hoy (1977) later published a modification of the Fuller and Hoy criterion for bored piles. Failure load is defined as the load at the intersection of the initial tangent and the point with a slope of 0.01 in/t (about 0.25 mm/1000 kg). At working load, the pile head settlement should also not exceed 13 mm.

Davisson (1972) developed a failure criterion using the results of loading tests on 12 in (about 300 mm) diameter driven piles in conjunction with wave equation analysis and dynamic measurements. It was intended for use with the quick load test. Failure load was defined as the load corresponding to a movement which exceeds the elastic compression of the pile by a value of 0.15 in (about 3.8 mm) plus 'quake'. Quake, designated Q_{tip}, is defined as the movement required to cause yielding of the soil at the pile tip and Davisson suggested it is equal to around 0.1 in (about 2.5 mm) for 12 in (about 300 mm) diameter piles in most

[2] Quick tests refer to constant rate of penetration tests. A constant strain rate of 0.25 to 1.25 mm/min and 0.75 to 2.5 mm/min is commonly used for clays and granular soils, respectively (ASTM, 1995).

soils. The criterion is given by the following equation:

$$\Delta_M < \frac{PL}{AE} + 0.15 + Q_{tip} \qquad \text{(in)} \qquad (1.5)$$

where P is the applied load, L is the pile length, A is the pile area and E is the pile shaft elastic modulus. However, Davisson (1972) also stated that the value of Q_{tip} depended on soil type. He gave another example of a pile in micaceous soil with a Q_{tip} of 0.3 in (about 7.6 mm).

An advantage of Equation (1.5) is that the interpretation of failure load is non-subjective, requiring only a simple mathematical calculation and not requiring graphical interpretation of the load–movement relationship. Also, the method can be included in contract specifications so that the engineer can determine in advance the maximum allowable movement for a load with consideration of the length and size of a pile (Fellenius, 1980). The Canadian Geotechnical Society (CGS, 1995) suggested an offset limit load method for interpreting failure load in pile load tests which slightly modified the Davisson criterion as given in Equation (1.5). This criterion can be used in pile load tests by the constant rate of penetration method (Davisson, 1972). The CGS (1995) suggests this criterion is intended for the quick maintained load test method but it can also be used for interpreting failure load for the slow maintained load test involving no unloading and loading cycles.

Weltman (1980) suggested that a movement of 10% of the pile diameter be used to determine failure load. For piles in granular soils, Weltman suggests that a unique ultimate load may not exist as the pile resistance may increase with penetration to larger movements than would be achieved in testing in practice. However, 10% of the pile diameter is still recommended on the basis that it is the minimum movement that could be associated with failure of the soil beneath the pile toe. The International Society for Soil Mechanics and Foundation Engineering (ISSMFE, 1985) also suggests that the failure load be taken as the load causing a gross settlement of 10% of the equivalent diameter. This is also suggested by BS 8004 (BSI, 1986) as well as by Lehane and De Cock (1999) and Eurocode 7 (BSI, 1997) for practical purposes unless the failure load can otherwise be defined by some clearly recognisable feature of the load–movement curve. Where appropriate, BS 8004 allows a correction to be made for the elastic shortening of the pile shaft under axial loading.

Fellenius (1980) recommended the use of four different criteria, including the Davisson criterion, the Brinch Hansen 80% criterion and the Butler and Hoy criterion mentioned above. He also suggested the use of other criteria depending on the circumstances. The CGS (1995) left the choice of which method to use as a failure criterion open to the designer and described the Davisson (1972) criterion, the Brinch Hansen (1963) 80% criterion and the Chin (1970) criterion as examples. Similarly, the American Society of Civil Engineers (ASCE, 1993) details a number of different criteria, but does not recommend anyone in particular.

Hirany and Kulhawy (1989) suggested the use of an interpreted failure load for drilled shafts (bored piles), which corresponds to 'failure threshold'. This is defined to be the point at which the load–movement curve becomes linear again after an initial linear region followed by a curved 'transition' region. By studying a number of loading tests on bored piles between 0.4 and 1.1 m in diameter and less than 17 m in length, they found that the failure threshold generally occurs at a pile head movement of 4% of the pile diameter and therefore effectively the method of interpretation is implemented as a limiting movement criterion.

The Australian AS-2159 standard (SAA, 1995) sets out a maximum movement criterion of 50 mm, and 30 mm upon unloading for 1.5 times the design action effect. The 'design action effect' (S) is defined under the AS limit state design system and is similar to the working load. At serviceability load (equal to $0.75S$), maximum movements of 15 mm, and 7 mm upon unloading apply. The criterion also allows a consideration of elastic shortening for long piles, although details of how to calculate the elastic shortening are not provided.

The Florida Department of Transportation (FDT, 1999) set out a maximum movement criterion as shown in Equation (1.6) for large-diameter bored piles; in any case where the pile head movement under maximum test load is less than the value given by the proposed method, the maximum test load can be taken as the failure load. For small-diameter piles, the Davisson criterion as given in Equation (1.5) can be used.

$$\Delta_M < \frac{PL}{AE} + \frac{d}{30} \quad \text{(in)} \tag{1.6}$$

Three definitions of failure load are given in the People's Republic of China Technical Code for Building Pile Foundations (JGJ, 1995). Firstly, failure load is defined as the load at which settlement continues to increase without any increase of that load. For large-diameter piles, the failure load is defined as the load corresponding to Δ_M less than 3% to 6% d, and for slender piles ($L/d > 80$) corresponding to Δ_M less than 60 to 80 mm. The last definition defines the failure load just before an obvious bend occurs in the Δ_M versus $\log t$ (t = time) plot.

A summary of the methods and criteria around the world for interpreting failure load from the literature is given in Table 1.3 and from Hong Kong experience is given in Table 1.4. It should be noted that the residual pile settlement criteria listed in the table should be abolished completely. Although the residual pile movement may indicate some degree of soil

Table 1.3 Settlement/failure load criteria for pile load tests from literature

	Criteria	Test method/pile type and size	Settlement rate	Key assumptions	Related references
Brinch Hansen 90% criterion (1963) (adopted by CED (1992); Norway)	Ultimate load, Q, is the load that gives twice the pile head settlement as obtained for 90% of that load, Q, i.e. $\Delta_Q = 2 \times \Delta_{0.9Q}$	Nil	Nil	The criterion was originally developed using laboratory measured stress–strain relationships of soil and defines failure as the stress for which the strain is twice the strain at a 10% smaller stress and assumes a hyperbolic stress–strain relationship. The criterion was not specifically intended for use with piles.	Brinch Hansen (1963) Fellenius (1980, 1990)
Brinch Hansen 80% criterion (1963) (suggested by CGS (1995))	Failure load is defined as the load that gives four times the pile head displacement as obtained for 80% of the failure load.	Not stated. Applicable for slow, quick maintained load test, constant rate of penetration test involved no unloading cycles (Fellenius, 1980, 1990).	Nil	The criterion was originally developed using laboratory measured stress–strain relationships and assumes a parabolic stress–strain relationship. The criterion was not specifically intended for use with piles. This is an extrapolation method.	Brinch Hansen (1963) CGS (1995) Fellenius (1980, 1990)
De Beer (1967)	This criterion defines failure load as the load at the intersection of two straight lines, which are plotted on the load–displacement relationship in double logarithmic scale.	Slow maintained load test. The criterion was applied to tests on Franki piles with overexpanded bases.	Nil	?	Fellenius (1980, 1990)
Hirany and Kulhawy (1989)	$\Delta_M < 4\%\ d$ Load Failure load Δ_M	The criterion is verified by bored piles with diameter between 0.4 m and 1.1 m and less than 17 m in length.	Nil	Failure load is defined to be the point at which the load–movement curve becomes linear again after an initial linear region followed by a curved transition region.	Hirany and Kulhawy (1989)

Table 1.3 Settlement/failure load criteria for pile load tests from literature (continued)

	Criteria	Test method/pile type and size	Settlement rate	Key assumptions	Related references
Chin (1970) (suggested by CGS (1995))	$\frac{\Delta}{P} = m\Delta + c$, ultimate load $= 1/m$, c is a constant (no particular meaning).	Slow and quick maintained load test provided that constant time intervals are used (Fellenius, 1980, 1990). Chin used 1.94 in diameter steel pipe piles and three 14 in reinforced concrete piles to verify his criterion, and he also applied his criterion to pile data from literature. The load increments were carried out in constant time intervals (48 h).	?	Failure load reached only after full mobilization of bearing capacity. Load–displacement relationship is approximately hyperbolic. This is an extrapolation method.	Chin (1970) CGS (1995) Fellenius (1980, 1990)
Fuller and Hoy (1970) (adopted by FHWA 1970)	Failure load should be load at the maximum slope of the load–displacement curve or where the load–displacement curve is sloping 0.05 in/t. Δ at WL should be less than the allowed displacement of the structure.	Slow and quick test method and CRP. Verified by driven pile (size not stated).	Nil	Fuller and Hoy gave an example which was 0.05 in/t for pile-head displace- ment and 0.03 in/t for plastic displace- ment of which they took the lowest. This is still an arbitrary definition of failure but is a more generalized approach.	Fuller and Hoy (1970) Fellenius (1980, 1990)
Butler and Hoy (1977) (adopted by FHWA (1978) (for drilled shaft)	Failure load is the load at the intersection of initial tangent and the line of tangent with slope of 0.01 in/t of the load–displacement curve. Proven max. safe static load is equal to $\frac{1}{2}$ of the failure load, provided that, $\Delta_{WL} < 13$ mm (for drilled shaft). For driven pile use the same criterion as Fuller and Hoy (1970).	Quick maintained load test method (constant time interval). Verified by 36 in (900 mm) diameter, 60 ft (18.3 m) long drilled shaft.	Nil	Required to test to plunging failure. Failure load < plunging failure load.	Butler and Hoy (1977) Fellenius (1980, 1990)

Table 1.3 Settlement/failure load criteria for pile load tests from literature (continued)

	Criteria	Test method/pile type and size	Settlement rate	Key assumptions	Related references
Davisson criterion (1972) (suggested by CGS (1995), BD (1997), NAVFAC (1982), Norway)	$\Delta_M < \frac{PL}{AE} + 0.15\,\text{in} + Q_{tip}$, where PL/AE is the elastic shortening; Δ_M is the pile head settlement, 0.15 in – based on observation and experience; Q_{tip} is the displacement required at pile toe to cause soil 'quake' (displacement at which yielding of soil at the pile base occurs).	Static load test. Verified by driven pile of dimension 1 ft. Intended for quick maintained load test with no unloading cycle (Fellenius, 1980; CGS, 1995).	Nil	Developed for point bearing driven pile, considering the displacement required to cause yielding of soil at pile base plus a movement of 0.15 in. Q_{tip} can vary for different pile types (examples given in Davisson, 1972)	Fellenius (1980, 1990) CGS (1995) Davisson (1972) Fellenius (1980, 1990)
10% diameter (adopted by BSI (1986); EC7 (BSI, 1997); ISSMFE (1985); Norway)	Failure load is the load causing pile-head settlement equal to 10% of pile-base diameter. For long pile, the 10% may be required to adjust for elastic shortening (BSI, 1986).	?	?	?	BSI (1986) ISSMFE (1985) Fellenius (1980, 1990)
AS-2159 (SAA, 1995)	$\Delta_M < 50\,\text{mm}$ at 1.5 design action effect (D.A.E). $\Delta_R < 30\,\text{mm}$ after unloaded from 1.5 design action effect. $\Delta_{WL} < 15\,\text{mm}$ at serviceability load ($= 0.75S$). $\Delta_{R,WL} < 7\,\text{mm}$ after unloaded from serviceability load.	Quick test method. Pile type and size not stated.	Nil	Usually depend on structural consideration. Correction for elastic shortening should be considered for long piles.	AS-2159 (SAA, 1995)
JGJ (1995) (Technical Code for Building Pile Founda-tions, People's Republic of China)	Ultimate load is the load at which settlement continues to increase without any increase of load. For large-diameter pile, the failure load is defined as the load corresponding to	Quick maintained load test with requirement of rate of settlement. Applicable for both driven and bored piles	Before increments: $<0.1\,\text{mm/hr}$ for at least 1.5 hours.	Nil	JGJ (1995)

Table 1.3 Settlement/failure load criteria for pile load tests from literature (continued)

	Criteria	Test method/pile type and size	Settlement rate	Key assumptions	Related references
JGJ (1995) *continued*	$\Delta_M < 3\text{–}6\%\ d$ and for slender pile ($L/d > 80$) corresponding to $\Delta_M < 60\text{–}80$ mm Based on Δ–log t plot, the failure load is the load just before an obvious bend occurs.				
FDT (1999)	$\Delta < \frac{PL}{AE} + \frac{d}{30}$ in Factor of safety = 2. Smaller piles use Davisson criterion.	Quick maintained load test. Diameter > 600 mm.	Nil	?	FDT (1999)

yield at the pile tip, the magnitude of the residual movement is strongly affected by 'locked-in movement' of the pile as a result of side shear acting downward against the pile upon unloading thus preventing it from rebounding to its original position. This locked-in effect is almost impossible to assess accurately and reliably. In addition, the residual movement is affected by any creep in the concrete pile. These two phenomena are likely to be more prominent in long and large-diameter bored piles and barrettes than short and small-diameter driven piles (GEO, 1996; Ng *et al.*, 2001c). The residual pile movement criterion is therefore irrational, unreliable and unsuitable for use as an acceptance criterion, especially for long piles and barrettes, and it always leads to unnecessarily conservative design (GEO, 1996).

Table 1.4 Settlement/failure load criteria for pile load tests in Hong Kong

	Criteria	Test method/pile type and size	Settlement rate	Key assumptions	Related references
Buildings Department, HKSAR (BD)	$\Delta_M < \frac{Pl}{AE} + \frac{D}{120} + 4$ (mm) $\Delta_R < \frac{D}{120} + 4$ (mm) where PL/AE is the elastic shortening, D is the pile diameter, Δ_M is the pile head settlement, Δ_R is the residual settlement.	Slow maintained load test in 2 cycles Max. test load: $2 \times$ design working load. Unloading at end of each cycle. 72 h hold at peak load of cycle 2. Pile type and size not stated but intended for bored piles.	Before increments/ decrements: <0.05 mm in 10 min. Before hold: <0.1 mm/h.	Based on Davisson criterion (1972) which are based on driven pile 1 ft wide (0.3 m). $D/120$ is directly transformed from 0.1 in on 1 ft pile, given in Davisson (1972). Reasoning behind residual settlement criterion not stated.	GEO (1996) PNAP 66, BD (1997a) Davisson (1972)
Housing Authority, HKSAR (HA)	$\Delta < \frac{PL}{AE} + \frac{D}{30}$ (in mm) $\Delta_R^f < \frac{D}{50}$ or 10 mm whichever is smaller. $\Delta_R^b < \frac{D}{100}$ or 5 mm whichever is smaller where Δ_R^f is the residual settlement for pile embedded in soil, Δ_R^b is the residual settlement for pile bearing on rock.	HA specification required test carried out in slow maintained load test method. Load applied in 3 cycles. Test load includes allowance for negative skin friction (NSF) if this is expected in long term. Max. test load: $2 \times$ design working load $+ 3 \times$ NSF. Pile type and size not stated.	Before increments/ decrements: <0.05 mm in 10 min. Before hold: <0.1 mm/h.	Based on Yiu and Lam 1990; the criterion verified by five driven piles; it seems that the test method used was CRP (constant rate of penetration) with 1 to 2 mm/min penetration. Verified by 5 driven pile of diameter less than 500 mm.	HA (1998) GEO (1996) Yiu & Lam (1990)
Civil Engineering Department, HKSAR (CED/GEO)	Ultimate load, Q, is the load that produces double the pile head settlement as 90% of that load, Q, i.e. $\Delta_Q < 2 \times \Delta_{0.9Q}$ $\Delta_{WL} < 20$ mm for buildings $\Delta_{WL} < 10$ mm for other structures. Preliminary piles shall have factor of safety of at least 2. Δ_{WL} is the pile head settlement under working load. [Chart labels: Ultimate load; Load; Q; $0.9Q$; $0.5\Delta_{ult}$; Δ_{ult}; Δ]	Slow maintained load test in 3 cycles. Load increments/ decrements to be in 25% of the design working load. Unloading at end of each cycle. Max. test load for preliminary piles: $2 \times$ design working load. Max. test load for working piles: $1.8 \times$ design working load. 24 h hold at peak load of each cycle. Pile type and size not stated.	Before increments/ decrements: <0.1 mm in 20 min. Before hold: <0.1 mm/h.	Based on Brinch Hansen 90% criterion (1963) with additional settlement requirements at design working load. Additional working load settlement, possibly required for structural requirement. The criterion was originally developed using laboratory measured stress–strain relationships of soil and defines failure as the stress for which the strain is twice the strain at a 10% smaller stress and assumes a hyperbolic stress–strain relationship. The criterion was not specifically intended for use with piles.	GEO (1996) HKG (1992) Brinch Hansen (1963)

CHAPTER FOUR

Establishment of a new failure load criterion for large-diameter bored piles and barrettes

Overview: the problem of large-diameter, long bored piles

Ng *et al.* (2001c) have observed that often in the literature, piles used to verify design criteria have been small-diameter driven piles; hence, there is uncertainty as to how suitable these criteria are for large-diameter long bored piles which are common in the Far East in general and in Hong Kong in particular. Because the load-carrying capacities of large-diameter bored piles are relatively large, high-capacity loading-reaction systems are required and the targeted failure load may not always be practical to achieve. A total of 38 large-diameter test bored (drilled) piles founded in weathered rocks and saprolites were critically examined and investigated. 'Saprolites' refer to soils that retain various degrees of the original texture, fabric and structure of the parent rock from which they were formed by decomposition (GCO, 1988). The test piles under review range from 0.6 to 1.8 m in diameter and from 12 to 66 m in length. The rock sockets were generally around 2 m in depth. Details of all of the tests are given in Table 1.5.

Among the 38 piles, 31 tests were examined for toe resistance and shaft shortening and incorporated in an investigation of pile behaviour related to the development of a failure load criterion. Based on the study of the mobilization of toe resistance and on a consideration of shaft shortening, a new semi-empirical method was proposed for determining an approximate moderately conservative failure load for large-diameter bored piles. The interpreted failure loads using the new criterion were compared for the five pile tests and further verified independently for two additional piles that were not used in the formation of the criterion. Moreover, pile movements at factored loads were reported and discussed using global and partial mobilization factors.

A comparative study of failure loads using some existing criteria is shown in Table 1.6. Pile-head movements corresponding to failure are not reached according to the failure criteria of Brinch Hansen (1963) and Butler and Hoy (1977) and failure load as suggested by the ISSMFE (1985). However, criteria such as the modified Davisson criterion and

Table 1.5 Database of large-diameter bored piles in weathered geomaterials reviewed by Ng et al. *(2001c)*

Pile	d (m)	L (m)	S/R	MC	MLTM	P_{max} (kN)	Δ_{max} (mm)	${}^{\psi}\Delta_{PS}$ at P_{max}	Notes	Reference
$1^{\#}$	1.80	45.0	GS	GW	SR	24 000	87.0	7.0	A, B, C, Ω	De Silva *et al.* (1998)
2	1.50	30.2	MS	GWP	SR	24 480	125.8	5.5	A, C, Ω	Ng *et al.* (2001c)
3	0.65	30.1	GS	Auger	SR	2 450	38.5	?	A, Λ	Chang and Wong (1995)
4	0.80	66.0	RS	?	SR	7 000	57.4	?	A, Λ	Liu *et al.* (1997)
$5^{\#}$	1.20	31.7	WVR	RCD	RT	10 770	61.0	3.5	A, B, C, Ω	Ng *et al.* (2001c)
$6^{\#}$	1.32	23.0	WGR	RCD	RT	26 000	79.8	14.0	A, B, C, Ω	Ng *et al.* (2001c)
7	0.62	12.0	WR	?	RT	3 460	36.0	?	A, B, Λ	Webb (1976)
8^{ϕ}	1.20	38.7	VS	GCW	SR	13 630	35.1	6.5	C, Ω	Ng *et al.* (2001c)
9^{ϕ}	1.20	38.7	VS	GCW	SR	5 200	26.7	4.5	C, Ω	Ng *et al.* (2001c)
10	1.20	31.8	VS	GCW	SR	19 100	47.0	8.1	C, Ω	Ng *et al.* (2001c)
11^{ϕ}	1.20	25.6	VS	GCW	SR	11 150	23.3	7.9	C, Ω	Ng *et al.* (2001c)
12	1.20	31.2	WVR	RCD	RT	30 220	31.5	28.0	C, Ω	Ng *et al.* (2001c)
13	1.46	52.1	GS	GCW	SR	11 500	19.6	10.8	C, Ω	GEO (1993)
14	1.46	40.6	GS	GCW	SR	11 600	14.4	11.0	C, Ω	GEO (1993)
15	1.46	42.2	GS	GCW	SR	11 200	20.4	11.1	C, Ω	GEO (1993)
16	1.00	36.9	GS	GRCD	SR	8 700	45.0	15.9	C, Λ	Holt *et al.* (1982)
17	1.31	50.9	MS	GCB	SR	5 400	16.3	3.5	C, Ω	Ng *et al.* (2001c)
18	1.32	50.2	MS	GCB	SR	5 500	15.4	3.8	C, Ω	Ng *et al.* (2001c)
19	0.60	25.0	MS	?	SR	6 180	~42.0	?	B, Λ	Tan *et al.* (1998)
20	1.05	23.0	MS	?	SR	8 250	~50.0	?	B, Λ	Tan *et al.* (1998)
21	1.00	63.0	MS	?	SR	15 000	~49.0	?	B, Λ	Tan *et al.* (1998)
22	0.61	20.0	MS	?	SR	7 350	~22.0	?	B, Λ	Tan *et al.* (1998)
23	0.90	29.0	GS	?	SR	4 320	~48.0	?	B, Λ	Tan *et al.* (1998)
24	0.75	15.0	MS	Dry	SR	?	?	?	B, Λ	Balakrishnan *et al.* (1999)
25	0.60	15.0	MS	Dry	SR	?	?	?	B, Λ	Balakrishnan *et al.* (1999)
$26^{\#}$	1.20	40.4	WMR	RCD	RT	30 250	25.0	20.0	C, Ω	Ng *et al.* (2001c)
$27^{\#}$	1.30	60.2	WGR	RCD	RS	13 200	38.9	13.1	C, Ω	Ng *et al.* (2001c)
$28^{\#}$	1.30	60.2	WGR	RCD	RT	24 500	60.0	29.0	C, Ω	Ng *et al.* (2001c)
$29^{\#}$	1.30	25.8	WGR	RCD	RT	23 000	23.8	14.5	C, Ω	Ng *et al.* (2001c)
$30^{\#}$	1.30	24.7	WGR	RCD	RS	11 400	5.5	5.4	C, Ω	Ng *et al.* (2001c)

31#	1.30	24.7	WGR	RCD	RT	23 000	17.4	11.9	C, Ω	Ng *et al.* (2001c)
32#	1.00	43.2	WGR	RCD	RT	18 260	20.1	15.8	C, Ω	Ng *et al.* (2001c)
33#	1.02	35.6	WVR	RCD	RT	25 500	54.0	24.0	C, Ω	Ng *et al.* (2001c)
34#	1.20	29.1	WVR	RCD	RT	26 400	17.5	15.6	C, Ω	Ng *et al.* (2001c)
35#	1.20	40.6	M	RCD	TR	30 000	23.6	20.8	C, Ω	Ng *et al.* (2001c)
36	1.50	49.4	WGR	RCD	TR	19 600	20.6	17.6	C, Ω	Ng *et al.* (2001c)
37#	1.20	39.4	WMR	RCD	TR	30 270	18.4	16.4	C, Ω	Ng *et al.* (2001c)
38#	1.02	37.1	WVR	RCD	TR	18 000	49.0	19.7	C, Ω	Ng *et al.* (2001c)

Notes: A – load–movement relationships analysed: B – toe resistance–movement relationships analysed; C – shaft shortening analysed; Dry – excavation under dry conditions; GCB – grab and chisel under bentonite; GRCD – grab and RCD under water with temporary casing; GS – granitic saprolite; GW – grab under water with temporary casing; GWP – grab under water with temporary casing and post grouted; MC – method of construction; MLTM – major load transfer mechanisms; M – marble; MS – metasedimentary saprolite: RCD – grab and chisel under water with temporary casing and RCD without casing; GCW – grab and chisel under water with temporary casing; RS – rock socket resistance; RT – rock socket and toe resistance; S/R – soil/rock description: SR – shaft resistance; TR – toe resistance; VS – volcanic saprolite; RS – residual soil: WGR – weathered granitic rock; WMR – weathered metasedimentary rock; WVR – weathered volcanic rock; Δ_{PS} – shaft shortening; ϕ – pile with soft toe; # – pile with slip liner; ? – data unavailable; Λ – unknown test method; Ω – slow maintained load test method; ψ – measured by extensometer or strain gauges.

Table 1.6 Interpreted failure loads (kN) from relevant specifications and criteria: Piles 1, 2, 5, 6 and 7 (after Ng et al.*, 2001c)*

Pile	Max. test load	10% dia.	Butler and Hoy	De Beer	Brinch Hansen 80%	Chin	Brinch Hansen 90% (CED)	AS max.	AS res.	Modified Davisson (BD max.)	BD res.	HA max.	HA res.	Hirany and Kulhawy	O'Neill and Reese (1999)	Proposed criterion
1#	24 000	N/A	N/A	16 400	23 765	26 800^E	N/A	22 000	20 800	19 000	19 800	22 800	19 500	22 700	N/A	24 000
2	24 480	N/A	N/A	N/A	78 700^E	40 000^E	N/A	17 500	13 400	11 800	12 200	17 500	10 400	17 800	19 800	19 500
5#	10 770	N/A	8700	7 740	10 770	11 700^E	9600	10 600	9 800	9 500	9 100	10 200	7 500	10 200	10 700	10 700
6#	26 000	N/A	N/A	17 000	30 000^E	34 200^E	N/A	25 100	23 000	15 800	17 200	23 800	13 000	23 200	25 000	25 400
7	3 460	N/A	2320	1 390	3 640^E	4 130^E	N/A	N/A	3 450	2 600	2 500	3 100	2 140	3 100	3 350	3 300

Note: # = pile with slip liner; N/A = ultimate load defined by analysis method not reached in loading test; ^{E} = values extrapolated from load–movement relationship.

the failure threshold of Hirany and Kulhawy (1989) were achieved in all cases. The method of Hirany and Kulhawy (1989) seems to be attractive for use with large-diameter bored piles, as it was verified specifically by loading tests on bored piles, most of which were of large-diameter but founded at relatively shallow depths. The method does not explicitly include the shaft shortening, because the amount of shaft shortening for the short piles considered is small and so the pile-head movement is approximately equal to the pile-toe movement. However, inconsistencies in interpreted failure load could be expected for long piles of 30–70 m in length, if shaft shortening is not explicitly considered.

The modified Davisson criterion does include an estimation of the shaft shortening, but, from the comparisons, the criterion appears to be over-conservative. Possibly, the general form of the modified Davisson criterion could be adopted, but greater movement of the pile toe should be specified to provide less conservative results. Also, the estimation of shaft shortening in the modified Davisson criterion may not be accurate enough due to shaft resistance reducing the axial load in the shaft, and the estimation should be refined based on actual measured shaft shortening results from loading tests.

Method for establishing a new failure load criterion

Study of pile toe movement

Weltman (1980) and the ISSMFE (1985) suggest that a movement of 10% of the pile diameter corresponds to ultimate load, based on the suggestions of Terzaghi. However, Reese and O'Neill (1988) provide toe resistance–toe movement relationships of bored piles in clays suggesting that full mobilization of toe resistance capacity occurs at a movement of around 45% of the pile diameter. For bored piles in sands, they report that toe resistance may continue to increase at movements exceeding 15% of the pile diameter. For the purpose of interpreting an approximate failure load for moderately conservative designs, a moderately conservative movement for the mobilization of capacity is preferred. For piles displaying increasing toe resistance at movements greater than the moderately conservative movement, the interpreted failure load may not represent the actual total load-carrying capacity of the pile, but the interpreted failure load will be comparable with the failure load determined for other piles. Furthermore, as suggested by Reese and O'Neill (1988) and Weltman (1980), actual ultimate toe resistance for piles in granular materials effectively may never be achieved.

Figure 1.5 shows the toe resistance–toe movement relationships for 11 of the piles having reliable data available. Details of the piles are given in Table 1.5. The unit toe resistance, r_t, was generally calculated using the measured strain near the pile toe by strain gauges (or by extensometers)

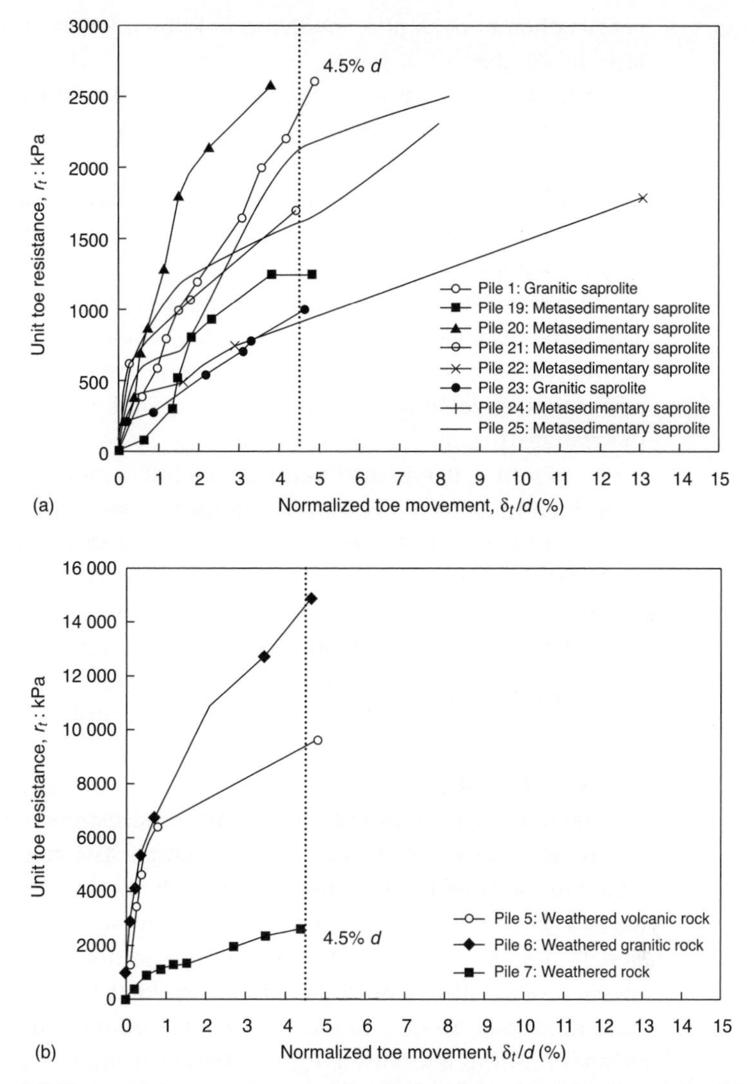

Fig. 1.5 Toe resistance–movement relationships for (a) piles embedded in saprolites; (b) piles socketed in and founded on weak rocks (after Ng et al., *2001c)*

and by estimating the pile modulus of elasticity. The movement δ_t is the 'local' toe movement, obtained by subtracting the shaft shortening (measured by extensometers or strain gauges) from the measured pile head movement and is normalized by d. Figure 1.5(a) shows the relationships for piles founded on saprolites. The figure indicates scatter in the rate of mobilization of resistance and varying magnitudes of r_t, at which yielding began. The variation may be due to the soil originating from different rock types, the varying density and confining pressure. The construction of each pile, involving unknown levels of disturbance, must also be a factor.

For most of the test piles, toe resistance capacity was not reached. An exception is Pile 20, which did achieve its capacity, and this occurred at a displacement of 4.5% of d. Figure 1.5(b) shows the relationships for three piles founded on weathered rocks. There is scatter in the magnitudes of mobilized toe resistance, but all three relationships are consistent in that significant yielding clearly occurs at a movement of 0.5% of d. However, resistance is observed to be still increasing after this point. The three tests were terminated at a toe movement of around 4.5% of d and in the case of Pile 5, the test was terminated due to the rate of pile movement exceeding the maximum allowed by the test requirement (0.5 mm in 10 mm).

From Fig. 1.5, it appears likely that the full mobilization of toe resistance is difficult to achieve, as it may require extremely large toe movements. However, a moderately conservative movement for the mobilization of toe resistance capacity has been identified to be 4.5% of d, and this is consistent for both piles founded on saprolites and piles founded on weathered rocks. The results are also consistent with the suggestions of Reese and O'Neill (1988). This approximate movement could be adopted in a new method of interpretation of moderately conservative failure load.

Study of shaft shortening

The modified Davisson criterion includes an estimation of the pile shaft shortening equal to PL/AE. The estimation assumes that no frictional resistance acts along the whole pile length. The load transferred as shaft friction depends on many factors, such as soil type and construction details. Whether the pile toe is 'floating', i.e. founded in soil or bearing on rock, also affects the total axial load in the shaft. A comparison of measured pile shaft shortening and calculated pile shaft shortening using PL/AE at measured P_{max} obtained from 28 pile loading tests is shown in Fig. 1.6. Details of the 28 test piles are given in Table 1.5.

All of the measurements in Fig. 1.6 lie on or above the line for calculated shortening equal to PL/AE, indicating that shortening is overestimated by the expression. The piles founded in soils generally have shortening measurements between values calculated by $\frac{3}{4}PL/AE$ and $\frac{1}{2}PL/AE$. The second expression is consistent with the theoretical shortening for piles with zero toe resistance and for which the average shaft resistance is constant along the pile length. For piles with toe resistance, the theoretical shortening will be greater than $\frac{1}{2}PL/AE$. This explains the measured values in soil, which are as high as $\frac{3}{4}PL/AE$. Figure 1.6 also shows results for 'end-bearing piles', piles with rock sockets and piles with slip liners. 'End-bearing piles' refer to piles founded on rocks and relying predomi-

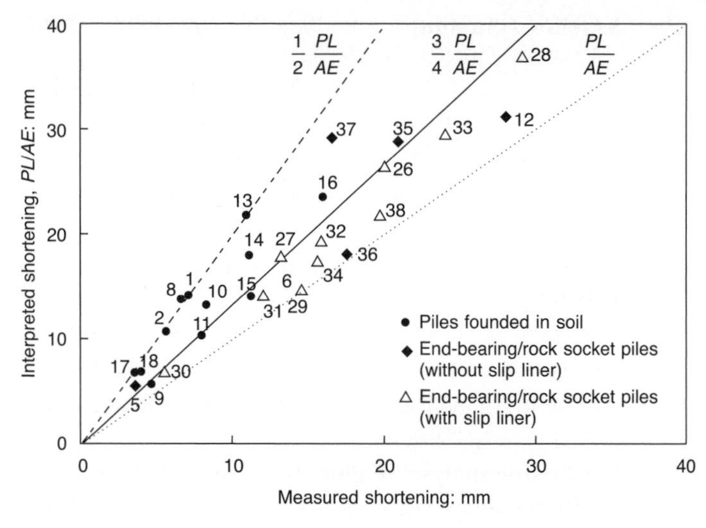

Fig. 1.6 Pile shaft shortening (after Ng et al.*, 2001c)*

nantly on toe resistance. 'Slip liners' refer to casings, often coated in a material such as bitumen and sometimes surrounded by a soft grout annulus. They are intended to allow minimal shaft resistance and thus allow maximum axial load to reach a test section below the liner. The piles with rock sockets and slip liners and the end-bearing piles may be expected to give similar results, due to the majority of the pile load being transferred at or close to the pile toe. The measured shortening would be expected to be higher than the piles founded in soils, due to shaft resistance being relatively lower, and this is supported by Fig. 1.6 with the results lying between PL/AE and $\frac{3}{4}PL/AE$.

For use in interpreting failure load, a moderately conservative estimate of shaft shortening may be appropriate. From the results in Fig. 1.6, the moderately conservative estimates are $\frac{1}{2}PL/AE$ and $\frac{3}{4}PL/AE$ for piles founded in soils and end-bearing piles, respectively. End-bearing piles are piles that rely predominantly on toe resistance, piles with rock sockets and piles with slip liners. Note that these estimations may also err on the conservative side for the reason that the maximum test load P_{max} may be less than failure load, as determined by the new criterion in which the estimation will be incorporated.

New proposed method of interpretation

From the previous discussions, 4.5% of the pile diameter (or $0.045d$) seems to be a moderately conservative toe movement for the mobilization of toe resistance capacity of large-diameter bored piles founded on saprolites and weathered rocks. Combining this movement with the observations of shaft shortening, a new method for interpreting failure load is given as follows (in mm):

$$\Delta_M \approx 0.045d + \frac{1}{2}\frac{PL}{AE} \tag{1.7}$$

for piles founded in soils, and

$$\Delta_M \approx 0.045d + \frac{3}{4}\frac{PL}{AE} \tag{1.8}$$

for piles founded on rocks and/or with short rock sockets and for piles with slip liners.

In any case that the pile-head displacement under maximum test load is less than the value given by the proposed method, the maximum test load can be taken as the conservative failure load for design.

Sensitivity analysis of global Factor of Safety

For practising civil engineers, pile-head movements at the interpreted failure load scaled down by factors of safety are often of interest. Movement limits of piles will vary with structural requirements, the design methodology, and the choice of factor safety. These issues will not be addressed here. However, the effectiveness of some of the failure criteria introduced in this chapter in controlling pile-head movements can be examined by studying the pile-head movements measured at the interpreted failure load and at the interpreted failure load divided by a factor of safety.

Figure 1.7 shows the relationship between pile-head movement, Δ_{Awork} (assumed working load movement), as measured during loading testing at

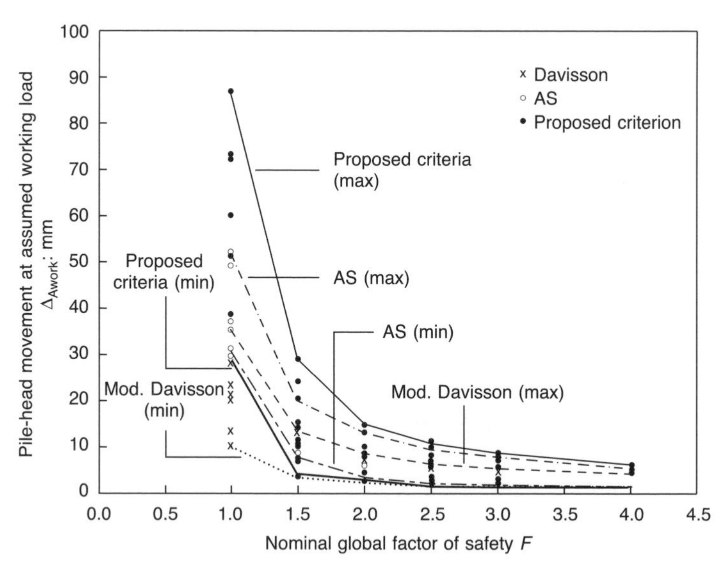

Fig. 1.7 Pile movement at assumed working loads (after Ng et al., *2001c)*

a load equal to the interpreted failure load divided by a nominal global factor of safety, F, defined as 'working load'. Values of F from 1.0 to 4.0 were applied to the interpreted failure loads for Piles 1–7, calculated using the new proposed criterion, the modified Davisson criterion by BD, and the AS criteria, which is dominated by its residual movement criterion. The 'max' and 'min' in Fig. 1.7 refer to the maximum and minimum measured movements for each criterion or set of criteria, respectively. At F equal to 1.0 (i.e. ultimate load) the scatter in Δ_{Awork} is considerable, but the scatter decreases as F increases. The modified Davisson criterion produces the lowest values of Δ_{Awork} of the three criteria, with the maximum line for this criterion on the figure being less than 12 mm for F equal to 1.5 and less than 8 mm for F equal to 2.0. For F equal to 2.0, the AS criteria and the new method provide similar results, with the maximum movement being around 15 mm.

Unlike the new method, the AS criteria are not dependent on pile diameter; hence, at greater pile sizes, the AS criteria and the new proposed criteria would be expected to produce a greater variation in interpreted results. The AS criteria should tend to limit the factored movements consistently to around the same values as those shown in Fig. 1.7, which could be considered to be an advantage by structural engineers. However, as mentioned in previous sections, the degree of mobilization at failure load for different pile diameters is likely to vary because the criteria are independent of diameter. Provided that structural tolerable movements are not exceeded, the new method will result in a more effective design. If structural requirements are demanding, for greater pile sizes the safety factor may need to be increased to a value greater than 2.0.

Summary

A new failure load criterion for large-diameter bored piles and barrettes is that the pile head displacement Δ_M at failure load in weathered geomaterials may be given as follows (in mm):

$$\Delta_M \approx 0.045d + \frac{1}{2}\frac{PL}{AE} \qquad (1.7)$$

for piles founded in soils, and

$$\Delta_M \approx 0.045d + \frac{3}{4}\frac{PL}{AE} \qquad (1.8)$$

for piles founded on rocks and/or with short rock sockets and for piles with slip liners.

CHAPTER FIVE

Analysis, design and parameters

Categories of analysis/design procedures

As pointed out by Poulos (1989), analysis and design procedures can be divided into three broad categories, depending on the level of sophistication and rigour. An extended classification system of these procedures has been proposed by Poulos and Hull (1989) and is shown in Table 1.7. Category 1 procedures probably account for most pile design done throughout the world. Category 2 procedures have a proper theoretical basis, albeit simplified, and are being increasingly used for pile deflection calculations.

Such procedures involve the use of simple computational methods or design charts, and generally do not demand the use of a computer. Category 3 procedures involve the use of a site-specific analysis based on relatively advanced analytical or numerical techniques such as the finite-element method or the boundary-element method. In most cases, such procedures require the use of a computer. Category 3 procedures are frequently used to carry out the necessary parametric solutions and develop the design charts which can then be used as category 2 solutions.

Typical examples of the various categories of procedures for axially loaded piles are shown in Table 1.8. In choosing an appropriate category of design for a practical problem, the following factors need to be considered:

- the significance and scale of the problem
- the available budget for foundation design
- the geotechnical data available
- the complexity of both the geotechnical profile and the design loading conditions
- the stage of the design process (i.e. whether a feasibility, preliminary or final design is being carried out)
- the experience of the designer with the methods being considered.

Estimation of geotechnical parameters

If soils can be idealised as elastic materials, the most significant parameters required for many of the categories 2A, 2B, 3A and 3B analyses

Table 1.7 Categories of analysis/design procedures (after Poulos, 1989)

Category	Subdivision	Characteristics	Method of parameter determination
1	–	Empirical – not based on soil mechanics principles	Simple in situ or laboratory tests, with correlations
2	2A	Based on simplified theory or charts – uses soil mechanics principles – amenable to hand calculation. Theory is linear elastic (deformation) or rigid plastic (stability)	Routine relevant in situ tests – may require some correlations
	2B	As for 2A, but theory is non-linear (deformation) or elasto-plastic (stability)	
3	3A	Based on theory using site-specific analysis, uses soil mechanics principles. Theory is linear elastic (deformation) or rigid plastic (stability)	Careful laboratory and/or in situ tests which follow the appropriate stress paths
	3B	As for 3A, but non-linearity is allowed for in a relatively simple manner	
	3C	As for 3A, but non-linearity is allowed for by way of proper constitutive models of soil behaviour	

of pile behaviour under static loading are as follows:

- the side resistance f_s
- the end-bearing resistance f_b
- Young's modulus of the soil E_s
- Poisson's ratio of the soil ν_s.

For calculation of axial pile load capacity, f_s and f_b must be estimated as accurately as possible. For the calculation of settlement resulting from direct axial loading, the theoretical solutions reveal that the choice of an appropriate value of E_s is generally crucial, unless the piles are long and compressible. For piles in soil subjected to external movement, the pile behaviour is generally much less dependent on E_s and, provided that the soil movement is known,[3] an approximate estimate of E_s may be

[3] The soil movement is treated here as an independent variable, although it will be influenced to some extent by the soil modulus.

Table 1.8 Examples of categorization of methods for evaluation of axial pile response (after Poulos, 1989)

Category	Axial pile capacity	Settlement
1	Correlations with CPT (e.g. Schmertmann, 1975; De Ruiter and Beringen, 1979). Correlations with SPT (Meyerhof, 1956; Thorburn and McVicar, 1971) Total stress (α) method (Tomlinson, 1957; Semple and Rigden, 1984)	Approximate correlations with pile diameter (Meyerhof, 1959; Frank, 1985) Column deflection multiplied by a factor (Focht, 1967)
2A	Effective stress (β) method (Burland, 1973; Meyerhof, 1976; Stas and Kulhawy, 1984)	Elastic solutions (Randolph and Wroth, 1978a; Poulos and Davis, 1980)
2B	Effective stress method (Fleming *et al.*, 1985)	Elastic solutions modified for slip (Poulos and Davis, 1980)
3A	Plasticity solutions for end-bearing capacity (Meyerhof, 1963; Giroud *et al.*, 1973)	Elastic finite-element analysis (e.g. Valliappan *et al.*, 1974)
3B	Non-linear load transfer analysis (e.g. Coyle and Reese, 1966; Kraft *et al.*, 1981) Non-linear boundary-element analysis (e.g. Poulos and Davis, 1980) Non-linear finite-element analysis (e.g. Desai, 1974; Jardine *et al.*, 1986)	
3C	Finite-element analysis, including simulation of pile installation (e.g. Randolph *et al.*, 1979; Withiam and Kulhawy, 1979; Nystrom, 1984)	

adequate, although reasonable estimates of shaft and end-bearing resistance are desirable.

Methods for determining parameters

For evaluating the parameters for static pile response, a number of methods can be contemplated, including

- laboratory testing
- appropriate interpretation of field pile load tests
- empirical correlations with laboratory determined parameters
- empirical correlations with the results of in situ test data.

Conventional laboratory tests, such as triaxial or oedometer tests, are generally not suitable for direct measurement of the soil's Young's modulus as they do not follow, even approximately, the stress path which the soil adjacent to the pile follows. Laboratory model pile tests may overcome this deficiency to some extent, but may not accurately reflect the behaviour of prototype piles because of the presence of scale effects, particularly for piles in sand. Some potential exists for more sophisticated tests such as the constant normal stiffness (CNS) direct shear test (Johnston and Lam, 1984; Ooi and Carter, 1987), and this type of test has been used in the design of grouted piles in offshore carbonate sediments (Johnston *et al.*, 1988). However, the direct utilisation of laboratory tests for pile design is infrequent in practice, and still requires further research before it can be applied with confidence.

The most reliable means of determining f_s and E_s is by back-analysing from the results of pile load tests. Methods for interpreting the pile load test data have been detailed by Poulos and Davis (1980) and Stewart and Kulhawy (1981), among others. Such methods are particularly effective if the pile is instrumented so that details of the load transfer along the pile shaft are available; it is then possible to determine detailed distributions of soil modulus and limiting pile–soil resistance along the pile shaft (Ng *et al.*, 2001b).

Side resistance f_s

Tables 1.9 and 1.10 summarize available methods for determining side resistance, f_s, from laboratory strength data, for both driven and bored piles. Effective stress approaches can be used for all soil types, whereas a total stress approach is still adopted commonly for piles in clay. The parameters α and β (or K and δ) are usually obtained from empirical correlations, despite the fact that the effective stress β approach is fundamentally sound and falls into category 2.

A summary of some suggested correlations between f_s and the standard penetration resistance N are given in Table 1.11. Considerable variations occur in these correlations, particularly for bored and cast-in-place piles.

Figures 1.8 and 1.9 show values of f_s correlated with the static cone resistance, q_c. These relationships have been developed by Poulos (1989) from the correlations suggested by Bustamante and Gianeselli (1982) and cover a wide range of pile types in both clay and silica sand. The classification of these pile types is shown in Table 1.12. It should be emphasized that several other correlations have been proposed and that wide variations exist between some of these. Figure 1.10 shows an example of this variability, for driven piles in silica sand. The potential inaccuracy of shaft capacity prediction using category 1 correlations, especially for loose sands, is clearly demonstrated.

Schmertmann (1975; 1978) proposes a different approach to the utilization of cone data, whereby the pile side resistance is related to the measured sleeve resistance of the penetrometer. Corrections are applied,

Table 1.9 Shaft resistance f_s for driven piles, determined from laboratory strength data (after Poulos, 1989)

Soil type	Equation	Remarks	Reference
Clay	$f_s = \alpha c_u$	$\alpha = 1.0 (c_u \leq 25\,\text{kPa})$ $\alpha = 0.5 (c_u \geq 70\,\text{kPa})$ Linear variation in between	API (1984)
		$\alpha = 1.0 (c_u \leq 35\,\text{kPa})$ $\alpha = 0.5 (c_u \geq 80\,\text{kPa})$ Linear variation in between. Length factor applies for $L/d > 50$	Semple and Rigden (1984)
		$\alpha = \left(\frac{c_u}{\sigma'_v}\right)^{0.5}_{nc} \left(\frac{c_u}{\sigma'_v}\right)^{-0.5}$ for $\left(\frac{c_u}{\sigma'_{vo}} \leq 1\right)$ $\alpha = \left(\frac{c_u}{\sigma'_v}\right)^{0.5}_{nc} \left(\frac{c_u}{\sigma'_v}\right)^{-0.25}$ for $\left(\frac{c_u}{\sigma'_{vo}} \geq 1\right)$	Fleming *et al.* (1985)
	$f_s = \beta \sigma'_v$	$\beta = (1 - \sin\phi') \tan\phi' (\text{OCR})^{0.5}$	Burland (1973 Meyerhof (1976)
Silica sand	$f_s = \beta \sigma'_v$ $(f_s \not> f_{slim})$	$\beta = 0.15$–0.35 (compression) 0.10–0.24 (tension)	McClelland (1974)
		$\beta = 0.44$ for $\phi' = 28°$ 0.75 for $\phi' = 35°$ 1.2 for $\phi' = 37°$	Meyerhof (1976)
		$\beta = (K/K_0) \cdot K_0 \tan(\phi \cdot \delta/\phi)$ δ/ϕ depends on interface materials (range 0.5–1.0); K/K_0 depends on installation method (range 0.5–2.0). K_0 = coefficient of earth pressure at rest, and is a function of OCR	Stas and Kulhawy (1984)
Uncemented calcareous sand	$f_s = \beta \alpha'_v$	$\beta = 0.05$–0.1	Poulos (1988d)

depending on soil type, pile type, relative pile length and depth below the surface. Robertson *et al.* (1985) have found that the method proposed by Schmertmann provides a more reliable prediction than the direct correlation to q_c when applied to piles in a clayey silt.

Correlations such as those outlined must always be employed with caution, as a number of other factors may also influence side resistance, e.g. the presence of overlying layers (Tomlinson, 1977).

End-bearing resistance f_b

Table 1.13 summarizes the two methods usually used for assessment of the end-bearing resistance of piles using laboratory data. A total stress approach is almost invariably used for piles in clay, whereas an effective

Table 1.10 Shaft resistance f_s *for bored piles, determined from laboratory strength data (after Poulos, 1989)*

Soil type	Equation	Remarks	Reference
Clay	$f_s = \alpha c_u$	$\alpha = 0.45$ (London clay)	Skempton (1959)
		$\alpha = 0.7$ times value for driven displacement pile	Fleming *et al.* (1985)
	$f_s = K \tan \delta \sigma'_v$	K is lesser of K_0 or $0.5(1 + K_0)$	Fleming *et al.* (1985)
		$K/K_0 = 2/3$ to 1; K_0 is function of OCR; δ depends on interface materials	Stas and Kulhawy (1984)
Silica sand	$f_s = \beta \sigma'_v$	$\beta = 0.1$ for $\phi' = 33°$ 0.2 for $\phi' = 35°$ 0.35 for $\phi' = 37°$	Meyerhof (1976)
		$\beta = F \tan(\phi' - 5°)$ where $F = 0.7$ (compression) and 0.5 (tension)	Kraft and Lyons (1974)
Uncemented calcareous sand	$f_s = \beta \sigma'_v$ ($f_s \ngtr f_{slim}$)	$\beta = 0.5$ to 0.8 $f_{slim} = 60$ to 100 kPa	Poulos (1988d)

stress approach is used for piles in sand. Two main problems arise in the latter case:

- some experimental evidence suggests that a limiting value of f_b may occur when the pile is embedded more than 10 to 20 diameters; no entirely satisfactory method of theoretical analysis has been developed to take this into account and an empirical upper limit to f_b is usually specified
- the theoretical bearing capacity factor, N_q, is very sensitive to the angle of shearing resistance ϕ' for values of ϕ' in excess of about 35°, thus small changes in ϕ' can theoretically lead to large changes in N_q although the effects of soil compressibility are then more important and may reduce the dependence of N_q on ϕ'.

Table 1.14 shows some empirical correlations between f_b and the standard penetration resistance in the vicinity of the pile tip. These correlations indicate that bored or cast-in-place piles develop a significantly smaller end-bearing resistance than do driven piles.

Bustamante and Gianeselli (1982) suggested correlations between f_b and the average cone penetration resistance value near the pile tip. The correlation factor to the latter value is between 0.3 and 0.55, depending on soil and pile type. These correlations contrast with procedures such as those proposed by Belcotec (1985) and De Ruiter and Beringen (1979), in which a factor of unity is applied to the average value (computed

Table 1.11 Correlations between shaft resistance f_s *and SPT value, with* $f_s = \alpha + \beta N$ *kPa (after Poulos, 1989)*

Pile type	Soil type	α	β	Remarks	Reference
Driven displacement	Cohesionless	0	2.0	f_s = average value over shaft	Meyerhof (1956)
				N = average SPT along shaft Halve f_s for small-displacement pile	Shioi and Fukui (1982)
	Cohesionless and cohesive	10	3.3	Pile type not specified $50 \geq N \geq 3$ $f_s \not> 170$ kPa	Decourt (1982)
	Cohesive	0	10.0		Shioi and Fukui (1982)
Cast-in-place	Cohesionless	30	2.0	$f_s \not> 200$ kPa	Yamashita *et al.* (1987) Shioi and Fukui (1982)
		0	5.0		
	Cohesive	0	5.0	$f_s \not> 150$ kPa	Yamashita *et al.* (1987) Shioi and Fukui (1982)
		0	10.0		
Bored	Cohesionless	0	1.0		Findlay (1984) Shioi and Fukui (1982)
		0	3.3		Wright and Reese (1979)
	Cohesive	0	5.0		Shioi and Fukui (1982)
	Cohesive	10	3.3	Piles cast under bentonite $50 \not> N \not> 3$ $f_s \not> 170$ kPa	Decourt (1982)
	Chalk	−125	12.5	$30 > N > 15$ $f_s \not> 250$ kPa	After Fletcher and Mizon (1984)

differently than in the Bustamante and Gianeselli approach). However, the latter approaches are confined to driven piles, whereas the Bustamante and Gianeselli approach is more general, simpler to apply, and probably more conservative.

Young's modulus E_s of the soil

Ideally, for piles in clay, a distinction should be made between the undrained Young's modulus, used for calculations of immediate or undrained settlement, and the drained Young's modulus, used for calculations of total settlement of a pile. However, for many clays, the difference between the drained and undrained modulus values is not great and the approximate nature of most correlations makes such a distinction impractical. It is therefore suggested that the correlations presented here should be considered to apply to the drained Young's modulus. For all soil types,

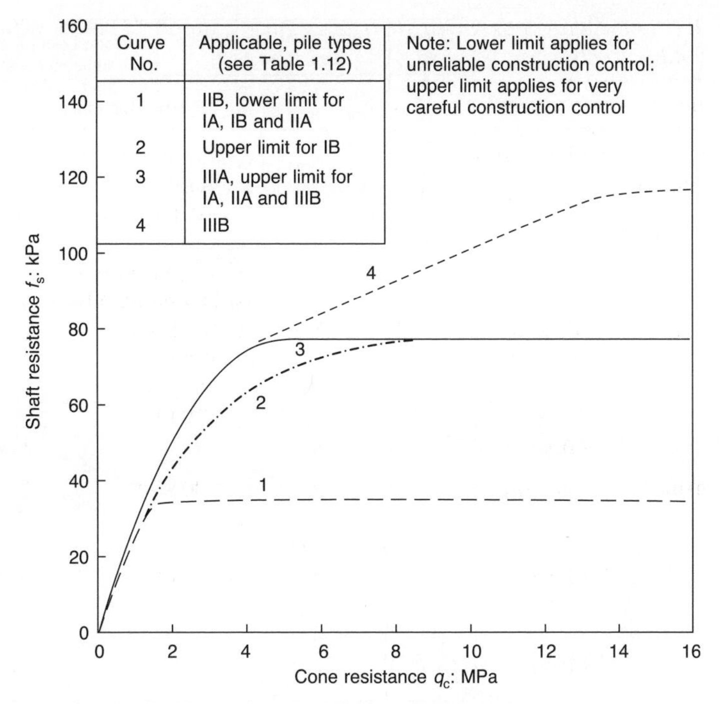

Fig. 1.8 Design values of shaft resistance for piles in clay (based on Bustamante and Gianeselli, 1982; after Poulos, 1989)

a further distinction needs to be made, between the tangent value of the Young's modulus (if a non-linear interface model is being used) and the secant value of the Young's modulus (if a purely linear analysis is being used). Again, it is often difficult to make such distinctions with rough empirical correlations and, unless specified, the Young's modulus referred to here will be a secant value, relevant for normal working load levels of between one-third to one-half of the ultimate load capacity.

For piles in clay, the Young's modulus has often been correlated with laboratory-measured undrained shear strength, c_u. Some of these correlations are shown in Fig. 1.11, and a feature of this figure is the wide spread of correlations. Possible reasons for this spread might include differences in the method of determining c_u, differences in the method of determining the modulus values, differences in the load level at which the modulus was determined, differences in the overconsolidation ratio of the clay between different tests, and differences between the clay types. Callanan and Kulhawy (1985) find that values of E_s/c_u generally range between 200 and 900, with an average value of about 500. These values apply to piles with a length-to-diameter ratio in excess of about 15. For shorter piles, the upper range of E_s/c_u may be greater because of the possible effects of fissuring, desiccation and overconsolidation of the clay near the surface.

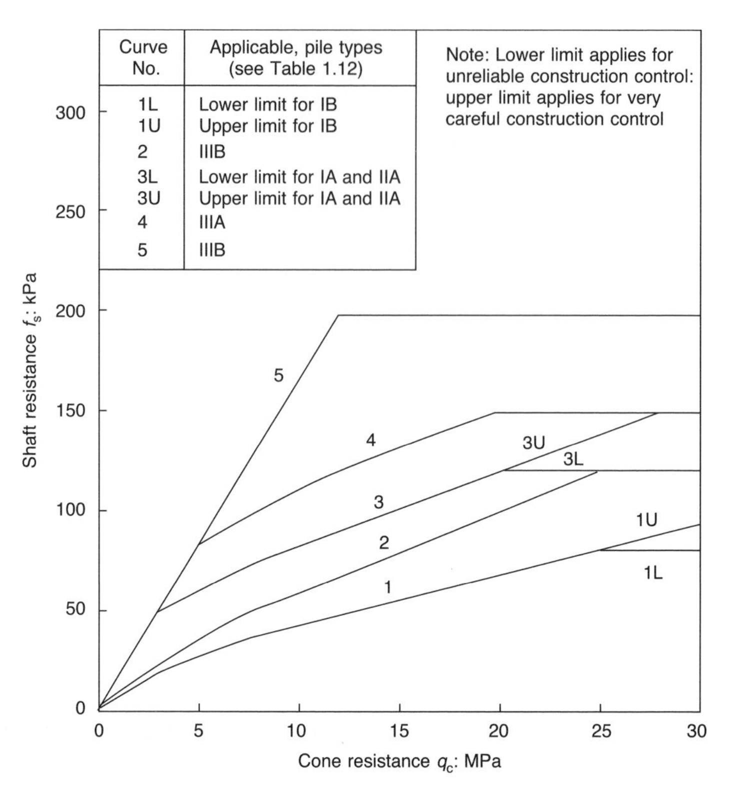

Fig. 1.9 Design values of shaft resistance for piles in sand (based on Bustamante and Gianeselli, 1982; after Poulos, 1989)

Table 1.12 Classification of pile types (Bustamante and Gianeselli, 1982; after Poulos, 1989)

Pile category	Type of pile
IA	Plain bored piles, mud bored piles, hollow auger bored piles, cast screwed piles Type I micropiles, piers, barrettes
IB	Cased bored piles Driven cast piles
IIA	Driven precast piles Prestressed tubular piles Jacked concrete piles
IIB	Driven steel piles Jacked steel piles
IIIA	Driven grouted piles Driven rammed piles
IIIB	High-pressure grouted piles ($d > 0.25$ m) Type II micropiles

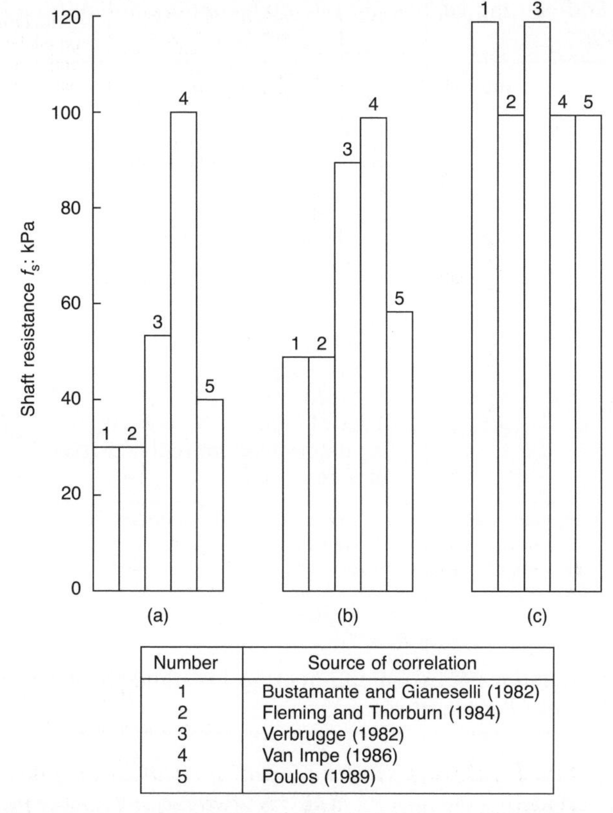

Number	Source of correlation
1	Bustamante and Gianeselli (1982)
2	Fleming and Thorburn (1984)
3	Verbrugge (1982)
4	Van Impe (1986)
5	Poulos (1989)

Fig. 1.10 Example of variations between correlations for shaft resistance against CPT-driven piles in sand: (a) $q_c = 6\,MPa$*; (b)* $q_c = 10\,MPa$*; (c)* $q_c = 20\,MPa$ *(after Poulos, 1989)*

Some correlations between the Young's modulus and the standard penetration test number are summarized in Fig. 1.12, and show alarming variability. At least some of this variability may be attributed to differences in the determination or definition of the SPT value, but it is clear that the potential for selection of inappropriate values of E_s is great.

Table 1.15 shows some suggested correlations between E_s and the cone penetration resistance, q_c, and, as with most of the other correlations, the range is large. Two correlations for initial tangent modulus, E_{st}, are shown, both being derived from dynamic triaxial tests but believed to be relevant for piles.

Poisson's ratio ν_s

The Poisson's ratio of the soil is a necessary input parameter into analyses that involve elastic continuum theory, but its effect is generally quite minor when the solutions are expressed in terms of the Young's modulus

Table 1.13 End-bearing capacity of pile tip f_b: determination from laboratory data (after Poulos, 1989)

Soil type	Equation	Remarks	Reference
Clay	$f_b = N_c c_{ub}$	$N_c = 9$ for $L/d \geq 3$ c_{ub} = value of c_u in vicinity of pile tip	Skempton (1959)
Silica sand*	$f_b = N_q \sigma'_v \not> f_{blim}$ †	$N_q = 40$	API (1984)
		N_q plotted against ϕ'	Berezantzev *et al.* (1961)
		N_q related to ϕ', relative density and mean effective stress	Fleming *et al.* (1985)
		N_q from cavity expansion theory, as a function of ϕ' and volume compressibility	Vesic (1972)
Uncemented calcareous sand	$f_b = N_q \sigma'_v \not> f_{blim}$	$N_q = 20$	Datta *et al.* (1980)
		Typical range of N_q = 8–20	Poulos (1988d)
		N_q determined for reduced value of ϕ' (e.g. 18°)	Dutt and Ingram (1984)

* For silica and calcareous sands, the above expressions apply for driven piles only.

† Typical limiting values f_{blim} range from 10–15 MPa for silica sand, and 3–5 MPa for calcareous sand; the latter value depends on soil compressibility (Nauroy *et al.*, 1986).

Table 1.14 Correlations between end-bearing resistance f_b and SPT value, with f_b = KN MPa (after Poulos, 1989)

Pile type	Soil type	K	Remarks	Reference
Driven displacement	Sand	0.45	N = average SPT value in local failure zone	Martin *et al.* (1987)
	Sand	0.40		Decourt (1982)
	Silt, sandy silt	0.35		Martin *et al.* (1987)
	Glacial coarse to fine silt deposits	0.25		Thorburn and MacVicar (1971)
	Residual sandy silts	0.25		Decourt (1982)
	Residual clayey silts	0.20		Decourt (1982)
	Clay	0.20		Martin *et al.* (1987)
	Clay	0.12		Decourt (1982)
	All soils	0.30	For $L/d \geq 5$ If $L/d < 5$; $K = 0.1 + 0.04L/d$ (closed-end piles) or $K = 0.06L/d$ (open-ended piles)	Shioi and Fukui (1982)
Cast-in-place	Cohesionless		$f_b = 3.0$ MPa	Shioi and Fukui (1982)
		0.15	$f_b \not> 7.5$ MPa	Yamashita *et al.* (1987)
	Cohesive	–	$f_b = 0.09(1 + 0.16z)$ where z = tip depth (m)	Yamashita *et al.* (1987)
Bored	Sand	0.1		Shioi and Fukui (1982)
	Clay	0.15		Shioi and Fukui (1982)
	Chalk	0.25	$N < 30$	Hobbs (1977)
		0.20	$N > 40$	

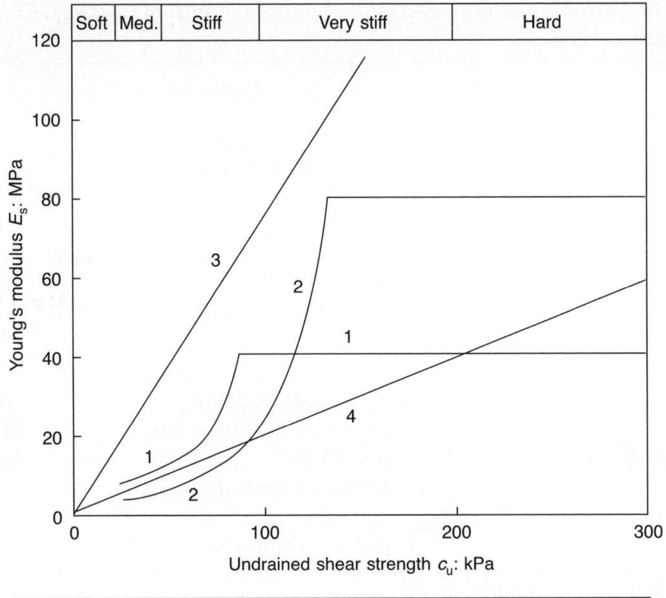

Legend	Remarks	Reference
1	Driven piles	Poulos (1968)
2	Bored piles	Poulos (1968)
3	Driven piles (E_s = 750 c_u)	Aschenbrenner and Olsen (1984)
4	Bored piles, lower bound (E_s = 200 c_u)	Callanan and Kulhawy (1985)

Fig. 1.11 Correlations for soil modulus for piles in clay (after Callanan and Kulhawy, 1985; Poulos, 1989)

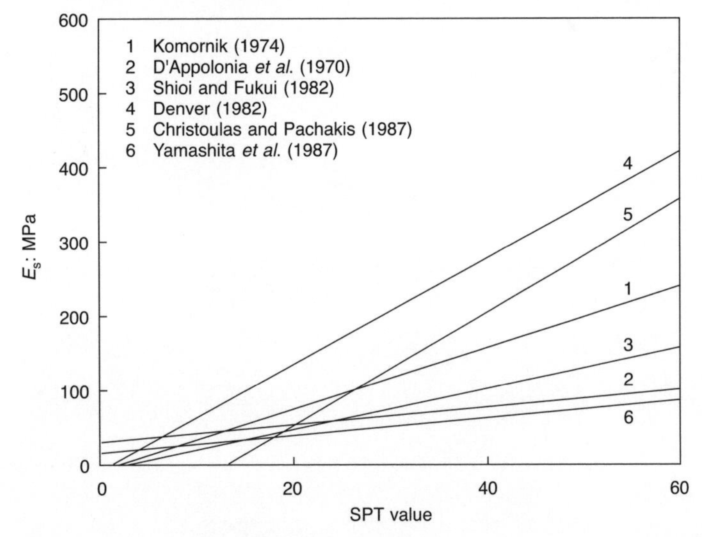

Fig. 1.12 Comparison between correlations for soil modulus and SPT value for driven piles in sand (after Poulos, 1989)

Table 1.15 Correlations between soil's Young's modulus E_s *and CPT value for driven piles (after Poulos, 1989)*

Soil type	Correlation	Remarks	Reference
Clay and silts	$E_s{}^* = 21.0q_c^{1.09}$	Various pile types E_s and q_c in MPa	Christoulas (1988)
	$E_s = 15q_c$		Poulos (1988c)
Silica sands	$E_s = \alpha q_c$	$\alpha = 20$–40	Milovic and Stevanovic (1982)
		$\alpha = 5$ (normally consolidated sands) $\alpha = 7.5$ (over-consolidated sands)	Poulos (1988c)
	$E_{st}{}^{\dagger} = 53q_c^{0.61}$	E_{st} and q_c in MPa Dynamic modulus value	Imai and Tonouchi (1982)
Unspecified	$E_{st} = \alpha q_c$	$\alpha = 24$–30 Dynamic modulus value	Holeyman (1985)
	$E_s = 10.8 + 6.6q_c$	E_s and q_c in MPa (for $q_c > 0.4$ MPa)	Verbrugge (1982)

* E_s = secant Young's modulus.
† E_{st} = initial tangent Young's modulus.

of the soil. For saturated clays under undrained conditions, ν_s can be taken as 0.5. For clays under drained conditions, ν_s generally lies within the range 0.35 ± 0.05, whereas for silica sands, ν_s is usually within the range 0.3 ± 0.1. Lower values, within the range 0.15 ± 0.1, are applicable for many marine calcareous sediments.

CHAPTER SIX

Dynamic formulae

Overview

As pointed out by Poulos and Davis (1980), perhaps the oldest and most frequently used method of estimating the load capacity of driven piles is to use a driving formula, or dynamic formula. All such formulae relate ultimate load capacity to pile set (the vertical movement per blow of the driving hammer) and assume that the driving resistance is equal to the load capacity of the pile under static loading. They are based on an idealised representation of the action of the hammer on the pile in the last stage of its embedment. There are a great number of driving formulae available, of varying degrees of reliability.

Smith (1960) states that the editors of the *Engineering News Record* have on file 450 such formulae. The derivation of some of these formulae is discussed by Whitaker (1970), while details of some of the parameters required are available in Chellis (1961).

The primary objectives in using a pile-driving formula are usually either to establish a safe working load for a pile by using the driving record of the pile, or to determine the driving requirements for a required working load. The working load is usually determined by applying a suitable safety factor to the ultimate load calculated by the formula. This safety factor, however, varies considerably depending on the formula used and the type of pile being driven. Also, because pile driving formulae take no account of the nature of the soil, the appropriate safety factor may vary from one site to another.

An improvement in the estimation of load capacity by dynamic methods has resulted from the use of the wave equation to examine the transmission of compression waves down the pile, rather than assuming that a force is generated instantly throughout the pile, as is done in deriving driving formulae. The main objective in using the wave-equation approach is to obtain a better relationship between ultimate pile-load and pile-set than can be obtained from a simple driving formula. As well as providing a means of load capacity estimation, this relationship allows an assessment to be made of the driveability of a pile with a particular set of equipment. Moreover, this approach also enables a rational analysis to be made of the stresses in the pile during driving and can therefore be useful in the structural design of the pile.

Pile-driving formulae

Derivation of general formulae

Pile-driving formulae attempt to relate the dynamic to the static resistance of a pile. They have been established on an empirical or a theoretical basis. Several of the latter are based on Newton's law of impact, modified in some cases for energy losses during impact and stress propagation.

The assumed relationship between pile resistance and downward movement of the pile is shown in Fig. 1.13. The materials of the pile and the driving cushion are assumed to be perfectly elastic, and inertia forces in the soil and energy losses stemming from irreversible deformations (except of the soil) are disregarded.

Dynamic equations

Dynamic equations can be categorized into three groups: theoretical equations, empirical equations and those which consist of a combination of the two. The theoretical equations are formulated around analyses which evaluate the total resistance of the pile, based on the work done by the pile during penetration. These formulations assume elasto-plastic force–displacement relations, as shown in Fig. 1.13. The total work is therefore:

$$W = R_u\left(S + \frac{Q}{2}\right) \tag{1.9}$$

where R_u is the yield resistance, Q is the quake denoting the combined elastic deformation of the pile and the soil, and S is the set, denoting the

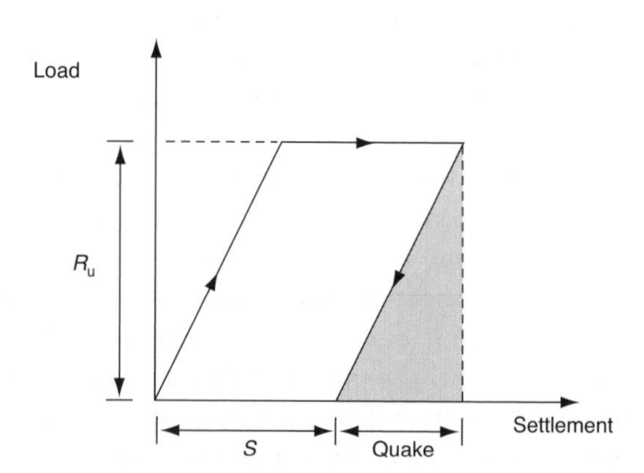

Fig. 1.13 Assumed load–settlement curve for pile (after Poulos and Davis, 1980)

plastic deformation (permanent displacement – also called the 'final set value') under each blow.

When the work of the resisting forces W is equated to the energy delivered to the pile, that is, $W = E_{after}$ (where E_{after} is the energy transferred to the pile–soil system after impact), then:

$$R_u = \frac{E_{after}}{S + \frac{Q}{2}} \qquad (1.10)$$

The low reliability of dynamic equations has been reported (Housel, 1965). There are several reasons for this:

- the parameters, such as the efficiency of energy transfer and the pile/soil quake, are assumed, and therefore may not reflect the high variability of the field conditions
- the theoretical analysis of the 'rational' pile formula (see for example Bowles, 1988) relates the energy transfer mechanism to a Newtonian analysis of ram–pile impact. This formulation is theoretically invalid for representing the 'elastic' wave propagation mechanism which actually takes place, and
- no differentiation is made between static and dynamic soil resistances.

A clear distinction is therefore required between the underlying valid energy analysis and additional estimations of the different parameters, many of which are either invalid theoretically or practically limited in their accuracy.

If the anvil–pile cushion is included as part of the pile–soil system, energy will also be stored during elastic compression of the pile cushion. The elastic compression should therefore comprise three components, namely, $Q = C_c + C_p + C_q$ where C_c, C_p and C_q are the temporary compression of pile cushion, pile and soil respectively.

If the anvil–pile cushion is not considered as part of the pile–soil system, the energy transferred to the pile–soil system may be simplified to energy stored in the pile and soil only. In this case, $Q = C_p + C_q$ and E_{after} may be approximated by the total energy transferred to the pile head, E_d, where E_d can be measured directly using a PDA (Pile Driving Analyser). E_d can also be expressed in terms of the energy transfer ratio X, defined as $E_d = XWh$, where W is the hammer weight and h is the drop height. The values of E_d and X are normally standard output in a PDA test report.

In some pile-driving formulae, E_d is related to the kinetic energy of the hammer before impact E'. It is common to express E' in terms of the hammer efficiency α defined as $E' = \alpha Wh$. The hammer efficiency measures the energy loss during descent of the hammer. Some hydraulic hammer systems impart energy to the hammer during descent of the hammer and the input energy of the hammer for pile driving is actually higher than Wh. This may give a hammer efficiency higher than 1.0 even if there is energy loss before impact. As there is further energy loss after impact, E_d can be expressed in terms of the efficiency of hammer blow η. This gives $E_d = \eta E' = \eta\alpha Wh$.

Hiley formula

By considering the physics of impact of two spheres, Hiley (1922) derived the following expression to calculate η:

$$\eta = \frac{W + (W_p + W_r)e^2}{W + (W_p + W_r)} = e^2 + \frac{W(1 - e^2)}{W + (W_p + W_r)} \tag{1.11}$$

where W_p is the weight of pile, W_r is the weight of anvil and e is the coefficient of restitution.

The Hiley formula considers the anvil as part of the pile–soil system. By substituting $E_d = \eta\alpha Wh$ and $Q = C_c + C_p + C_q$ into Equation (1.10), then

$$R_u = \frac{\eta\alpha Wh}{S + \frac{1}{2}(C_c + C_p + C_q)} \tag{1.12}$$

Major deficiencies in using the Hiley formula for long piles have been pointed out and explained by Triantafyllidis (2001) and hence it should not be used blindly.

Reliability of dynamic formulae

Poulos and Davis (1980) point out that several investigations have been made to determine the reliability of the various pile-driving formulae by comparing the load capacity computed from the appropriate formula with the measured capacity from a pile loading test. Some of the most comprehensive investigations have been reported by Sorensen and Hansen (1957), Agerschou (1962), Flaate (1964), Housel (1966) and Olsen and Flaate (1967).

Sorensen and Hansen used data from 78 load tests on concrete, steel and wooden piles, most of these having their points bearing on sand (a few were founded on hard moraine clay). The results of their comparisons are shown in Fig. 1.14, in which the ratio, μ, of the measured to the computed load-capacity is plotted against the percentage of load tests smaller than μ. This plot is a probability plot, and a straight line on this plot represents a normal or Gaussian distribution of results. Figure 1.14 shows that all the formulae considered, with the exception of the Eytelwein formula, follow approximately a Gaussian distribution. There is very little difference in the accuracy of the Danish, Hiley and Janbu formulae, and the theoretical curves derived from the wave equation, but the Eytelwein formula is very inaccurate.

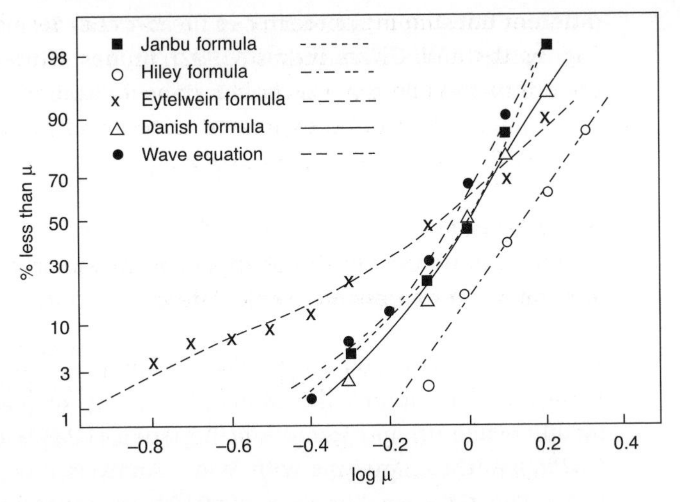

Fig. 1.14 Statistical distribution of load test results (after Sorensen and Hansen, 1957; Poulos and Davis, 1980)

Agerschou's investigation concentrated on the *Engineering News* formula but also broadly confirmed the conclusions of Sorensen and Hansen regarding the Danish, Hiley and Janbu formulae. Agerschou showed that despite its popularity, the *Engineering News* formula is unreliable. It has the highest standard deviation, and 96% of the allowable loads determined by this formula will have actual safety factors ranging between 1.1 and 30. Flaate (1964) investigated the accuracy of the Janbu, Hiley, and *Engineering News* formulae for 116 tests on timber, concrete and steel piles in sand. The conclusions reached by Agerschou regarding the unreliability of the *Engineering News* formula are reinforced by Flaate's results. There is relatively little difference between the Janbu and Hiley formulae, although the former is perhaps the more reliable overall and gives good results for timber and concrete piles. Hiley's formula is also reasonable for timber piles.

Modified pile driving formula for long piles

For relatively long piles the η factor in the Hiley formula suggests very low energy transfer coefficients from the hammer onto the pile, because the whole mass of the pile is taken into account (Triantafyllidis, 2001). If the pile is relatively long it is worth considering only that portion of the pile that is affected during the duration of the impact. Therefore a more realistic approach is to consider the impact between two rods instead of the impact between two spheres.

Alternatively, due to the availability of field equipment for measuring the forces and accelerations in a pile during driving by using a PDA, a different but simple approach can be adopted. Broms and Lim (1988) and Paikowsky and Chernauskas (1992) proposed an energy equation to determine the pile-driving resistance as follows:

$$R_u = \frac{E_d}{S + \frac{1}{2}(C_p + C_q)} \tag{1.13}$$

where E_d can be directly measured by a PDA.

In this modified formula the anvil–pile cushion is not considered as part of the pile–soil system and hence C_c is not included in the equation.

Accuracy of the modified pile-driving formula

Figure 1.15 compares the ultimate capacity of piles predicted by the modified pile-driving formula of Equation (1.13) with that predicted by CAPWAP (Case Analysis with Wave Analysis Program) or pile loading tests. The data smaller than 3000 kN are taken from Paikowsky and Chernauskas (1992) for which the failure loads were obtained from pile loading tests. The other data points are obtained from Li *et al.* (2003). A total of seven test piles of Grade 55C $305 \times 305 \times 224$ kg/m H-piles constructed in completely decomposed granite (CDG) were loaded to failure or 3.3 times the design pile capacity. The length of the piles ranged from 35 to 46 m. Each test pile was initially driven to final set by hydraulic hammer followed by drop hammer. As shown in Fig. 1.15 it can be seen that the modified formula predicts the test loads for long piles reasonably well. However, further verifications of this modified formula are still needed.

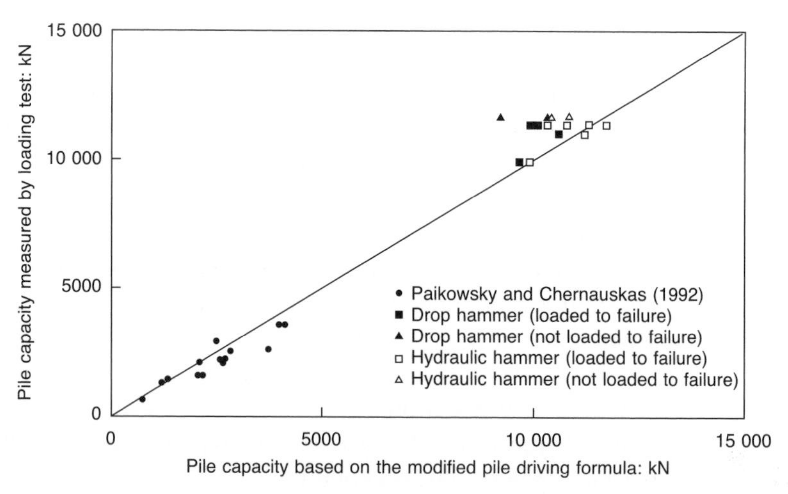

Fig. 1.15 Pile capacity based on the modified pile-driving formula correlated with measurements from loading tests

The Wave Equation

Stress propagation

Stress propagation in a pile during driving can be described by the following equation of motion:

$$E_p \frac{\partial^2 u}{\partial x^2} - \frac{S_p}{A_p} f_s = \rho_p \frac{\partial^2 u}{\partial t^2} \tag{1.14}$$

where $u(x, t)$ = longitudinal displacement of an infinitesimal segment; A_p, S_p = pile area and circumference, respectively; E_p, ρ_p = modulus of elasticity and unit density of the pile material, respectively.

The friction stresses f_s are activated by the pile movement and under free wave motion ($f_s = 0$) Equation (1.14) becomes the familiar one-dimensional wave equation.

Equation (1.14) may be solved for the appropriate initial and boundary conditions to determine the relationship among displacement, time and position in the pile, from which the stress variation in the pile may be determined. Because of the complications involved in practical piling problems, analytical solutions to Equation (1.14) generally are not feasible, and resort must be made to numerical means of solution. A convenient numerical method has been described by Smith (1960).

Smith (1960) proposed an idealization of the hammer–pile–soil system which would be capable of representing the passage of the stress wave down the pile (Fig. 1.16). The pile is modelled as a series of rigid masses connected by springs which can act in both compression and tension. The hammer ram and the anvil or pile cap are also modelled as rigid masses, but interfaces between ram and anvil and between anvil and top of pile are idealized by springs capable of sustaining compression but not tension. Side resistance and end-bearing actions of the soil into which the pile is being driven are represented by bi-linear springs acting on the embedded rigid masses. Side resistance springs can exert forces either upward or downward while the end-bearing spring can act only in compression. Soil springs act linearly up to a limiting displacement termed the 'quake'. In addition, increased resistance of the soil due to viscous damping is represented by a series of dashpots. Smith expresses the instantaneous soil resistance force, R, acting on an adjacent rigid mass as

$$R = R_s(1 + JV) \tag{1.15}$$

where R_s = static soil resistance, J = a damping constant, V = instantaneous velocity of the adjacent mass.

As can be seen in Fig. 1.16, the basic Smith idealization is representative of a pile driven by a drop hammer or a single acting hammer. Diesel

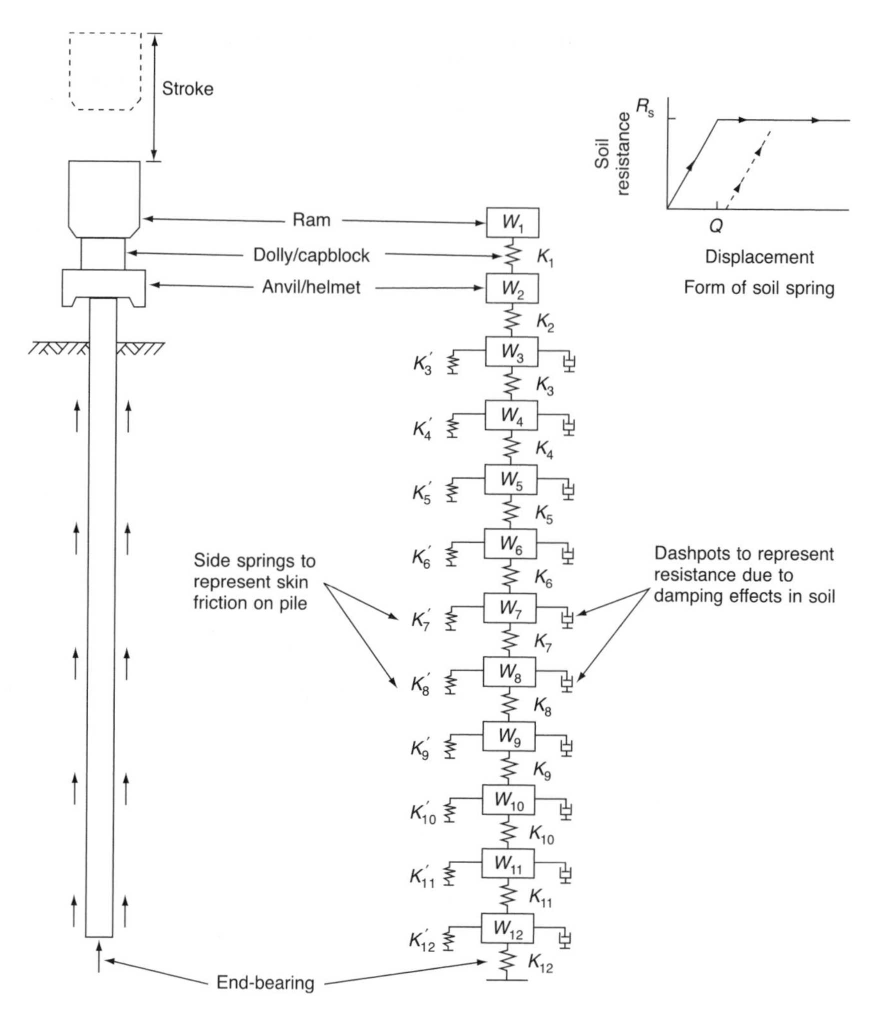

Fig. 1.16 Smith (1960) idealization for Wave Equation analysis of pile during driving (after Tomlinson, 2001)

hammers have to be considered in a rather different manner since the actual energy generated by such a hammer will vary with the resistance of the pile that is being driven. For low resistance there will be low energies per blow at a high number of blows per minute, while for high resistances the energy per blow will increase and the number of blows per minute decreases. Manufacturers can provide charts of energy versus rate of striking for diesel hammers, and when drivability predictions are being made the range of energies at which a particular hammer may operate should be considered. An example of this is shown in Fig. 1.17 in which back-analyses of static load tests have been used to generate blow count versus driving resistance curves for closed-ended tubular

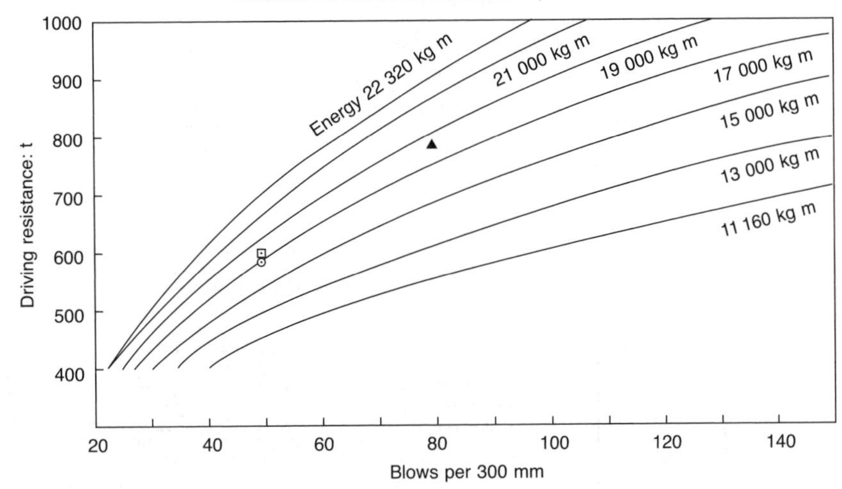

Fig. 1.17 Pile drivability analysis for diesel hammer (after Tomlinson, 2001)

steel piles being driven into calcareous sands and corals. In these ground conditions time-dependent effects are negligible, and so the driving resistance is considered to be the same as the static resistance. Although the original Smith idealization is incapable of modelling the interaction between driving resistance and energy generated for a diesel hammer, advances have been made. Goble and Rausche (1976) published details of a computer program which will model diesel hammer behaviour realistically; this program proceeds by iterations until compatibility is obtained between the pile–soil system and the energy–blows per minute performance of the hammer.

Data required to undertake a pile drivability analysis include values for the stiffness and coefficient of restitution of the dolly/capblock and the quakes Q and damping constants J for the soil. Typical values of Q and J are given in Table 1.16. These may be used in pre-driving studies, but it must be remembered that they can be subject to significant variations and whenever possible they should be back-figured to fit load test results. Since the late 1970s the exact form of damping of soil for Wave Equation analyses has been the subject of much investigation and Heerema (1979) and Litkouhi and Poskitt (1980) have published findings indicating that for end-bearing in clays and sands and for side resistance in clays, damping is dependent on the fifth root of velocity, that is, Equation

Table 1.16 Soil property values typically used in Wave Equation analyses (after Tomlinson, 2001)

Soil type	Damping constant		Quake side and end Q (mm)
	Skin friction J_s (s/m)	End-bearing J_p (s/m)	
Clay	0.65	0.01–1.0	2.5
Sand	0.15	0.33–0.65	2.5
Silt	0.33–0.5	0.33–1.5	2.5

(1.15) is modified to

$$R = R_s(1 + JV^{0.2}) \tag{1.16}$$

Capacity evaluation in the field using Wave Equation

The procedure of monitoring pile driving by dynamic measurements is well established. Early large-scale studies (e.g. Michigan State Highway Commission, 1965; Texas Highway Department, 1973; and Ohio Department of Transportation, 1975; and Goble *et al.*, 1970) led to the development of commercial systems which enable complete and relatively easy acquisition of dynamic measurements and analysis during driving. The PDA (Pile Driving Analyser, see Pile Dynamics Inc., 1990) which is the most commonly used device in the USA, utilizes a simplified pile-capacity evaluation method, known as the Case method (see Goble *et al.*, 1970). The formulation of the method is based on a simplification of the Wave Equation and employs the force and velocity measurements taken at the pile top in order to obtain the total resistance. The static resistance is then evaluated based on a dimensionless damping coefficient J_c (Case damping) which was correlated to the soil type at the pile tip (Goble *et al.*, 1967, 1975; Zhang *et al.*, 2001a). The method encountered two fundamental difficulties:

- the total resistance is time dependent and different variations of the method produce different results
- the dimensionless damping coefficient was found to have questionable correlation to soil type (e.g. Paikowsky, 1982; Thompson and Goble, 1988) and therefore had to be calibrated for the specific pile, soil and site conditions.

Reliability of dynamic methods

McVay *et al.* (2000) evaluated the parameters for load and resistance factor design (LRFD) of driven piles using dynamic methods. They considered a database of 218 pile cases in Florida. Eight dynamic methods were studied: ENR, modified ENR, FDOT and Gates driving formulae, and the CAPWAP, Case method for PDA, Paikowsky's energy method and Sakai's energy method. It was demonstrated that the modern methods based on wave mechanics, such as CAPWAP, PDA and Paikowsky's energy method, are roughly twice as cost-effective in reaching the target reliability indices of 2.0 to 2.5 (failure probability $p_f = 0.62$ to 2.5%) as the ENR and modified ENR driving formulae. They concluded that the past designs with driving formulae reveal a large redundancy of pile groups against failure and correspondingly underestimate pile group effects. They therefore suggested that use of a relatively low reliability target index, $\beta_T = 2.0$ ($p_f = 2.5\%$), is sufficient for single-pile design.

CHAPTER SEVEN

Design of rock-socketed piles

Research and practice in North America

Ng *et al.* (2001a) report that Rosenberg and Journeaux (1976) studied the results of a limited number of field load tests on small-diameter piles (200–610 mm). They proposed a tentative relationship between the unconfined compressive strength (UCS) of rock σ_r (MPa) and the ultimate side resistance q_s (MPa) as follows:

$$q_s = 0.375(\sigma_r)^{0.515} \tag{1.17a}$$

Horvath and Kenney (1979) reviewed the results of load tests on rock-socketed piles and anchors from, mainly, Australia, Canada, the United Kingdom and the United States, including 49 tests on piles with diameters between 410 and 1220 mm. The majority of the piles were socketed into sedimentary rock types, mostly shale or mudstone. They reported socket friction generally became fully mobilized at a displacement of around 6 mm (0.5–1.5% of the pile diameter). A correlation was suggested between q_s (MPa) and the compressive strength of the weaker socket material (concrete or rock) σ_{rc} (MPa) for the piles given by

$$q_s = b\sqrt{\sigma_{rc}} \tag{1.17b}$$

where $b = 0.2$–0.25 for piles >410 mm in diameter. Horvath *et al.* (1983) later revised b to be equal to 0.2–0.3 for 'large'-diameter piles, presumably >410 mm, and proposed a method for estimating the effects of socket wall roughness on the capacity of side resistance using a 'roughness factor', which is dependent on asperity height, socket radius and socket length. However, in practice it is difficult to determine the roughness factor in the field accurately, particularly for piles constructed under water or bentonite.

Rowe and Armitage (1987) recommended a correlation of $q_s = 0.45\sqrt{\sigma_{rc}}$ for regular 'clean' sockets and $q_s = 0.6\sqrt{\sigma_{rc}}$ for clean 'rough' sockets for an initial estimate of side resistance from the results of a large number of field load tests. 'Rough' sockets were defined as having grooves or undulations of depth and width >10 mm at a spacing of 50–200 mm. A more detailed design procedure taking into account socket dimensions was provided to refine the prediction. Rowe and Pells (1980) studied theoretical considerations of rock-socket behaviour. That paper is significant in that it predicted variations in side resistance capacity as a result of varying stress distributions, related to varying ratios of socket length to diameter and pile modulus to rock modulus.

Reese and O'Neill (1988) recommended that the first of the following equations be used for side resistance for rock (or concrete) with a UCS up to 1.9 MPa and the second equation be used for higher strengths (MPa):

$$q_s = 0.15\sigma_{rc} \tag{1.18a}$$

$$q_s = 0.2\sqrt{\sigma_{rc}} \tag{1.18b}$$

The first equation was taken from Carter and Kulhawy (1987) and the second equation from Horvath and Kenney (1979). The recommendations are adopted by the US Department of Transportation. Recently, Zhang and Einstein (1998) suggested relationships based on a review of the results of Rosenberg and Journeaux (1976), Meigh and Wolski (1979), Williams and Pells (1981), Horvath (1982) and Kulhawy and Phoon (1993). These are as follows for 'smooth' and 'rough' sockets (MPa), respectively:

$$q_s = 0.4\sqrt{\sigma_{rc}} \tag{1.19}$$

$$q_s = 0.85\sqrt{\sigma_{rc}} \tag{1.20}$$

Research and practice in Australia

In Australia, most experience in the design of rock-socket piles has been gained in mudstone, shale and sandstone. One example is the Melbourne mudstone, which Chiu and Johnston (1980) described as having characteristics very similar to over-consolidated clays. Williams *et al.* (1980b) described the results of a number of field load tests in Melbourne mudstone. The socket friction–displacement relationships exhibited peak, followed by residual, friction behaviour, although the residual values were only around 5% less than the peak values for 'rough' sockets. 'Rough' sockets were defined by minimum statistical requirements for asperity height and angle and were achieved with excavation by conventional bucket augers. Williams *et al.* (1980a) also described the results of tests in Melbourne mudstone of which at least one reached peak friction at a displacement of 6 mm.

Williams and Pells (1981) suggested a design method based on the results of Williams *et al.* (1980b), Pells *et al.* (1980) and Horvath (1978) but with emphasis placed on test results in Australia. For cases in which the socket wall is sufficiently rough to prevent brittle side shear behaviour, a curve for the ratio of q_s to σ_{rc} termed the 'side resistance reduction factor' α is used. The side resistance factor decreases with increasing σ_{rc}. A further factor β is applied to take into account a variation of side resistance with the rock mass factor (ratio of jointed rock modulus to intact rock modulus) giving

$$q_s = \alpha\beta\sigma_{rc} \tag{1.21}$$

As mentioned above, Rowe and Pells (1980) predicted the relationship between socket side resistance capacity and the socket dimensions and modulus ratio.

Design practice in Hong Kong

In Hong Kong design practice, for large-diameter bored piles bearing on granitic and volcanic rocks, allowable values of toe resistance may be used as specified in PNAP 141 (BOO, 1990). The presumptive end-bearing values range from 3 to 7.5 MPa, depending on the rock 'category'. The category is defined in terms of the rock decomposition grade, strength (UCS) and total core recovery. Generally, the presumed values may be used without a settlement check for structures which are not unduly sensitive to settlement. The details of the presumed end-bearing values from PNAP 141 are summarized in Table 1.17. Presumptive values of rock-socket side resistance are not defined. However, PNAP 141 allows

Table 1.17 Presumed safe vertical bearing stress for foundations on horizontal ground in Hong Kong, simplified from PNAP 141 (BOO, 1990; after Ng et al., *2001a)*

Category	Granitic and volcanic rock	Presumed bearing stress (kPa)
1(a)	Fresh to slightly decomposed strong rock of material weathering grade II or better, with total core recovery of >95% of grade and minimum uniaxial compressive strength of rock material σ, not less than 50 MPa (equivalent point load index strength PLI_{50}* not less than 2 MPa)	7500
1(b)	Slightly to moderately decomposed moderately strong rock of material weathering grades II or III or better, with total core recovery of >85% of grade and minimum uniaxial compressive strength of rock material σ, not less than 25 MPa (PLI_{50}* not less than 1 MPa)	5000
1(c)	Moderately decomposed moderately strong to moderately weak rock of material weathering Grades III or IV or better, with total core recovery of >50% of grade	3000

* Point load index strength (i.e., PLI_{50} values) of rock quoted is equivalent value for 50 mm diameter cores (ISRM, 1978).

the use of a 'rational design' approach and rock-socket side resistance can generally be incorporated in a design, provided that the parameters and assumptions are verified by field testing.

For caissons in granite, the Geotechnical Engineering Office (GEO, 1991) reported various findings from studies including those by Ladanyi (1977), Pells and Turner (1979) and Horvath *et al.* (1980) and acknowledges that socket side resistance can account for a significant portion of the load support capacity. For the preliminary design of side resistance in widely jointed rock, the GEO (1996) suggested the use of the empirical correlation proposed by Horvath *et al.* (1983) to estimate side resistance capacity. Horvath's correlation was developed using the results of load tests in mostly sedimentary rocks, however, whereas Hong Kong conditions are dominated by granitic, volcanic and metasedimentary rocks. The GEO (1996) also suggested that the allowable working load should be estimated using a minimum mobilization factor of 1.5 on the ultimate side resistance. The mobilization factor is intended to take into account debris left at the base of the rock socket, which would cause the base resistance to be unreliable at working load. It is implicit in the GEO's recommendation that the contribution of toe resistance is only taken into account at the ultimate limit state. Therefore the contribution of toe resistance may potentially become irrelevant altogether if the working load requirement governs the design.

CHAPTER EIGHT

Case studies: rock-socketed piles

Overview

Ng *et al.* (2001a) report that a number of empirical relationships have been published for estimating the side resistance of rock-socketed piles. All are based on studies of field load test results and laboratory tests and relate socket friction capacity to the UCS of rock or concrete, generally whichever is the weakest. Most of the load tests considered were carried out in sedimentary rocks having lower strengths than typical rock types encountered in Hong Kong.

Table 1.18 summarises details of the 13 tests carried out in Hong Kong. The tests were generally carried out in the preliminary stages of the respective projects. They were intended to investigate the feasibility of using rock-socket side resistance in design, rather than using the presumed end-bearing stresses detailed in PNAP 141 (BOO, 1990), and to establish design rock-socket side resistance values. Also, some of the rock types tested were not volcanic or granitic rocks and therefore could not be designed using the presumed values in PNAP 141. The chosen methodology falls under the PNAP 141 clause permitting the use of a 'rational' approach to foundation design. Based on the loading test results, the use of rock-socket side resistance in design for each of the specific projects was approved by the relevant checking bodies and detailed design and construction proceeded accordingly. Of the buildings being supported by the piles, the settlements at this stage are, to the writers' knowledge (Ng *et al.*, 2001a), within acceptable structural limits.

With the exception of the hand-dug caisson reported by Lam *et al.* (1991), excavation was by RCD (Rotary Core Drilling) under water or bentonite. The geological descriptions including the rock decomposition grades and the rock quality designations (RQDs) were assigned by geologists. Rock types have also been classified generally as being granitic, volcanic or metasedimentary. The UCS σ_r values were either measured directly by laboratory tests or estimated from a correlation with the point load index test (PLIT) results. In the PLIT, rock specimens in the form of core, cut blocks, or irregular lumps are broken by application of a concentrated load through a pair of spherically truncated conical platens with a

Table 1.18 Database of Hong Kong load tests of large-diameter bored piles socketed into decomposed rocks (after Ng et al.*, 2001a)*

Pile No.	Reference	Pile length (m)	Socket diameter (mm)	Socket length (m)	Method of excavation	Rock type	Average UCS (MPa)	RQD (%)	$\bar{q}_{max}$ (kPa)	$\bar{q}_{1\%}$ (kPa)	DI
1	This study[a]	29.2	1200	2.0	RCD under water	Grade II fine ash tuff (volcanic)	105 (50)[b]	92	2900	–	*C*
2	This study[a]	32.5	1060	6.0	RCD under water	Grades II–V fine ash tuff (volcanic)	156 (50)[b]	64	1740	–	*C*
3	This study[a]	35.6	1020	2.0	RCD under water	Grade III coarse ash tuff (volcanic)	40	68	2860	–	*B*
4	This study[a]	40.3	1200	1.5	RCD under water	Grade III metasandstone (meta-sedimentary)	28.8	41	5100	–	*C*
5	This study[a]	25.1	1320	2.1	RCD under water	Grade IV/III granodiorite (granitic)	6	3	480	420	*A*
6	This study[c]	38.6	1200	1.2	RCD under bentonite	Grade II granite (granitic)	82.5 (50)[b]	100	1700	1700	*B*
7	This study[c]	60.2	1200	3.6	RCD under bentonite	Grade III/II granite (granitic)	10	69	610	500	*A*
8	This study[c]	41.7	1000	0.9	RCD under bentonite	Grade II granite (granitic)	125 (50)[b]	90	2865	–	*C*
9	This study[c]	48.3	1000	2.5	RCD under water	Grade III granite (granitic)	28.8	83	960	730	*B*
10	This study[c]	24.2	1200	1.5	RCD under bentonite	Grade III/II granite (granitic)	230 (50)[b]	?	1000	1000	*B*
11	This study[c]	23.9	1200	3.0	RCD under water	Grade III granite (granitic)	38	?	1210	–	*C*
12	This study[a]	30.2	1200	1.1	RCD under water	Grade IV/III breccia (volcanic)	–	20	360	–	*A*
13	Lam *et al.* (1991)	38.0	1000	0.8	Hand-dug caisson	Grade III/II granite (granitic)	120 (7)[b]	88	695	695	*A*

Note: $\bar{q}_{max}$ = mean maximum rock-socket side resistance, averaged over socket length. $\bar{q}_{1\%}$ = rock-socket side resistance at 1% of socket displacement. DI = Displacement Index (see p. 73)

[a] Tests carried out and data interpreted by writers.

[b] Value in parenthesis is approximate concrete UCS (only cases with concrete UCS less than rock UCS are shown).

[c] Test data interpreted by writers.

tip radius of 5 mm. Little or no specimen preparation is needed (International Society for Rock Mechanics (ISRM), 1985). A factor of 24 was applied to the PLIT results, as suggested for weathered granite by the GEO (1991). The suggestions of the ISRM Commission on Testing Methods were adopted for the PLIT tests (ISRM, 1985) and also the UCS tests (ISRM, 1978). The UCS values reported in Table 1.18 are the average values over the depth of the rock socket. Generally three or four tests in total were performed for each socket. It is suggested that, for purposes of comparison, the same or similar testing methods should be adopted in the future by other researchers. An anomaly may be observed in the results in Table 1.18. Pile 12 was tested in rock, which was too decomposed to allow a UCS test to be conducted, and the test may be considered to be not typical of the remaining tests in Table 1.18; the differences should be kept in mind in the analyses to follow.

In the load tests, the axial load distribution in the rock socket was determined using strain gauges installed within the pile shaft and by estimating the pile shaft modulus. The average strain across the socket was taken to be equal to the average reading of the gauges at each level. For the piles with which the writers were directly involved, the average strain gauge readings were checked by calculating the total socket shortening from the readings and comparing these with the shortening measured by rod extensometers. The difference in the results varied between 5 and 13%. The modulus was generally deduced using laboratory test results on concrete core samples taken from the pile shaft and by considering the variation of modulus with strain over the duration of the loading test using gauges located near the top of each pile. With the axial load distribution known, the unit socket side resistance was calculated accordingly. For most of the piles, the calculated toe resistance accounted for only up to 17% of the socket side resistance. The side resistance value q_{max} shown is the maximum recorded or extrapolated for each test and is the 'mean' value, averaged over the whole socket length.

Table 1.19 summarizes key details of the 66 other tests from the literature. Some contributed to the various empirical correlations for side resistance in the literature detailed earlier. The details and measurements shown in Table 1.19 are consistent with Table 1.18, although the level of detail is generally lower. The RQD is not shown and whether or not the excavation was 'dry' or 'wet' (constructed under bentonite or water) is unknown in many cases. For a number of tests in granitic rock in Singapore, the pile diameters are also unknown. Possibly some of these tests were on small-diameter rather than large-diameter piles. Note that, in a few cases, the sockets were artificially roughened. The UCS values shown were obtained using a range of methods, including empirical correlations with PLIT results and standard penetration test results.

Table 1.19 Database of non-Hong Kong load tests of large-diameter bored piles socketed into decomposed rocks (after Ng et al.*, 2001a)*

Pile No.	Reference	Pile length (m)	Socket diameter (mm)	Socket length (m)	Method of excavation	Rock type	Average UCS (MPa)	$\bar{q}_{max}$ (kPa)	DI
14	Leung (1996)	30.00	1000	1.00	Flight auger and chisel; wet bore	Granite (granitic)	12.50	800	*B*
15	Leung (1996)	12.40	810	10.00	Flight auger and chisel; dry bore	Siltstone (sedimentary)	6.00	560	*C*
16	Leung (1989)	14.00	1350	6.80	Flight auger and chisel; dry bore	Siltstone (sedimentary)	7.00	600	*C*
17	Leung (1989)	11.50	1500	11.50	Flight auger and chisel; dry bore	Siltstone (sedimentary)	9.00	800	*C*
18	Leung (1989)	7.30	710	7.30	Flight auger and chisel; dry bore	Siltstone (sedimentary)	9.00	700	*C*
19a	Leung (1996)	16.00	1400	2.50	Flight auger and chisel; dry bore	Siltstone (sedimentary)	3.50	390	*C*
19b	Leung (1996)	16.00	1400	3.00	Flight auger and chisel; dry bore	Siltstone (sedimentary)	6.50	620	*C*
20	Carrubba (1997)	18.50	1200	7.50	Drilling bucket and rock bit; casing	Marl (sedimentary)	0.90	140	*B*
21	Carrubba (1997)	19.00	1200	2.50	Drilling bucket and rock bit; casing	Diabasic breccia (volcanic)	15.00	490	*B*
22	Carrubba (1997)	37.00	1200	11.00	Drilling bucket and rock bit; casing	Gypsum (sedimentary)	6.00	120	*C*
23	Carrubba (1997)	20.00	1200	2.00	Drilling bucket and rock bit; casing	Diabase (sedimentary)	40.00	890	*C*
24	Carrubba (1997)	13.50	1200	2.50	Drilling bucket and rock bit; casing	Limestone (sedimentary)	2.50	400	*B*
25a	Walter *et al.* (1997)	22.00	900	1.00	Downhole jack	Mudstone (sedimentary)	3.20	600	*B*
25b	Walter *et al.* (1997)	22.00	900	2.10	Downhole jack	Siltstone (sedimentary)	8.90	1110	*B*
25c	Walter *et al.* (1997)	22.00	900	2.20	Downhole jack	Sandstone (sedimentary)	11.60	2160	*B*
26	Horvath *et al.* (1983)	1.97	710	1.40	Auger	Shale (sedimentary)	5.40	1110	*A*
27	Horvath *et al.* (1983)	1.97	710	1.40	Auger	Shale (sedimentary)	11.10	1110	*A*
28	Horvath *et al.* (1983)	1.97	710	1.40	Auger*	Shale (sedimentary)	5.60	2000	*A*
29	Horvath *et al.* (1983)	1.97	710	1.40	Auger*	Shale (sedimentary)	5.50	1750	*B*
30	Horvath *et al.* (1983)	1.97	710	1.40	Auger	Shale (sedimentary)	10.40	1090	*B*
31	Horvath and Kenney (1979)	1.97	635	0.90	?	Shale (sedimentary)	15.20	830	*B*
32	Vogan (1977)	?	610	2.90	?	Shale (sedimentary)	7.00	932	*C*
33	Osterberg and Gill (1973)	?	1220	1.20	?	Shale (sedimentary)	11.10	1040	*B*
34	Mason (1960)	?	610	1.80	?	Shale (sedimentary)	1.50	417	*B*
35	Horvath and Kenney (1979)	?	1070	3.20	?	Shale (sedimentary)	22.10	2600	*C*
36	Pells *et al.* (1978)	?	1090	1.50	?	Mudstone (sedimentary)	2.30	800	*B*
37	Buttling (1976)	27.30	1050	8.00	Bored under bentonite	Chalk (sedimentary)	1.00	190	*B*
38	Davis (1974)	?	760	4.00	?	Marl (sedimentary)	1.30	230	*B*
39	Webb (1976)	?	615	12.20	?	Diabase (sedimentary)	0.40	122	*B*
40	Williams *et al.* (1980a)	?	660	1.80	Bucket auger	Mudstone (sedimentary)	2.30	965	*A*
41	Johnstone and Donald (1979)	?	1200	1.00	?	Mudstone (sedimentary)	3.10	1050	*B*
42	Johnstone and Donald (1979)	?	1200	1.00	?	Mudstone (sedimentary)	1.90	940	*B*
43	Williams (1980)	?	660	1.50	?	Mudstone (sedimentary)	0.80	660	*B*
44	Williams (1980)	?	1120	2.60	?*	Mudstone (sedimentary)	0.60	510	*B*
45	Williams (1980)	?	1220	2.00	Bucket auger	Mudstone (sedimentary)	2.50	600	*B*
46	Williams (1980)	?	1300	2.00	Bucket auger*	Mudstone (sedimentary)	2.30	640	*B*

Table 1.19 Database of non-Hong Kong load tests of large-diameter bored piles socketed into decomposed rocks (after Ng et al., *2001a) (continued)*

Pile No.	Reference	Pile length (m)	Socket diameter (mm)	Socket length (m)	Method of excavation	Rock type	Average UCS (MPa)	$\bar{q}_{max}$ (kPa)	DI
47	Williams (1980)	?	1230	2.00	Bucket auger	Mudstone (sedimentary)	2.30	710	*B*
48	Williams (1980)	?	1350	2.00	Bucket auger*	Mudstone (sedimentary)	2.30	620	*B*
49	Williams (1980)	?	750	2.00	?	Claystone (sedimentary)	5.50	1300	*B*
50	Matich and Kozicki (1967)	13.60	610	6.00	Auger*	Shale (sedimentary)	0.48	310	*C*
51	Williams and Pells (1981)	?	690	3.40	Drilled, cast under bentonite	Shale (sedimentary)	3.10	1100	*B*
52	Williams and Pells (1981)	?	6600	5.20	Drilled, cast under bentonite	Shale (sedimentary)	0.50	300	*B*
53	Williams and Pells (1981)	?	790	8.90	Drilled, cast under bentonite	Shale (sedimentary)	2.70	720	*B*
54	Thorne (1980)	7.00	900	1.30	Flight auger	Shale (sedimentary)	21.00	1260	*C*
55	Pells *et al.* (1980)	?	710	0.90	Auger	Sandstone (sedimentary)	6.00	650	*B*
56	Thorburn (1966)	5.75	915	3.35	Bucket auger	Shale (sedimentary)	12.20	242	*C*
57	Wilson (1976)	4.00	900	1.00	Hochstasser rig	Mudstone (sedimentary)	1.10	120	*B*
58	Wilson (1976)	4.00	900	1.00	Hochstasser rig	Mudstone (sedimentary)	1.10	184	*B*
59	Buttling (1976)	31.00	900	10.00	?	Chalk (sedimentary)	2.40	120	*C*
60	Leung (1996)	?	?	?	?	Granite (granitic)	1.10Ψ	180Ψ	*C*
61	Leung (1996)	?	?	?	?	Granite (granitic)	1.10Ψ	80Ψ	*C*
62	Leung (1996)	?	?	?	?	Granite (granitic)	1.30Ψ	260Ψ	*C*
63	Leung (1996)	?	?	?	?	Granite (granitic)	1.30Ψ	130Ψ	*C*
64	Leung (1996)	?	?	?	?	Granite (granitic)	1.30Ψ	90Ψ	*C*
65	Leung (1996)	?	?	?	?	Granite (granitic)	2.00Ψ	600Ψ	*C*
66	Leung (1996)	?	?	?	?	Granite (granitic)	2.00Ψ	400Ψ	*C*
67	Leung (1996)	?	?	?	?	Granite (granitic)	2.00Ψ	125Ψ	*C*
68	Leung (1996)	?	?	?	?	Granite (granitic)	2.30Ψ	375Ψ	*C*
69	Leung (1996)	?	?	?	?	Granite (granitic)	2.80Ψ	210Ψ	*C*
70	Leung (1996)	?	?	?	?	Granite (granitic)	4.00Ψ	60Ψ	*C*
71	Leung (1996)	?	?	?	?	Granite (granitic)	4.10Ψ	310Ψ	*C*
72	Leung (1996)	?	?	?	?	Granite (granitic)	4.00Ψ	250Ψ	*C*
73	Leung (1996)	?	?	?	?	Granite (granitic)	5.00Ψ	1250Ψ	*C*
74	Leung (1996)	?	?	?	?	Granite (granitic)	13.00Ψ	300Ψ	*C*
75	Leung (1996)	?	?	?	?	Granite (granitic)	14.00Ψ	525Ψ	*C*
76	Leung (1996)	?	?	?	?	Granite (granitic)	20.00Ψ	725Ψ	*C*
77	Leung (1996)	?	?	?	?	Granite (granitic)	40.00Ψ	500Ψ	*C*
78	Leung (1996)	?	?	?	?	Granite (granitic)	45.00Ψ	900Ψ	*C*
79	Leung (1996)	?	?	?	?	Granite (granitic)	50.00Ψ	625Ψ	*C*

Note: DI = Displacement Index (see p. 73). Ψ = scaled-off figure in publication.
* Socket was artificially roughened.

Displacement Index (DI)

The database contains side resistance values recorded at varying displacements and degrees of mobilization. Therefore, a DI is assigned. This is an approximate measure of the pile 'local' displacement and degree of mobilization of side resistance for a given maximum side resistance value. The 'local' displacement is the pile displacement deduced at the vertical centre of the rock socket using the measured pile-head movement and shaft shortening at the rock socket midpoint. The DI values are assigned by observation of the field side resistance–displacement relationships. An index *A* is assigned to test results, which are observed to be close to the fully mobilized value, and is defined by the displacement reaching a minimum of 1% of the socket diameter and the gradient of the resistance–displacement relationship becoming <30 kPa/mm at the end of the curve (at the completion of the test). An index *B* is assigned to test results not achieving an index of *A* but recorded at a displacement of at least 0.4% of the pile diameter and for which the gradient of the resistance–displacement relationship is <200 kPa/mm at the end of the curve. An index *C* is assigned to the remaining results. The various proposed displacement indices are illustrated graphically in Fig. 1.18.

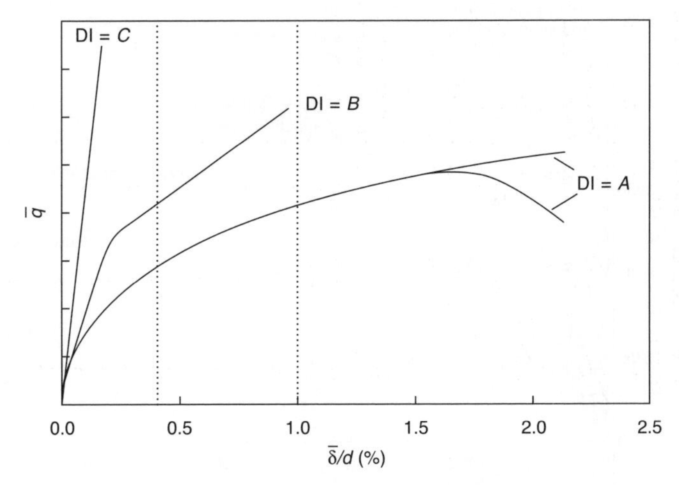

Fig. 1.18 Proposed criteria for Displacement Index (DI) (after Ng et al.*, 2001a)*

Capacity of side resistance and relationship with UCS and decomposition grade

Relationship with unconfined compressive strength, UCS

All of the various empirical relationships reviewed in the literature relate rock-socket side resistance to the UCS of rock or concrete, generally

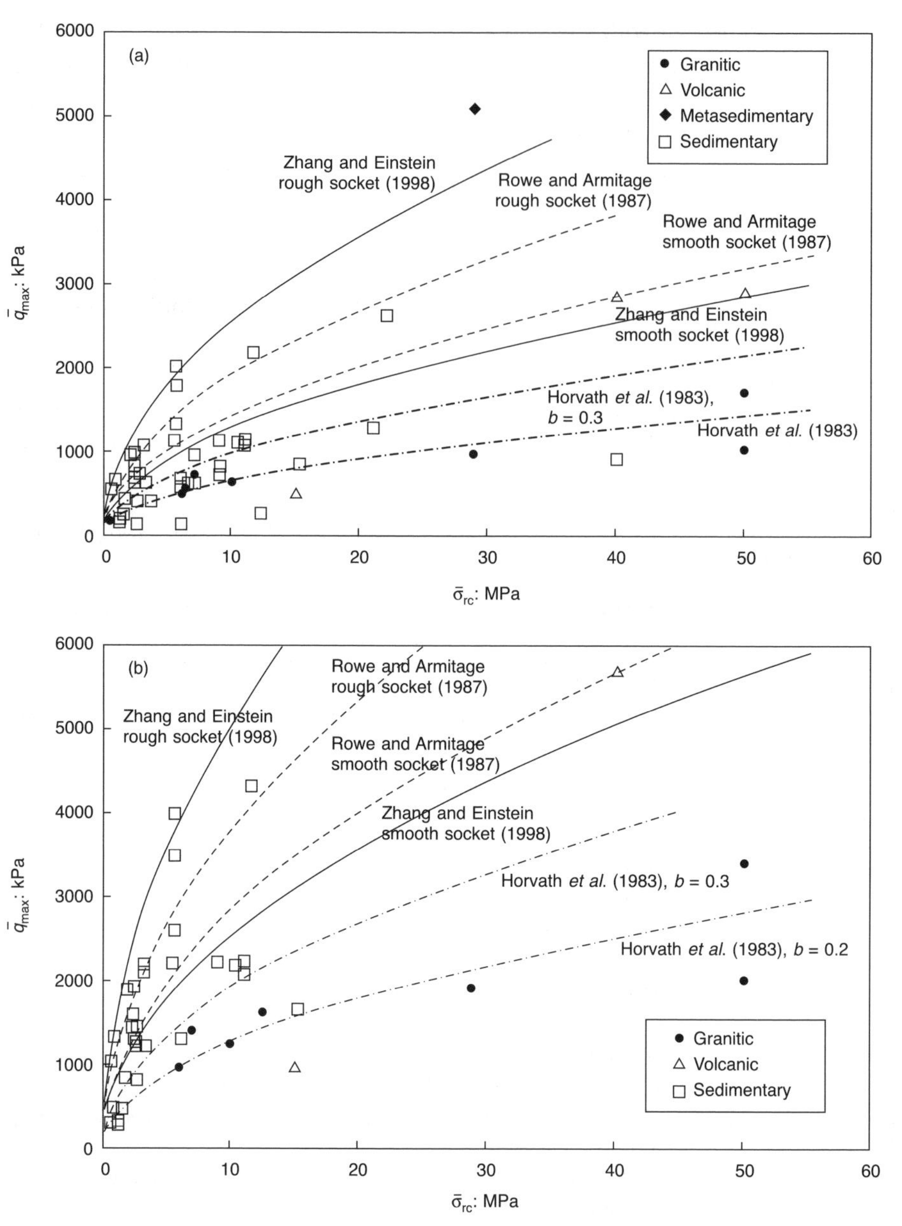

Fig. 1.19 Maximum achieved side resistance versus UCS: (a) all results, DI = A, B *or* C*; (b) DI =* A *or* B *only (after Ng* et al.*, 2001a)*

whichever is weaker. In Fig. 1.19 the mean maximum socket side resistance q_{max} has been plotted against the mean UCS σ_{rc} for all test results. Also shown are the suggested correlations of Horvath *et al.* (1983), Rowe and Armitage (1987) and Zhang and Einstein (1998). Overall, almost all the results lie below the correlation of Zhang and Einstein (1998) for 'rough' sockets but with a high degree of scatter. None of the correlations from the literature appear to be particularly suitable for predicting side resistance for all results. However, if the results in only sedimentary rock are examined, these are mostly higher in magnitude than results produced by the correlation of Horvath *et al.* (1983) for $b = 0.2$. Generally the results in granitic rock are lower than in sedimentary rock and tend to be lower than the empirical correlations. However, the UCS is also generally higher for these tests. The results in volcanic rock are few and scattered. The single result in metasedimentary rock indicates extremely high side resistance.

Relationship with decomposition grade

The Hong Kong tests were conducted in rocks varying from decomposition grade IV/III to II. From the descriptions of the rock decomposition grades in Table 1.20, it would be expected that rock strength and hence

Table 1.20 Classification of rock material decomposition grades of Hong Kong (simplified from GCO, 1988)

Descriptive term	Grade symbol	General characteristics for granite and volcanic rocks and other rocks of equivalent strength in fresh state
Residual soil	VI	Original rock texture completely destroyed Can be crumbled by hand and finger pressure into constituent grains
Completely decomposed	V	Original rock texture preserved Can be crumbled by hand and finger pressure into constituent grains Easily indented by point of geological pick Slakes when immersed in water Completely discoloured compared with fresh rock
Highly decomposed	IV	Can be broken by hand into smaller pieces Makes a dull sound when struck by geological hammer Not easily indented by point of geological pick Does not slake when immersed in water Completely discoloured compared with fresh rock
Moderately decomposed	III	Cannot usually be broken by hand; easily broken by geological hammer Makes a dull or slight ringing sound when struck by geological hammer Completely stained throughout
Slightly decomposed	II	Not broken easily by geological hammer Makes a ringing sound when struck by geological hammer Fresh rock colours generally retained but stained near joint surfaces
Fresh	I	Not broken easily by geological hammer Makes a ringing sound when struck by geological hammer No visible signs of decomposition (i.e. no discoloration)

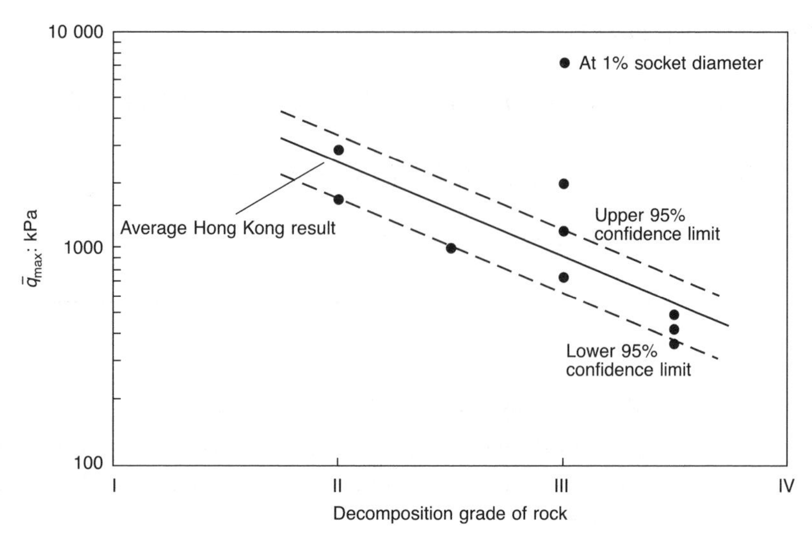

Fig. 1.20 Maximum achieved side resistance versus decomposition Grade for Hong Kong granitic and volcanic rocks (after Ng et al., *2001a)*

side resistance would decrease with the decomposition grade increasing from II to IV/III. Figure 1.20 shows the relationship between decomposition grade and side resistance. The side resistance results are as recorded at a displacement of 1% of the pile diameter in granitic and volcanic rocks. Each increment of decomposition grade is spaced evenly along the horizontal axis, and materials having two decomposition grades (identified by geologists) are shown halfway between the two increments. It is recognized that the evaluation of decomposition grade is subjective and is by no means accurate enough to provide rock strength, particularly between various rock types. Not surprisingly for some of the decomposition grades, there is a range of results. This may be due to a variation in the UCS, rock type and construction method. However, a trend line can be drawn as shown in the figure. This could be used for preliminary prediction of side resistance capacity without having information such as UCS available. The 95% confidence limits of the trend line are also included. Side resistance capacity in rock of higher decomposition grades is particularly useful to designers if there is uncertainty over rock quality over the whole socket length and diameter.

CHAPTER NINE

Pile instrumentation: a case study of a well-instrumented barrette in Kowloon Bay, Hong Kong

Overview: key questions

As reported by Ng *et al.* (2000b), a well-instrumented barrette was tested in order to answer the following key questions about the behaviour of barrettes:

- How are lateral stresses and pore water pressures distributed and how do they change during a load test?
- What is the importance of degree of mobilization of side resistance in 'floating' barrettes (i.e. that rely on side resistance for their load capacity)?
- What are realistic design values of side resistance and settlement?
- What is the influence of construction factors such as duration, use of bentonite and grouting?

Introduction

Limited space and high demand have made land in Hong Kong extremely expensive. Tall buildings are built to optimize the floor area to land area ratio. Many of the tall buildings located along the Victoria Harbour on Hong Kong Island and the Kowloon peninsula are commonly founded on reclaimed land. Thus deep foundations are required to resist both vertical and horizontal loads due to the weight of the building and wind, respectively. The prevailing deep foundation types for tall buildings on these reclaimed lands are large bored and excavated piles, which are very long, normally in excess of 50 m. These piles can be circular (bored piles/drilled shafts) or rectangular (barrette) in shape and must extend through the fill, underlying soft marine clay, sandy clay and alluvial sand deposit down into the deep weathered granite soil (saprolite), which is typically less weathered with an increase in depth. The thickness of the weathered granite can be up to 80 m in some places, and its depth can extend to more than 100 m from the ground surface.

Over the past 15 years, barrette foundations have become increasingly popular in parts of Asia such as Hong Kong and Malaysia for many civil engineering structures and tall buildings. The construction method for barrettes is very similar to that adopted for diaphragm walls, where a rectangular trench is excavated under bentonite by heavy grabs or hydrofraise[4] and filled with tremie concrete. In Hong Kong, single barrettes up to a size of 1.5 m wide by 6.6 m long (on plan) have been constructed (Pratt and Sims, 1990). Due to their rectangular shape, barrette foundations are particularly suited to resisting large vertical and significant horizontal loads in a chosen direction.

Site location and ground conditions

The test site is located on the Kowloon peninsula of Hong Kong, to the east of the runway of the old Kai Tak international airport, at the Kowloon Bay area. Figure 1.21 shows the geology and some relevant borehole information obtained at Kowloon Bay. The site is on marine reclaimed land and the ground level is at approximately 4.48 m above Principal Datum (PD). The groundwater level is about 3 m below ground surface. The ground conditions consist of about 6 m fill material overlying a succession of approximately 9.5 m marine clay deposits, 1.5 m of sandy clay (probably alluvial), 4.8 m alluvial sand of Quaternary age, and about 12 m of weathered granitic saprolites that overlie granitic rocks of the Upper Jurassic to Lower Cretaceous age (Strange, 1990). Detailed descriptions and measured N values (number of hammer blows per unit penetration) by Standard Penetration Tests for each type of material are given in Fig. 1.21.

Details of construction

The top 20 m of the barrette consisted of a reduced-section sheathing zone (Fig. 1.21) built with the intention of minimizing the interface side resistance developed between the barrette and the upper soil layers. This sheathing layer consisted of four layers: a 3 mm steel plate welded onto the reinforcement cage, a coating of bitumen, a flexible and weak 'voltex' layer (geotextile infilled with sodium bentonite), and a thin sheet of plywood. However, the final result was that the plywood was unfortunately attached to the steel plate with a dense matrix of high-strength screws, precluding the possibility of shear between the intermediate 'soft' layers. As a consequence, the theoretical gap of about 80 mm between the plywood and the surrounding soil was not back-filled with gravel as planned, so that a 'weak' friction zone would hopefully exist. However, steel rods inserted into this suspected bentonite-filled gap about two weeks after concreting

[4] A 'hydrofraise' is a drilling machine powered by three down-hole motors, with reverse mud circulation. A heavy metal frame serving as a guide is fitted at its base with two cutter drums carrying tungsten carbide tipped cutters which rotate in opposite directions and break up the soil. The hydraulic cutting device is designed to give the cutter drums a high torque at low speed of rotation.

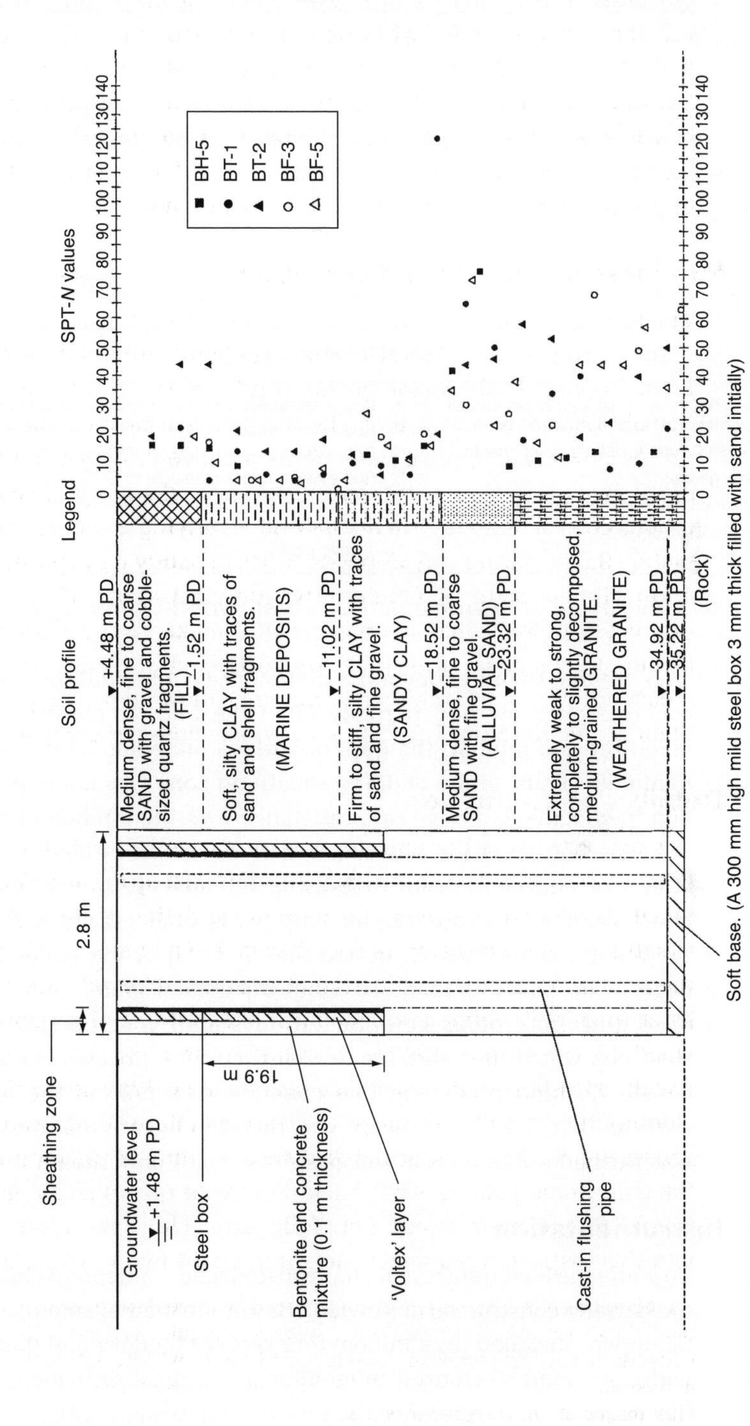

Fig. 1.21 Borehole logs and SPT values at Kowloon Bay, Hong Kong (after Ng et al., 2000b)

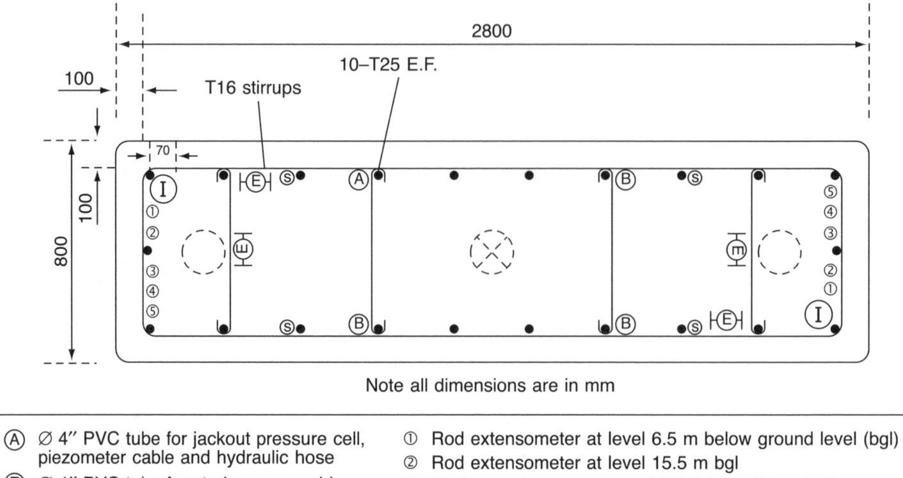

Ⓐ Ø 4″ PVC tube for jackout pressure cell, piezometer cable and hydraulic hose
Ⓑ Ø 4″ PVC tube for strain gauge cable
Ⓢ Strain gauges
Ⓔ Horizontal embedment strain gauges
Ⓘ In-place inclinometer
① Rod extensometer at level 6.5 m below ground level (bgl)
② Rod extensometer at level 15.5 m bgl
③ Rod extensometer at top of CDG layer (30 m bgl)
④ Rod extensometer at middle of CDG layer (36 m bgl)
⑤ Rod extensometer at bottom of CDG layer (41 m bgl)
Ø 6″ Cast-in flushing pipe
Tremie pipe

Fig. 1.22 Plan view layout of instrumentation of test barrette (after Ng et al.*, 2000b)*

were unable to probe beyond a metre or two all round the barrette. Either concrete overflow had partially filled the gap, or construction activities failed the soil infilling the gap, or surface materials mixed with solidified bentonite. Thus, in the end, the sheathing zone was not expected to function to effectively reduce side resistance over the top half of the barrette.

At the bottom of the barrette, a 'soft' base was formed to minimize the effects of end-bearing for mobilizing full side resistance at the soil–wall interface. This was done by placing a $2.8 \times 0.8 \times 0.3$ m high steel box to the bottom of the trench, before the lowering of the main reinforcement cages. The box was made of 3 mm thick steel plate, and it was initially filled with fine round sand. Seven days after concreting, the sand-filled steel box was drilled through and flushed with pressurized water via two cast-in flushing pipes and one concrete core hole in the middle of the barrette (Fig. 1.22). Great care was taken to ensure that most of the sand was flushed out to form a 'soft base' (i.e. void) underneath the barrette.

Instrumentation

To study the load transfer mechanism and load–settlement characteristics of the barrette constructed at Kowloon Bay, a substantial amount of instrumentation was installed. In addition, four sets of standard dial gauges together with surveying were used to monitor the vertical settlement of the top of the barrette and reference beams during testing. Strain gauges were

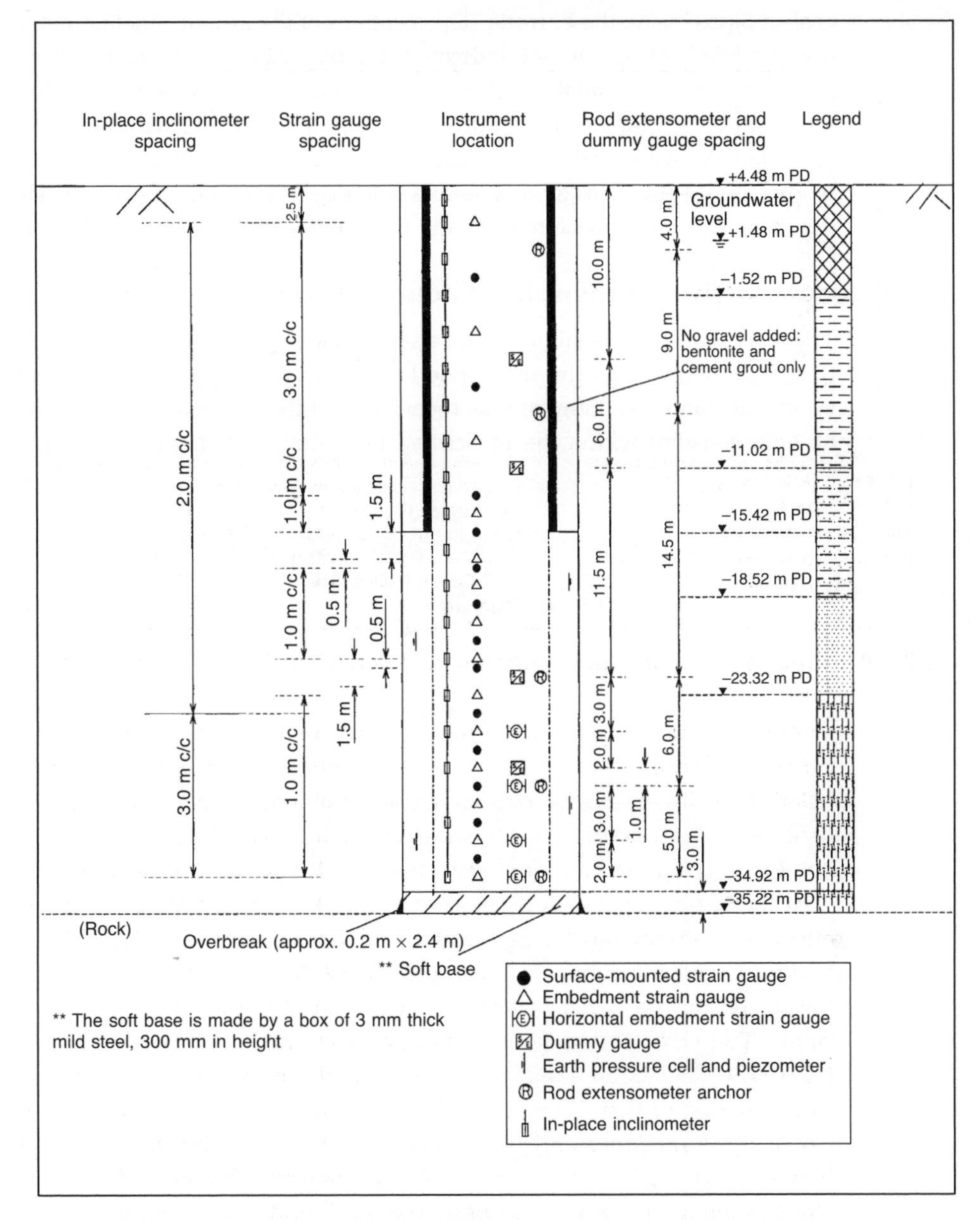

Fig. 1.23 Elevation layout of instrumentation of test barrette (after Ng et al., *2000b)*

placed at 27 levels on the reinforcement cages (Figs 1.22 and 1.23). Ten aluminium rod extensometers were sleeved individually in PVC tubes and installed to five depths at two different locations inside the barrette to monitor displacements between each depth and reference steel plate at the top of the barrette (Figs 1.22 and 1.23). A total of 38 biaxial servo accelerometer sensors were installed at 19 levels (most of them at 2 m intervals) in two cast-in pipes inside the barrette. The locations of the in-place inclinometers and the levels of sensors are indicated in Figs 1.22 and 1.23, respectively. The bottoms of the inclinometers were socketed in rock. A total of four vibrating wire total earth pressure cells, together with four vibrating wire piezometers, were installed at the barrette–soil interface at four levels within the layers of sandy clay, alluvial sand and weathered granite. The depths of the earth pressure cells and piezometers are shown in Fig. 1.23.

Load and displacement behaviour of barrette

The test programme originally comprised four loading and unloading cycles (Fig. 1.24). However, after the applied load reached 7455 kN at the second cycle, substantial settlement was recorded and the applied load could not be held constant within the prescribed maximum settlement tolerance of

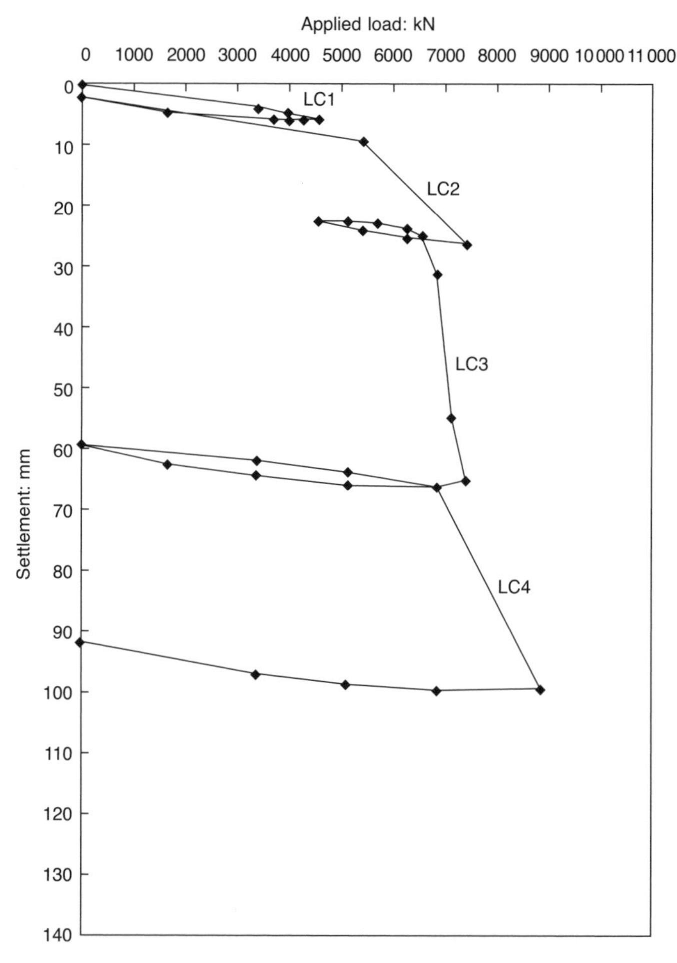

Fig. 1.24 Load–settlement response of test barrette (after Ng et al.*, 2000b)*

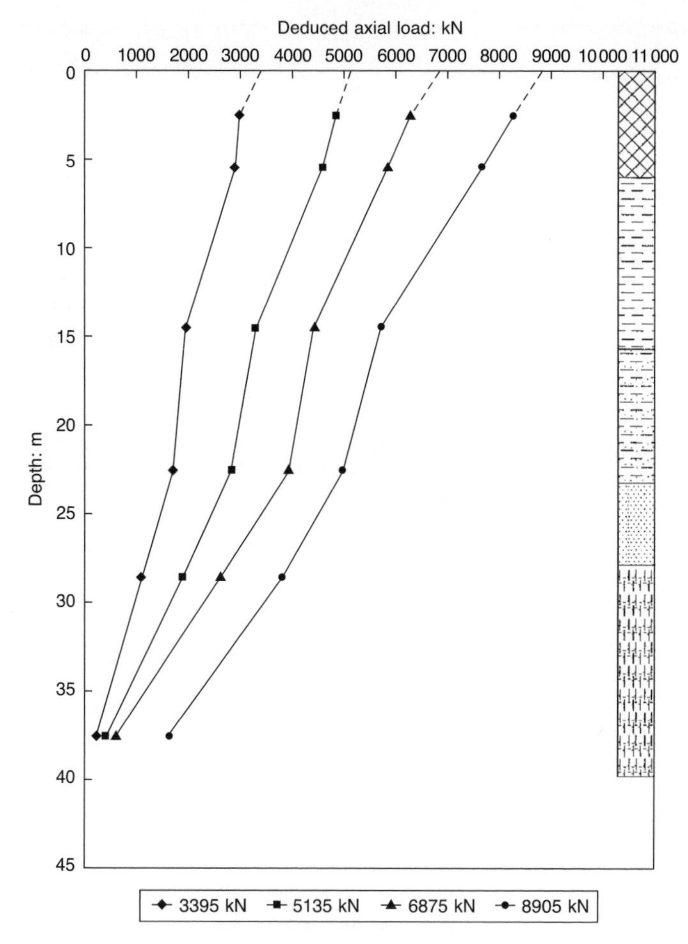

Fig. 1.25 Deduced axial load distribution at load test cycle 4 (after Ng et al., *2000b)*

0.05 mm/10 min. It was therefore decided to unload the barrette to 4555 kN and to hold it for 80 h (about three days). After the holding period, the testing programme resumed and two more loading cycles were performed. The barrette–soil interface appeared to gain strength as a result of consolidation, which is discussed later. Due to the presence of the 'soft base' (i.e. void) underneath the barrette, the barrette ultimately settled about 100 mm, enabling the side resistance to be fully mobilized along the shaft. Figure 1.25 shows the deduced axial load versus depth for the last loading cycle, LC4.

Barrette side resistance

From the gradient of the barrette normal stresses with depth, mobilized side resistance (interface shear stress, τ) was calculated and plotted

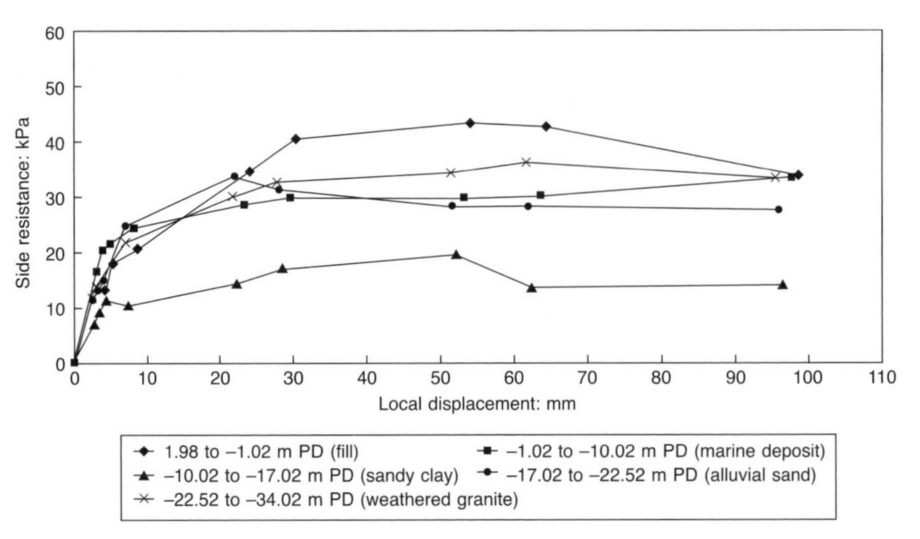

Fig. 1.26 Mobilization of side resistance with local displacement (after Ng et al., *2000b)*

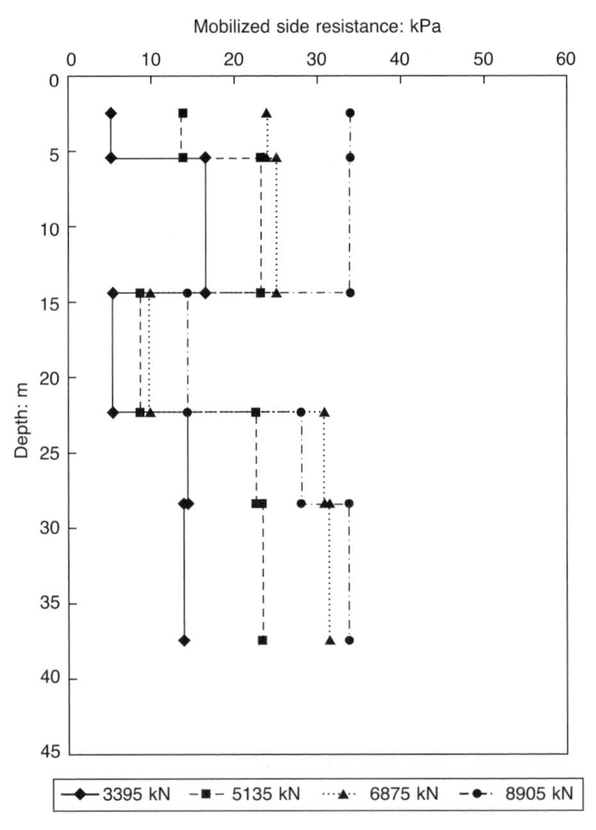

Fig. 1.27 Distribution of mobilized side resistance (after Ng et al., *2000b)*

against deduced local displacement of the barrette in each soil stratum as shown in Fig. 1.26. Figure 1.27 shows the distribution of mobilized side resistance with depth.

Pore pressure response at soil–barrette interface

After installation of the piezometers, readings were taken continuously to compare them with the initial hydrostatic pore water pressures in the ground (the initial groundwater table was located at about 1.3 m PD). These readings are shown in Fig. 1.28(a), (b) and (c). It was found that the measured pore water pressure at gauge P1 in the sandy clay layer (Fig. 1.28) was slightly higher than the corresponding hydrostatic value (1.3 m PD) before the loading test. Piezometric level (head) is defined as the sum of the pore water pressure head and the elevation head at each location. The measured piezometric heads recorded at P3 and P4 were a little lower than the hydrostatic conditions in the weathered granite before the commencement of the load test. Labels LC1–LC4 denote the commencement of the first to the fourth load cycles, respectively. Similarly, labels UC1–UC4 represent the start of the first to the fourth unloading cycles.

After the load test, dissipation of excess pore water pressures continued (Fig. 1.28(a)). As expected, the rate of dissipation was much slower in the sandy clay soil (P1) than in the weathered granite (P3 and P4). Dissipation at the alluvial sand interface was slow during loading, but generally the excess pore pressures generated returned to near their equilibrium value (around 2 m of head) fairly quickly after slip. Nearly all the excess

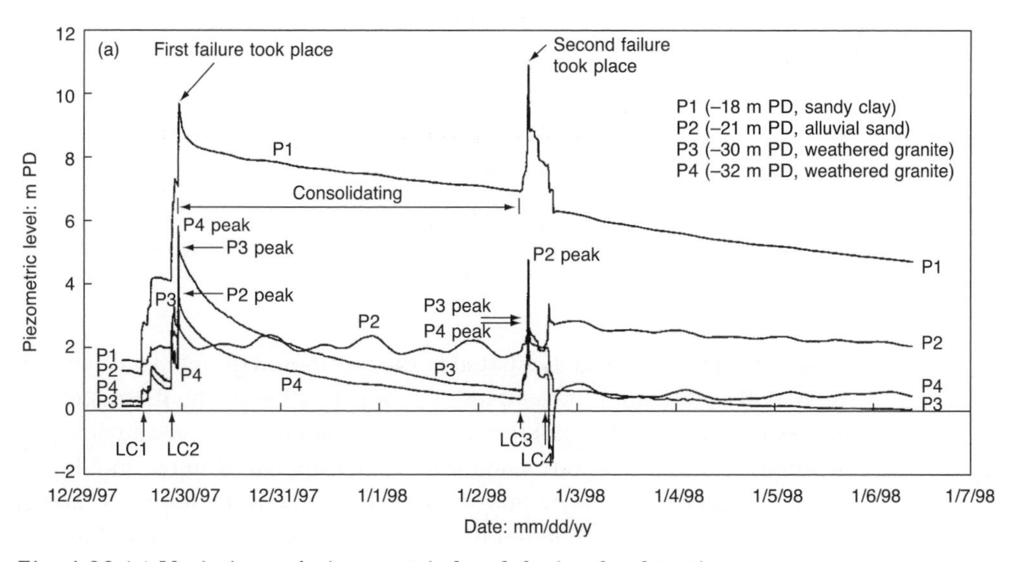

Fig. 1.28 (a) Variations of piezometric level during load testing.

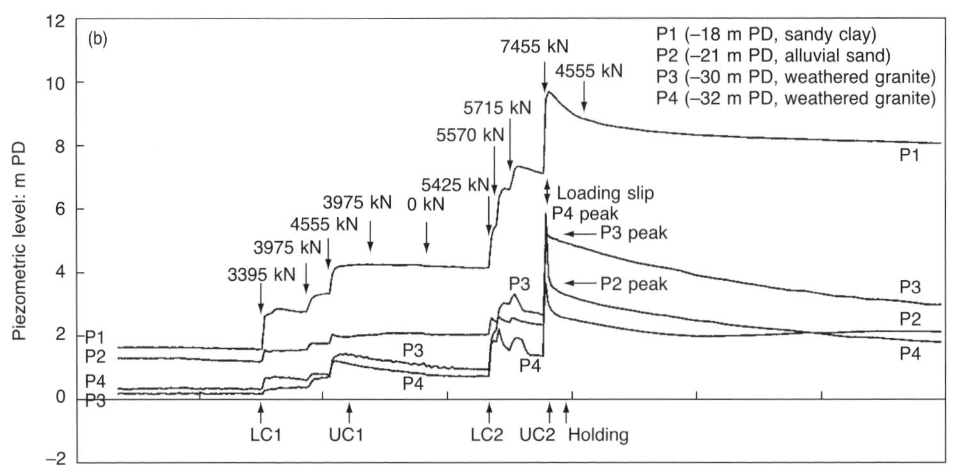

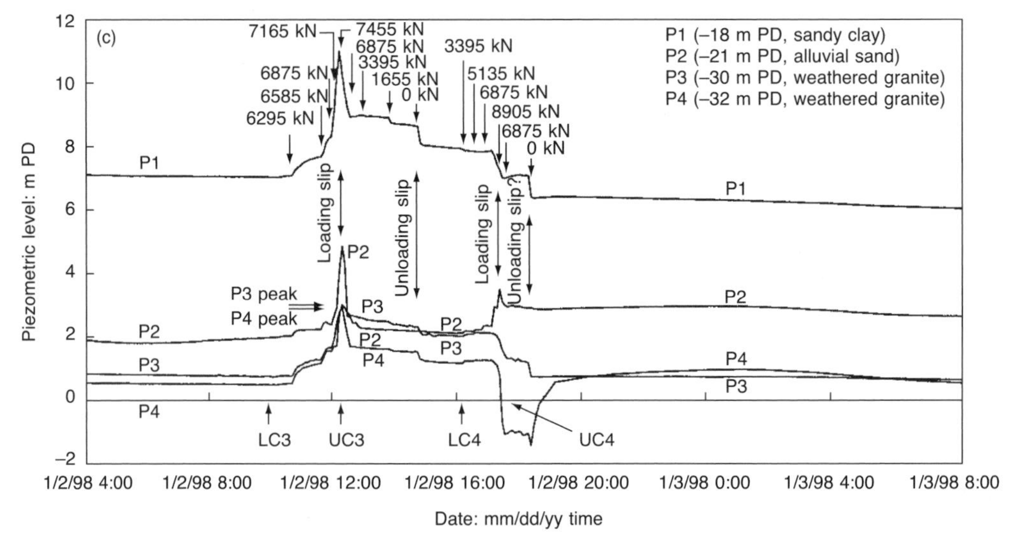

Fig. 1.28 (continued) (b) Variations of piezometric level during first two load test cycles. (c) Variations of piezometric level during last two load test cycles (after Ng et al., *2000b)*

pore water pressure was dissipated in the weathered granite by 6 January 1998, that is, 80 h after the load test. On the other hand, about 3 m and 1 m excess pore water head still remained in the sandy clay and alluvial sand layers, respectively, at this same time. What appeared to be tidal behaviour was picked up at P2 and P4, but it is difficult to see since actual peak tide magnitudes dropped by 50% and the scan frequency of instruments was reduced substantially after the pile test.

Changes of lateral stress at soil–barrette interface

During the subsequent three-week curing period before the vertical load test, a gradual, continuous decrease of lateral earth pressure was measured at all lateral earth pressure cells (about 30 kPa in the sandy clay and alluvial sand and about 60 kPa in the weathered granite). The observed reduction in the contact earth pressure may in large part be due to soil consolidation as indicated by the dissipation of excess positive pore water pressures generated during the construction of the barrette. In

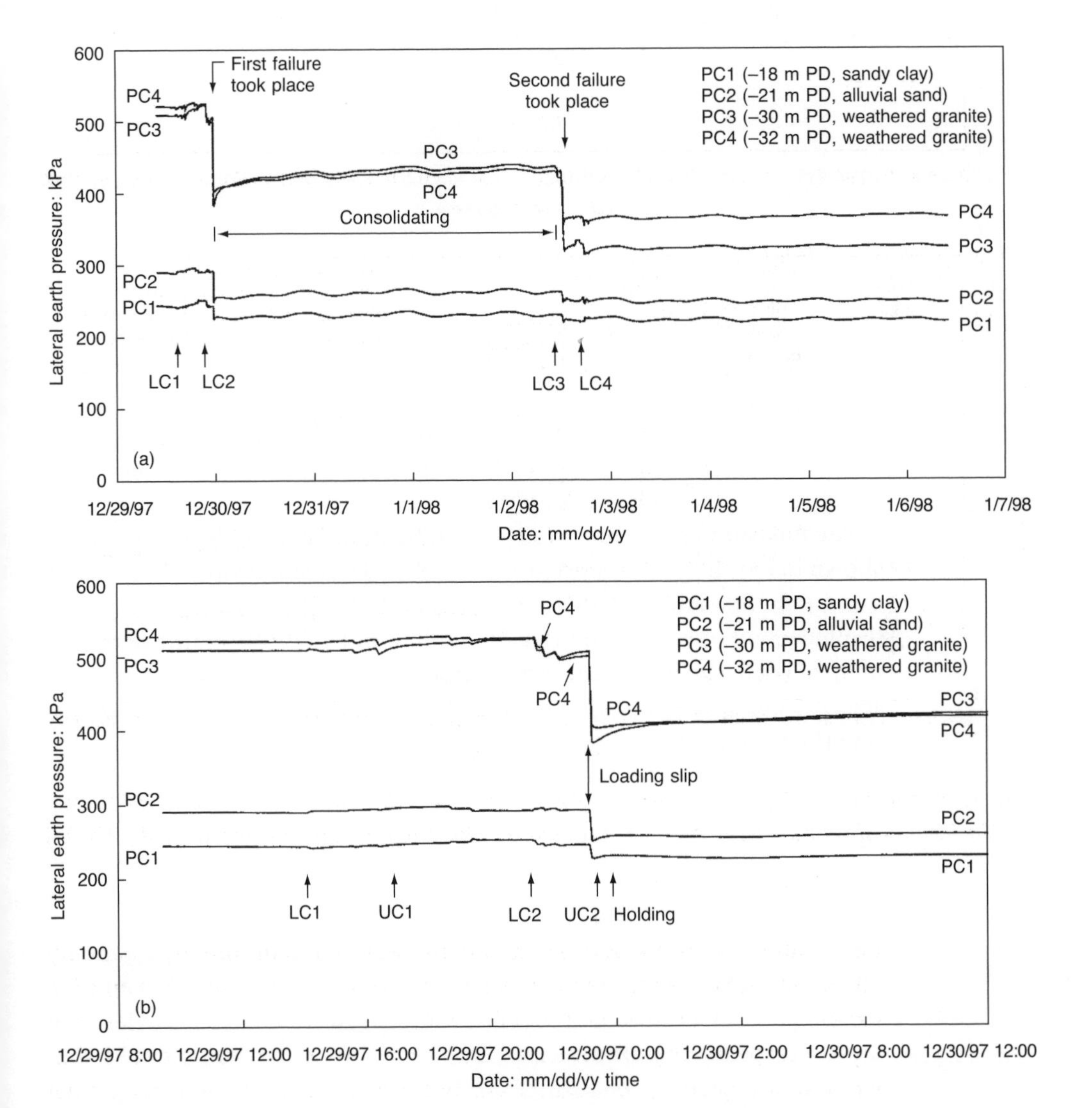

Fig. 1.29 (a) Variations of lateral earth pressure during load testing. (b) Variations of lateral earth pressure during first two load test cycles (after Ng et al., *2000b)*

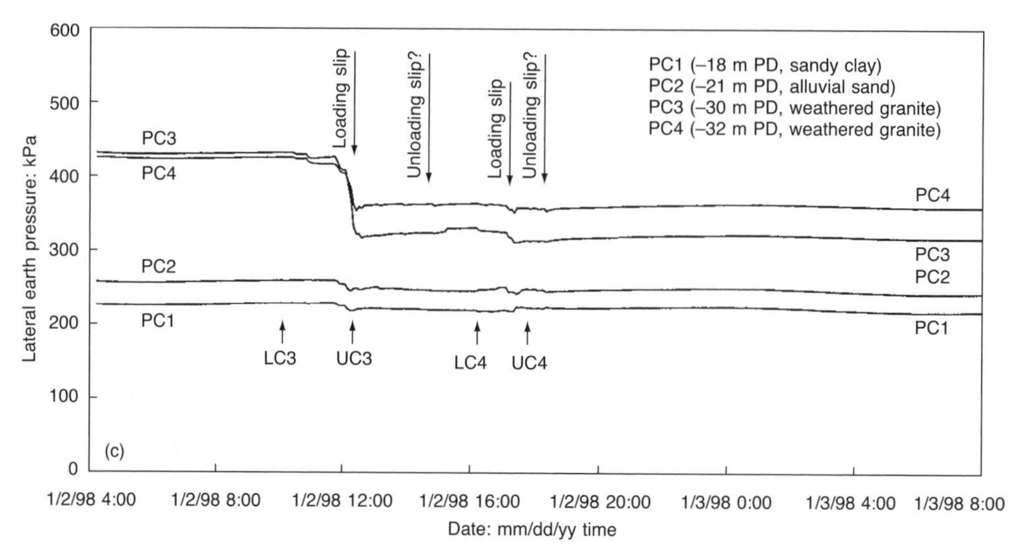

Fig. 1.29 (continued) (c) Variations of lateral earth pressure during last two load test cycles (after Ng et al.*, 2000b)*

addition, the reduction in earth pressure might be caused by a small shrinkage of the pressure cell units as a result of a fall in temperature after hydration of cement in the concrete. Lings *et al.* (1994) also reported reductions in earth pressure at the soil–diaphragm wall interface of a heavily over-consolidated stiff clay.

The measured earth pressures during the four cycles of load testing are shown in Fig. 1.29. The earth pressure cells all remained virtually constant during the first load and unload cycles (LC1 and UC1) when the barrette displacement was small (Figs 1.24, 1.29(a) and 1.29(b)). With the beginning of load cycle 2, some small drops of lateral earth pressure were seen in the weathered granite soil layer until the onset of pile failure (interface slip).

CHAPTER TEN

Case studies: 15 full-scale loading tests on barrettes in Hong Kong

Test barrettes reviewed

Tables 1.21 and 1.22 review 15 case studies of barrettes in Hong Kong of which 13 were plain (non-grouted) barrettes (identified as B1–B13) and two were shaft-grouted barrettes (G1 and G2), founded upon granitic saprolites (weathered/decomposed granitic soils). All 15 barrettes were constructed between 1987 and 1998. The design plan dimensions and depths of the barrettes ranged from 2.2 m × 0.6 m to 2.8 m × 1.0 m and from 36 to 63 m deep, respectively. The embedded depths of the barrettes into the granitic saprolites, d_g, ranged from 11.6 to 37.6 m.

Site locations and ground conditions

The typical ground conditions consist of fill material overlying a succession of marine deposits, alluvium and medium-to-coarse-grained granitic saprolites that overlie granitic rocks from the Jurassic to the Cretaceous age (see Fig. 1.30). Site investigations revealed that the typical thicknesses of the granitic saprolites range from about 10 to 40 m, corresponding to mean SPT $\bar{N}$ values ranging from about 60 to 120. The water table is located at about 1.5 m above the principal datum (PD) with a small fluctuation due to tidal action, as given in Table 1.23.

Methods of interpretation of test data

Interpretation of unit side resistance

In order to investigate the side resistance behaviour at the interface between the barrettes and the granitic saprolites, the unit side resistances were interpreted from the test data using a consistent method. The unit side resistance is defined as the side shear stress acting on the barrette shaft. For a non-sleeved barrette segment in granitic saprolite, the unit side resistance was calculated by dividing the axial load difference between the two ends of the segment by the perimeter surface area of the segment. The axial load distribution along the depth of the barrette was obtained by multiplying the average vertical strain measured from readings of the strain gauges at each level by a derived pile modulus.

Table 1.21 Summary of test barrettes in Hong Kong (after Ng and Lei, 2003a)

Identity	Depth (m)	Size (m)	d_g (m)	Toe level (m PD)	Water table (m PD)	d_s (m)	Void base	Concrete grade	Reinforcement ratio (%)	Reference
B1	53.0	2.2 × 0.6	29.5	−49.4	1.5	23.5	Yes	30/20	0.34	K. K. S. Ho (1993a)
B2	56.0	2.2 × 0.8	20.0	−51.5	?	36.0	Yes	?	?	Pratt (1989)
B3	35.9	2.2 × 0.6	13.5	−32.0	1.5	22.4	No	30/20	0.047	K. K. S. Ho (1993b)
B4	37.8	2.2 × 0.8	22.2	−27.3	1.5	15.6	No	30/20	0.39	K. K. S. Ho (1993c)
B5	55.3	2.2 × 0.8	20.0	−51.3	2.0	22.0	Yes	?	0.39	K. K. S. Ho (1994)
B6	55.6	2.8 × 1.0	37.6	−51.8	1.0	0.0	No	30/20	0.73 (30.3 m)[b] 0.29 (23.0 m)[b]	Lo (1997)
B7	39.4	2.8 × 0.8	11.6	−34.9	1.5	19.9	Yes	30/20	0.48	Ng and Lei (2003a)[d]
B8	39.5	2.2 × 0.8	12.5	−36.5	?	27.0	No	?	?	de Silva *et al.* (1998)
B9	52.5	2.2 × 0.6	23.0	−47.5	?	29.5	No	30/20	?	de Silva *et al.* (1998)
B10	41.0	2.3 × 0.6	22.6	−36.2	?	18.4	No	30/20	?	Ng and Lei (2003a)[e]
B11	63.2	2.8 × 0.8	18.0[a] 15.1[a]	−57.6	2.0	30.1	No	40/20	0.42 (30.1 m)[b] 0.66 (33.1 m)[b]	Ng and Lei (2003a)[e]
B12	49.0	2.8 × 0.8	17.0	−45.0	?	32.0	No	40/20	2.80	Ng and Lei (2003a)[e]
B13	42.7	2.8 × 0.8	15.0	−39.0	?	27.7	No	40/20	2.80	Ng and Lei (2003a)[e]
G1	51.4	2.8 × 0.8	12.3[c]	−46.3	1.6	39.1	No	45/20	0.86 (17.0 m)[b] 0.53 (12.0 m)[b] 0.34 (22.0 m)[b]	Ng and Lei (2003a)[d]
G2	39.7	2.8 × 0.8	16.0[c]	−34.6	1.6	23.7	No	45/20	0.44	Ng and Lei (2003a)[d]

Note: d_g = embedded depth of non-sleeved barrette shaft in granitic saprolite; d_s = sleeved length of barrette.
[a] Two layers of granitic saprolites with distinctly different $\bar{N}$ values.
[b] Different reinforcement ratios from top to bottom of barrette.
[c] Postgrouted.
[d] Tests carried out and data interpreted by the writers.
[e] Test data interpreted by the writers.

Table 1.22 Construction details of test barrettes (after Ng and Lei, 2003a)

Identity	Excavation			Trench opening				Concreting				Reaction system
	Tool	Total overbreak (%)	Overbreak in granitic saprolites (%)	t_e (h)	Side scraping	Base cleaning	t_s (h)	t_c (h)	Rate (m/h)	Temperature (°C)	Curing time (days)	
B1	Hydraulic grab	24.7	7.0	240.0	No	Yes	96.0	3.1	17.1	?	40	Kentledge
B2	Hydraulic grab	?	?	?	?	?	?	?	?	?	?	?
B3	Hydraulic grab	15.0	11.8	87.5	No	Yes	22.5	3.3	10.9	?	23	Reaction barrettes
B4	Hydraulic grab	9.0	11.2	40.5	No	Yes	53.5	3.3	11.5	?	29	Kentledge
B5	Hydraulic grab	?	?	?	No	Yes	?	?	?	?	?	Kentledge
B6	Mechanical grab	11.8	8.9	72.0	No	Yes	24.0	4.2	13.2	?	35	Reaction barrettes
B7	Hydraulic grab	11.8	8.4	62.0	No	Yes	43.0	4.5	10.3	27.6	21	Kentledge
B8	Mechanical grab	0.0	0.0	?	?	?	?	?	?	?	?	Kentledge
B9	Mechanical grab	19.6	5.6	40.3	?	?	27.3	3.9	13.1	?	?	Reaction barrettes
B10	?	?	?	?	?	?	?	?	?	?	?	Kentledge
B11	Hydrofraise (Mechanical grab[a])	?	?	15.5	Yes	Yes	22.5	6.0	10.5	?	32	Kentledge
B12	Hydrofraise (Mechanical grab[a])	5.7	0.0	19.0	Yes	Yes	31.5	5.5	8.9	?	40	Reaction barrettes
B13	Hydrofraise (Mechanical grab[a])	2.6	2.7	18.0	Yes	Yes	24.0	3.5	12.2	30.0	26	Reaction barrettes
G1	Hydrofraise (Mechanical grab[a])	7.3	2.0	25.0	Yes	Yes	22.0	4.0	12.9	29.0	21	Kentledge
G2	Hydrofraise (Mechanical grab[a])	16.0	4.7	20.0	Yes	Yes	13.0	3.4	11.7	29.0	28	Kentledge

Note: t_e = duration of excavation; t_s = standing time elapsed after excavation and before concreting; t_c = duration of concreting.
[a] Excavation tool for pretrenching.

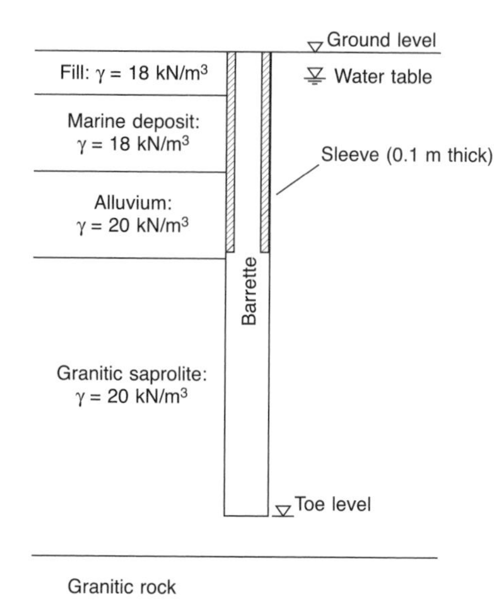

Fig. 1.30 Side view of typical test barrette (not to scale; after Ng and Lei, 2003a)

The vertical strains were calculated with the readings just before the loading test being taken as the reference datum. The derived pile modulus is the composite axial rigidity of concrete and steel in the barrette. The Young's modulus of the barrette was obtained by calibration against the applied load at the barrette head and the average vertical strain at the top set of gauges. In deriving the unit side resistance, the actual as-built cross-sectional area of the barrettes from the concreting record was considered.

To investigate the degree of mobilization of the barrette side resistance in granitic saprolites, the local displacements were also determined from the test data. The local displacement is the displacement deduced at the vertical centre of the non-sleeved barrette segment in the granitic saprolite. It was calculated by subtracting the measured barrette head movement from the shortening between the barrette head and the mid-height of the segment. The shortening was estimated using the measured vertical strains.

Mobilization rating system

Among many other factors, it is well recognized that the magnitude of the side resistance mobilized along the pile–soil interface is also a function of relative pile–soil movement, that is, the local displacement. A minimum amount of local displacement is needed to mobilize the side resistance fully. In many circumstances, the full capacity of the side resistance may not be achieved in full-scale barrette loading tests. In order to compare

Table 1.23 Summary of results from tested barrettes (after Ng and Lei, 2003a)

Identity	Concrete test		Loading test data			Unit shaft resistance in sleeved zone (kPa)	Interpreted result in granitic saprolite			Mobilization rating
	Average uniaxial compressive strength [MPa (days)]	Average Young's modulus [GPa (days)]	Maximum test load (kN)	Maximum pile head movement (mm)	Residual pile head movement (mm)		Maximum unit shaft resistance τ_s (kPa)	Mean SPT value $\bar{N}$	Mean effective vertical stress $\overline{\sigma'_v}$ (kPa)	
B1	?	26.3 (120)	9 800	19.0	7.3	10	51, 56[a]	66	377	*B*
B2	?	?	18 000	34.0	?	?	152	61	468	*C*
B3	?	?	18 000	36.7	13.8	36	127	132	290	*B*
B4	?	?	18 000	29.2	6.7	30	104	80	286	*B*
B5	?	?	20 000	34.2	19.1	15	89	61	449	*B*
B6	53.0 (28)	?	29 245	15.1	3.2	No sleeve	63, 78[a]	66	366	*A*
B7	37.5 (28)	26.1 (28)	8 905	99.3	91.4	25	37	24	335	*A*
B8	?	?	18 000	80.8	64.9	11	117	82	330	*A*
B9	37.7 (21)	?	24 282	46.9	20.0	30	156	119	386	*A*
B10	?	?	23 570	67.9	37.0	30	106	109	323	*B*
B11	?	?	29 558	32.5	8.8	34	78, 99[a]	50	367	*B*
							204	187	532	*C*
B12	55.5 (36)	?	9 211	51.5	44.7	12	49	43	382	*A*
B13	43.8 (23)	?	18 000	49.6	34.3	10	119	96	334	*A*
G1	57.0 (21)	?	30 500	64.2	25.9	15	232	166	447	*B*
G2	49.7 (21)	?	27 600	63.4	29.5	24	145	77	309	*A*

[a] Extrapolated maximum unit shaft resistance at 1% equivalent diameter of the barrette.

the interpreted values of the side resistance among the test barrettes at similar degrees of mobilization, a simple three-grade mobilization rating (MR) system (Ng *et al.*, 2001) was adopted to classify the degree of mobilization of the barrette side resistance.

The mobilization rating *A* (i.e. an MR = *A*) is assigned to test results that fall within 10% of the ultimate unit side resistance determined using Brinch Hansen's (1963) 80% criterion. The '80% criterion' defines failure as the stress (or the unit side resistance) at which the strain (or the local displacement) is equal to four times the strain at 20% less stress. The criterion assumes a hyperbolic relationship between stress and strain.

The rating *B* is assigned to results that do not achieve a rating of *A* but that achieve or have been extrapolated to a local displacement of at least 1% of the equivalent diameter of the barrette, assessed on an equal area basis. A displacement of 1% of the pile diameter is used as a benchmark because of the availability of test results recorded at and extrapolated to this displacement in the literature. It is also defined by the GEO (1996) as corresponding to the 'substantial' mobilization of the capacity of the side resistance, although not necessarily the full mobilization of capacity. The rating *C* is assigned to results for which the degree of mobilization is unknown or the displacement was less than 1% of the equivalent diameter of the barrette.

The mobilization ratings assigned to the test barrettes are shown in the last column of Table 1.23, together with the loading test data and the interpreted results. In addition, the concrete test results and the uncorrected mean SPT $\bar{N}$ values of granitic saprolites near the non-sleeved barrette shafts before barrette installation are given in Table 1.23.

Mobilization of side resistance in granitic saprolites

Figure 1.31 shows the mobilization of the unit side resistance with local displacement at the interface between the barrettes and the granitic saprolites. Figure 1.31b shows that, of the 13 plain barrettes, the lower portion of the shaft of barrette B11 (i.e. B11b) has the largest (yet seemingly far from fully) mobilized unit side resistance of 204 kPa, whereas barrette B7 has the smallest ultimate unit side resistance of 37 kPa (Fig. 1.31a). This is most likely because their corresponding mean SPT $\bar{N}$ values of granitic saprolites are, respectively, the largest and smallest ($\bar{N} = 187$ for B11b and $\bar{N} = 24$ for B7; see Table 1.23). Compared with most of the plain barrettes, the shaft-grouted barrettes Gl and G2 show a relatively larger unit side resistance response to a given local displacement and a higher ultimate unit side resistance, as expected.

In this section, the correlations are studied between the maximum unit side resistance τ_s and the $\bar{N}$ value and the mean effective vertical stress σ'_v.

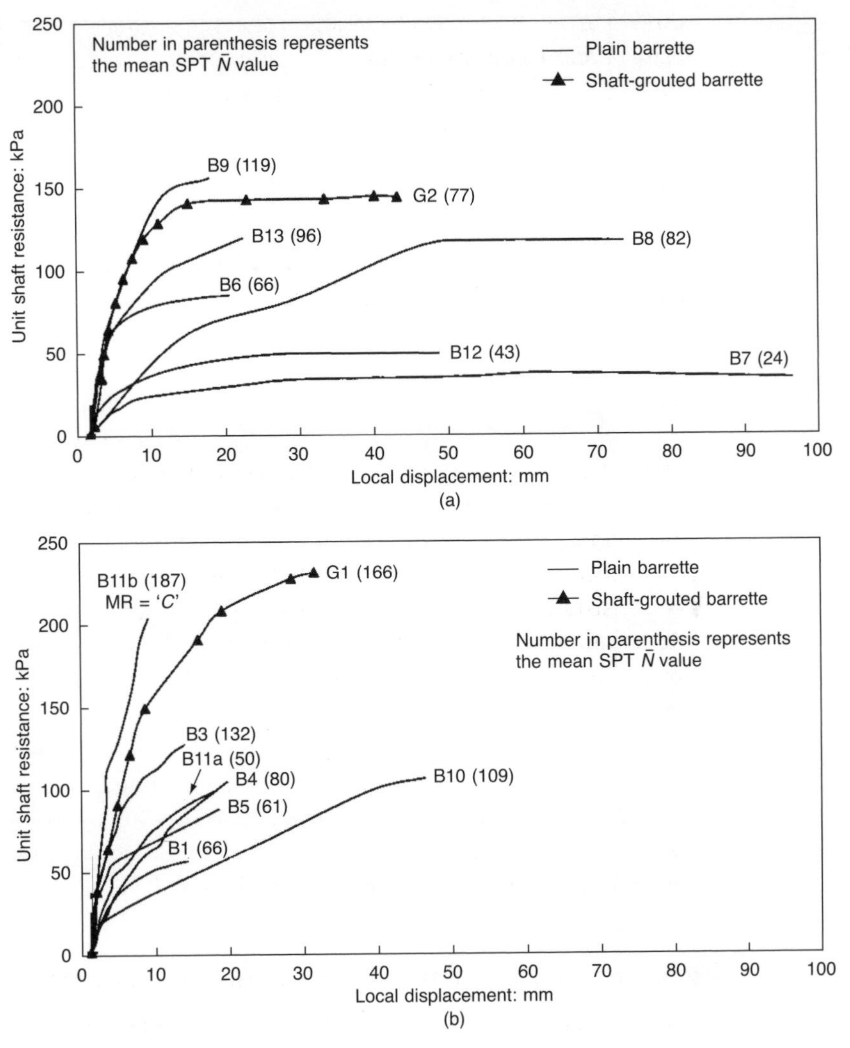

Fig. 1.31 Mobilization of unit shaft resistance with local displacement for barrettes with Mobility Rating (a) A *and (b)* B *and* C *(after Ng and Lei, 2003a)*

The traditional correlation between τ_s and $\bar{N}$ is purely empirical, whereas the correlation between τ_s and σ'_v is based on the principle of effective stress with an assumption of zero adhesion at the barrette interface and the ratio τ_s/σ'_v gives the well-known β value. Apart from these correlations, alternative empirical relationships are also studied, such as β versus $\bar{N}$, $\tau_s/\bar{N}$ versus $\bar{N}$, $\tau_s/\bar{N}$ versus pile depth and β versus pile depth.

Figure 1.32(a) shows the relationship between the maximum mobilized unit side resistance τ_s and the mean SPT $\bar{N}$ value of granitic saprolite for barrettes with an MR = A, that is, more than 90% of the ultimate unit

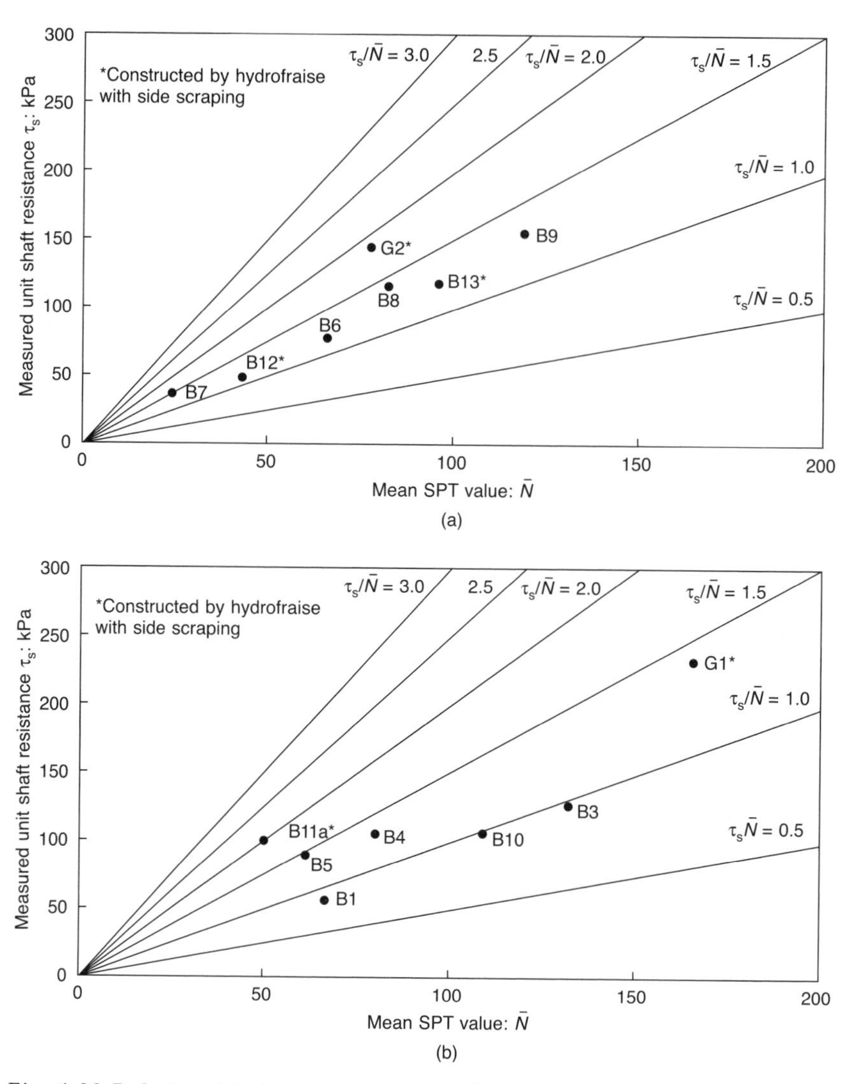

Fig. 1.32 Relationship between measured unit shaft resistance and mean Standard Penetration Test N̄ *value for barrettes with Mobility Rating (a)* A *and (b)* B *(after Ng and Lei, 2003a)*

side resistance has been mobilized. Tabulated values are presented in Table 1.23. Figure 1.32(b) shows the relationship between the maximum mobilized unit side resistance τ_s and the mean SPT $\bar{N}$ value of granitic saprolites for barrettes with an MR $= B$, that is, the maximum mobilized unit side resistance is less than 90% of the ultimate unit side resistance. Tabulated values are presented in Table 1.23.

Figure 1.33 shows the values of $\tau_s/\bar{N}$ plotted against $\bar{N}$ for plain barrettes with an MR $= A$. Although the correlation between the values

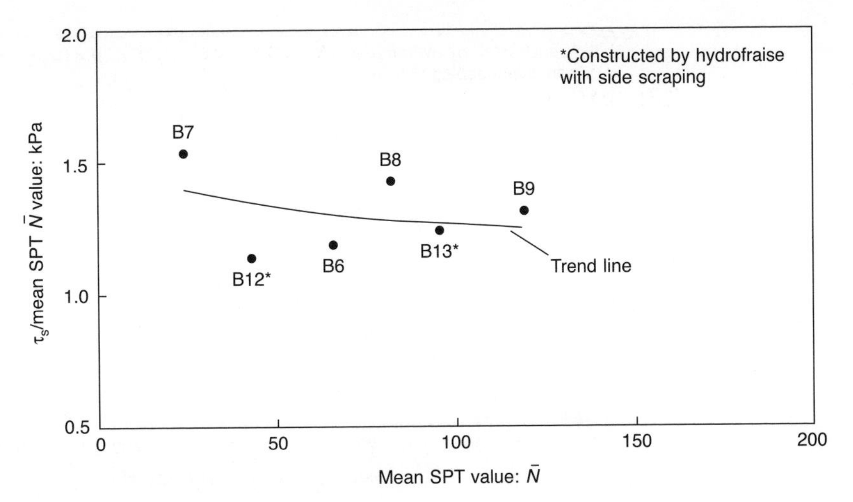

Fig. 1.33 $\tau_s/\bar{N}$ versus $\bar{N}$ for barrettes with Mobility Rating A (after Ng and Lei, 2003a)

of $\tau_s/\bar{N}$ and $\bar{N}$ is fairly weak, it appears that $\tau_s/\bar{N}$ decreases slightly with increasing $\bar{N}$ values, as shown by the trend line (the regression line). This is similar, although the similarity is relatively weak, to the trend observed for barrettes in old alluvium in Singapore (Ho and Lim, 1998).

Figure 1.34 shows the $\tau_s/\bar{N}$ values plotted against the embedded depths of plain barrettes with an MR $= A$. It is evident that the $\tau_s/\bar{N}$ value decreases with increasing pile depth. This might be due to the increase in the degree of construction disturbance to the granitic saprolites as the

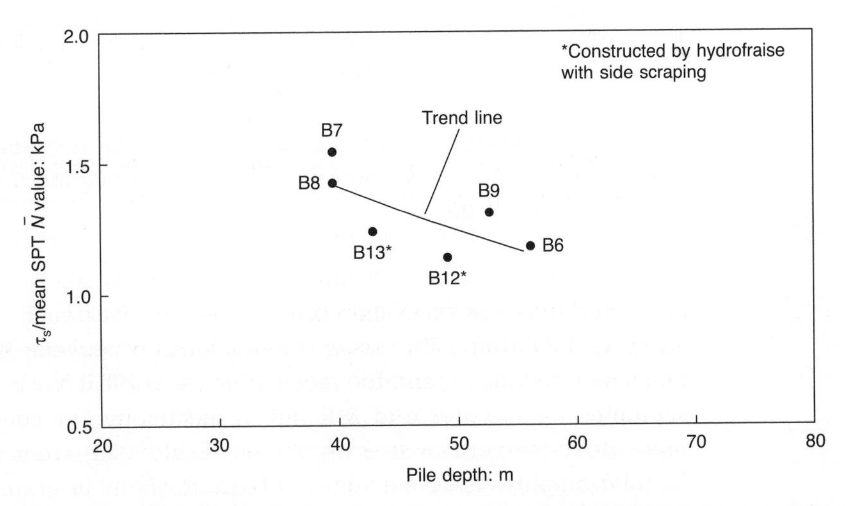

Fig. 1.34 $\tau_s/\bar{N}$ versus pile depth for barrettes with Mobility Rating A (after Ng and Lei, 2003a)

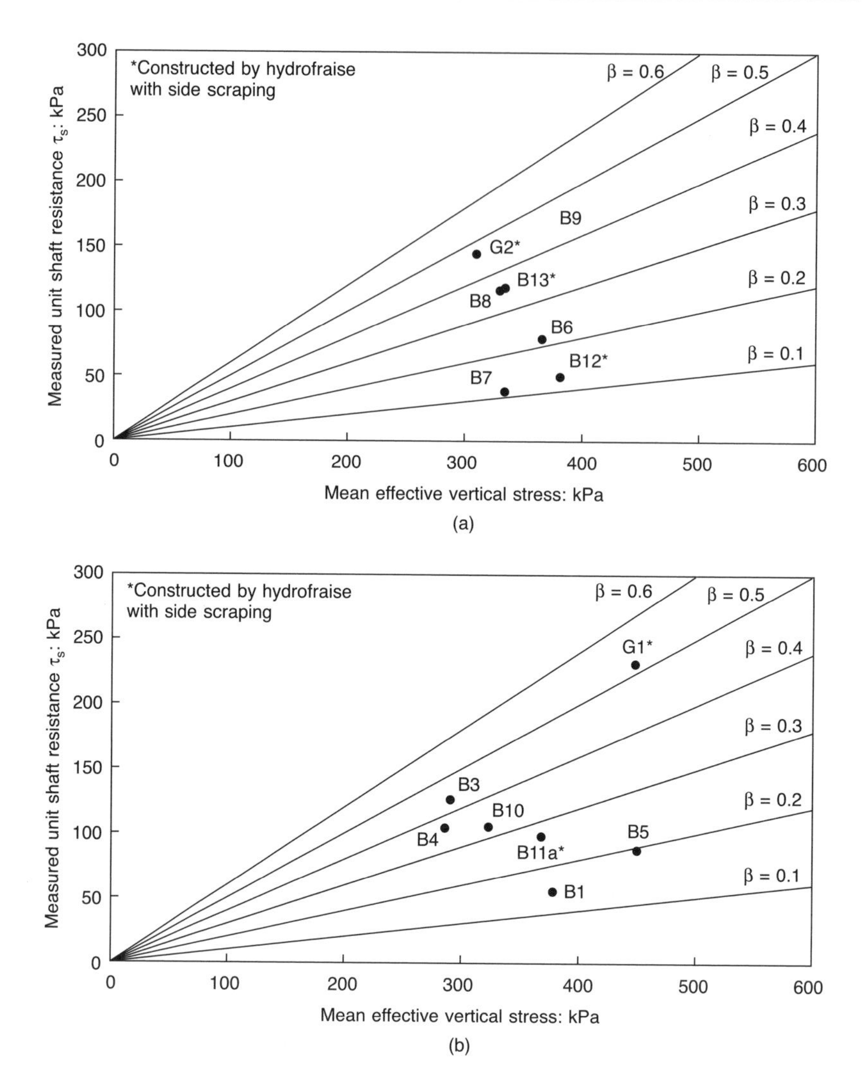

Fig. 1.35 Relationship between measured unit shaft resistance and mean effective vertical stress for barrettes with Mobility Rating (a) A *and (b)* B *(after Ng and Lei, 2003a)*

depth increases. Relatively larger stress relief is likely to have been induced during the excavation of a deeper barrette trench.

Figure 1.35(a) and (b) shows the relationship between the maximum unit side resistance τ_s and the mean effective vertical stress σ'_v in granitic saprolites for MR $= A$ and MR $= B$, respectively. For consistency, the mean effective vertical stresses σ'_v (see Table 1.23) were calculated at the mid-heights of the non-sleeved barrette shafts in granitic saprolites by assuming that the unit weights of the fill and marine deposits are $18.0\,kN/m^3$, the unit weights of the alluvium and granitic saprolites are

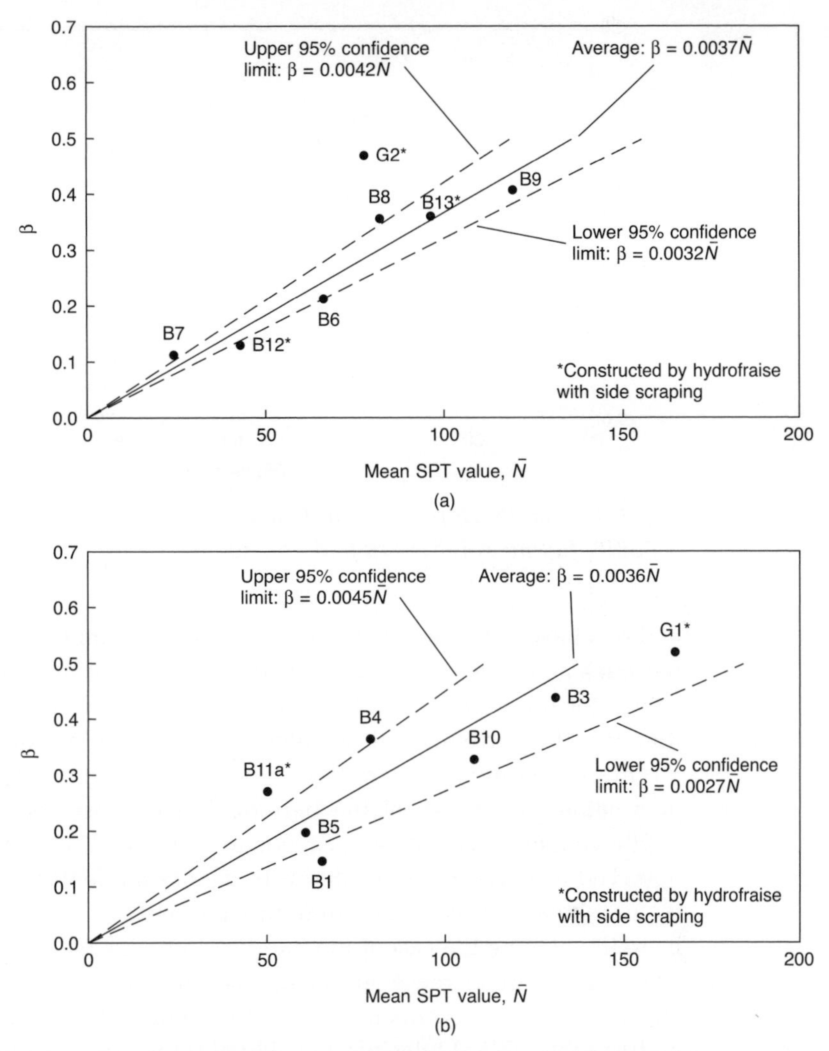

Fig. 1.36 Relationship between β value and mean Standard Penetration Test $\bar{N}$ *value for barrettes with Mobility Rating (a)* A *and (b)* B *(after Ng and Lei, 2003a)*

20.0 kN/m^3 (see Fig. 1.30), and the water table is at 1.5 mPD if the values are not known.

Figure 1.36(a) and (b) shows the correlation of β with $\bar{N}$ for barrettes with an MR = A and an MR = B, respectively. For plain barrettes, the β values lie between $0.0030\bar{N}$ and $0.0046\bar{N}$, with an average value of $0.0037\bar{N}$, a 95% confidence range of $0.0032\bar{N}$–$0.0042\bar{N}$, and a Coefficient of Variance (COV) of 0.17. Figure 1.37 shows the variation of β values with embedded depths of plain barrettes with an MR = A. Barrettes B8, B9 and B13 have $\bar{N}$ values greater than 80 (see Table 1.23), and barrettes B6, B7

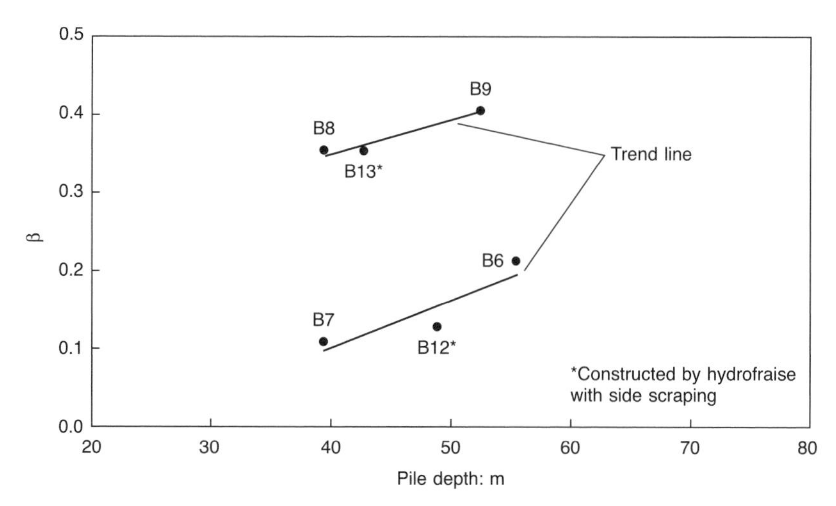

Fig. 1.37 Relationship between β value and depth for barrettes with Mobility Rating A (after Ng and Lei, 2003a)

and B12 have $\bar{N}$ values less than 70. The β values for these two groups of barrettes lie between two distinctly different ranges. It is clear that the β values for the barrettes with $\bar{N}$ values greater than 80 are higher than those for the barrettes with $\bar{N}$ values less than 70. Although the results for these two groups of barrettes are very limited, a trend of an increase in β value with depth of the barrette can be observed, but a critical depth cannot be identified, as pointed out by Kulhawy (1984). The observed increase in the β value with increasing depth might be caused by an increase in the shear strength with depth at the interface. This is more likely to be the case if the interface contains both bentonite and in situ soil, as an increase in shear strength with depth would not be expected if failure occurs in the bentonite only. This observation (Wates and Knight, 1975; Littlechild and Plumbridge, 1998) suggests that if a bentonite filter cake does exist, it consists of some amount of in situ soil.

Construction effects on side resistance

Effects of grouting

The above-interpreted test results have shown that, compared with plain barrettes, the side resistance response to local displacement for shaft-grouted barrettes is relatively stiffer. The average β and $\tau_s/\bar{N}$ values for shaft-grouted barrettes are about 1.3 and 1.45 times those for plain barrettes, respectively. The grouting effect in improving the side resistance for the barrettes in granitic saprolites is comparably as effective as that for a barrette in sand (Hamza and Ibrahim, 2000), which shows that the resistance of the grouted shaft is 1.5 times that of the non-grouted shaft. The stiffer side resistance response to local displacement and the higher maximum mobilized unit side resistance both result from the grouting which increases both the lateral pressure (i.e. the normal stress) on the barrette shaft and the shear strength of the barrette–granitic saprolite interface.

Effects of side scraping and excavation tools

After excavation and before base cleaning of a barrette trench, scraping the filter cake on the sidewalls of the barrette trench is sometimes carried out in practice, with the intention of minimizing the possible adverse effect of filter cake on the side resistance as reported in some references (O'Neill and Reese, 1970; Wates and Knight, 1975; Cernak, 1976; Baker *et al.*, 1994; Cheng and Haberfield, 1998). However, it is postulated that the side scraping action would smooth the sidewalls of the barrette trench, leading to a lower frictional angle at the barrette–soil interface, although the filter cake may be fully or partially removed. In this study, side scraping was carried out on shaft-grouted barrettes G1 and G2 and plain barrettes B11–B13, but not on barrettes B1 and B3–B7 (see Table 1.22). Figures 1.32(a), 1.35(a) and 1.36 show the relationships between τ_s and $\bar{N}$, τ_s and σ'_v, and β and $\bar{N}$, respectively, giving the $\tau_s/\bar{N}$, β and $\beta/\bar{N}$ values for barrettes with an MR $= A$. A superscript asterisk attached to the relevant symbols is used to identify the barrettes constructed with side scraping.

Effects of duration of excavation and trench standing time

Among many other factors, the maximum mobilized unit side resistance τ_s is also a function of soil density and confining pressure at the barrette–soil interface, which can be characterized roughly by the σ'_v and $\bar{N}$ values. In order to estimate the effects of duration of excavation and trench standing time on the maximum mobilized unit side resistance, the values of $\beta/\bar{N}$ ($= (\tau_s/\sigma'_v)/\bar{N}$) for the plain barrettes were compared, for the sake of minimizing the influence of soil density and confining pressure on the side resistance. Therefore, the $\beta/\bar{N}$ value is to some extent representative of a 'reference' capacity of the maximum unit side resistance mobilized under similar soil density and confining pressure conditions. The higher the $\beta/\bar{N}$ value, the greater the mobilization capacity of the maximum unit side resistance. Ideally, the duration of excavation or the rate of excavation in granitic saprolites should be used for this purpose. However, due to the absence of these data, only the duration of excavation for the entire barrette, as presented in Table 1.22, was used.

CHAPTER ELEVEN

Case studies: bored piles in weathered materials in Hong Kong

Overview: key questions

A total of 28 full-scale load tests are reviewed to answer the following key questions about the behaviour of bored piles:

- What is the importance of degree of mobilization of side resistance in 'floating' bored piles (i.e. that rely on side resistance for their load capacity)?
- What are realistic design values of side resistance and settlement?
- What is the influence of construction factors such as duration, use of bentonite and grouting?

Introduction

Deep piled foundations are commonly required to resist the large vertical and lateral wind loads acting on tall commercial and residential buildings in Hong Kong. Large-diameter circular bored piles (drilled caissons or piers), constructed using rope-operated grabs and/or RCD techniques, have often been used in saprolitic soils. 'Saprolites' refer to soils which retain various degrees of the original texture, fabric and structure of the parent rock from which they were formed by decomposition. In Hong Kong, with little exception, 'floating'-type bored piles which rely largely on side shear resistance for capacity, are founded within saprolites and 'toe bearing'-type piles, founding on rock, penetrate through saprolites. Compared with the number of toe bearing piles constructed, the number of floating piles has been small due to overregulated practice in Hong Kong.

In Hong Kong, the adopted design method for side shear resistance of large-diameter floating bored piles is usually a direct correlation of side shear resistance with the uncorrected Standard Penetration Test (SPT) blow count value, $\bar{N}$. The SPT blow count is termed 'uncorrected' due to the fact that no energy correction for the type of test apparatus or depth of bore is made. Calculations of capacity are made using a side shear correlation for the entire saprolite layer rather than for individual layers within the saprolite. Therefore the mean N value for the layer, $\bar{N}$, is used and the side shear resistance calculated is the 'mean side shear resistance.' Although GEO (1996) recommends a value for $\bar{\tau}/\bar{N}$ (mean side shear resistance normalized to the mean SPT $\bar{N}$ value) of 0.8 to 1.4 (kPa) be used for preliminary design, with the $\bar{N}$ value limited to 200, these design parameters must be verified by large-scale load testing.

Typical ground conditions and soil types in Hong Kong

The geology of Hong Kong is influenced heavily by subtropical weathering. Typically, superficial deposits of fill, alluvium and marine deposits overlie weathered rock of varied depth or colluvium on hillsides, beneath which lies fresh rock. The most common rock types include granite and volcanic tuff and rhyolite. The weathering profiles may include corestones and may be somewhat erratic but can also vary gradually from soil to rock. Table 1.20 sets out the classification of rock decomposition grades used in Hong Kong (GCO, 1988). In Table 1.20, saprolites are equivalent to 'rock' of decomposition grade IV and V. The physical properties of grade IV and V material are closer to those of soil than rock. Therefore, generally in Hong Kong, the term 'rock' refers only to material of decomposition grades I to III with I being fresh rock. Grade IV, V and VI materials are referred to as 'soil'. The occurrence of rock of weathering grade VI, referred to as 'residual soil' is uncommon in Hong Kong and is usually found only in thin layers. Since obtaining high-quality granular granitic and volcanic saprolites from great depths (more than 20 m) is extremely difficult if not impossible for laboratory testing, the SPT is the most common method adopted in Hong Kong to characterize their geotechnical properties empirically. Recently it has been found that the shear modulus of granitic saprolites is highly non-linear even at very small strains (Ng *et al.*, 2000a; Ng and Wang, 2001). However, for design purposes, a wide range of empirical correlations of Young's modulus with SPT $\bar{N}$ values such as $E = 200$ to $3000N$ (kPa) have been proposed (GEO, 1996). Moreover, empirical relationships of shear strength parameters (i.e. effective cohesion intercept, c' and angle of shearing resistance, ϕ') with SPT $\bar{N}$ values and fine contents are given by Pun and Ho (1996). Typically, ϕ' varies between 33° and 44° and c' changes from 0 to 6 kPa, depending on SPT $\bar{N}$ values and fine contents.

Load test procedures

For compressive static load testing of bored piles, hydraulic jacks are commonly used to apply load to the pile head with kentledge used as the reaction system. The kentledge is normally stacked concrete blocks, steel billets or Universal steel sections (see Fig. 1.38). The measurement of the applied load is achieved using load cells positioned under each

Fig. 1.38 Kentledge reaction stack for testing piles in saprolitic soils in Hong Kong

hydraulic jack. In two of 12 recent tests, an additional hydraulic jack was installed at the base of the pile. This was an Osterberg load cell (Osterberg, 1989), and was used to apply load directly to the pile at its base while applying an uplifting force to the pile shaft. Tests are generally maintained-load types in which the load is applied in increments, each being held until the rate of movement has reduced to an acceptably low value before the next load increment or decrement is applied. Two to five loading and unloading cycles are commonly used in each test, which consists of a number of increments or decrements (Li, 2000; Yau, 2000).

Load tests under review

Tables 1.24 and 1.25 summarize some details of the tests. For some cases from literature, the decomposition grade is not available and so cannot be reported in Table 1.25. However, it is known that the results are in saprolites (grade IV or V material) and the decomposition grade itself is not used as a basis for the analysis. Mean results are reported on the basis of each saprolite layer for each pile with the exception of a few piles. These have separate layers distinguished by decomposition grade or the inclusion or omission of post-grouting.

Table 1.24 Database of large-diameter floating bored piles in saprolitic soils in Hong Kong (after Ng et al., *2001b)*

Pile No.	References	Pile length (m)	Pile dia. (m)	Saprolite	Method of construction
1	Holt *et al.* (1982)	36.9	1.0	Granitic	RCD in water
2	De Silva *et al.* (1998)	45.0	1.8	Granitic	Grab, casing and water
3	GEO (1993)	52.1	1.5	Granitic	Grab, casing and water
4	GEO (1993)	40.6	1.5	Granitic	Grab, casing and water
5	GEO (1993)	42.2	1.5	Granitic	Grab, casing and water
6	GEO (1992)	65.0	1.0	Granitic	RCD in bentonite
7	GEO (1992)	75.0	1.0	Granitic	RCD in bentonite
8	Ho (1992)	32.8	1.2	Volcanic	Grab, casing and water
9	Ho (1992)	36.8	1.2	Volcanic	Grab, casing and water
10	Fraser and Kwok (1986)	30.0	1.5	Granitic	Grab, casing and water then RCD in water
11	Fraser and Kwok (1986)	22.6	1.5	Granitic	Grab, casing and water
12	Fraser and Kwok (1986)	22.0	1.5	Granitic	Grab, casing and water
13	GEO (1996)	48.2	1.5	Granitic	Grab, casing and water
14	GEO (1996)	48.0	1.0	Granitic	Grab, casing and water
15	GEO (1996)	42.6	1.5	Granitic	RCD in bentonite
16	GEO (1996)	59.1	1.5	Granitic	RCD in bentonite
17	Li (2000); Yau (2000)	40.6	1.2	Karstic surface deposits (metasedimentary)	Grab, casing and water
18	Li (2000); Yau (2000)	31.6	1.2	Rhyolite (volcanic)	Grab, casing and water then post-grouted
19	Li (2000); Yau (2000)	50.7	1.5	Granitic	Grab, casing and water then RCD in water
20	Li (2000); Yau (2000)	54.5	1.2	Granitic	Grab, casing and water then post-grouted
21	Li (2000); Yau (2000)	32.5	1.2	Tuff (volcanic)	Grab, casing and water
22	Li (2000); Yau (2000)	39.4	1.4	Metasiltstone (metasedimentary)	Grab, casing and water
23	Li (2000); Yau (2000)	30.2	1.5	Metasiltstone (metasedimentary)	Grab, casing and water then post-grouted
24	Li (2000); Yau (2000)	50.9	1.2	Metasiltstone/schist (metasedimentary)	RCD in bentonite
25	Li (2000); Yau (2000)	50.2	1.2	Metasiltstone/schist (metasedimentary)	RCD in bentonite
26	Li (2000); Yau (2000)	27.5	1.2	Rhyolite (volcanic)	Grab, casing and water
27	Li (2000); Yau (2000)	39.9	1.2	Rhyolite (volcanic)	Grab, casing and water
28	Li (2000); Yau (2000)	39.8	1.2	Rhyolite (volcanic)	Grab, casing and water

RCD = Reverse Circulation Drill

Table 1.25 Load test results for large-diameter floating bored piles in saprolitic soils in Hong Kong (after Ng et al., *2001b)*

Pile No.	Weathering grade	Mean SPT value $\bar{N}$	Mean effective vertical stress $\bar{\sigma}'_v$ (kPa)	Maximum side shear resistance				Maximum local displacement δ_{max} (% of pile diameter)	Estimated degree of mobilization (%)	Mobilization rating (MR)	Notes
				$\bar{\tau}_{s\,max}$ (kPa)	$\frac{\bar{\tau}_{s\,max}}{\bar{\sigma}'_v}$	$\frac{\bar{\tau}_{s\,max}}{\bar{N}}$ (kPa)	Achieved/ extrapolated				
1	?	>100	268	129	0.5	1.3	Achieved	3.9	94	*A*	*Ψ
2	V	74	310	84	0.3	1.1	Achieved	1.8	100	*A*	*Ψ
3	?	80	360	43	0.1	0.5	Extrapolated to 1% of diameter	0.5	100	*A*	*Ψ
4	?	107	302	77	0.3	0.7	Extrapolated to 1% of diameter	0.3	85	*B*	Ψ
5	?	65	270	96	0.4	1.5	Extrapolated to 1% of diameter	0.7	?	*B*	Ψ
6	?	92	624	155	0.3	1.7	Achieved	3.2	100	*A*	Ψ
7	?	68	627	72	0.1	1.1	Achieved	0.6	?	*C*	Ψ
8	?	35	194	30	0.2	0.9	Achieved	0.3	?	*B*	
9	?	78	205	50	0.2	0.6	Extrapolated to 1% of diameter	0.4	?	*B*	
10	?	55	184	22	0.1	0.4	Achieved	?	?	*C*	
11	?	80	134	80	0.6	1.0	Achieved	?	?	*C*	
12	?	65	120	23	0.2	0.4	Achieved	?	?	*C*	
13	?	104	319	45	0.1	0.4	Achieved	~0.1	?	*C*	
14	?	140	398	77	0.2	0.6	Achieved	1.0	?	*B*	
15	?	97	250	19	0.1	0.2	Achieved	1.3	?	*B*	
16a	?	77	223	28	0.1	0.4	Achieved	1.2	?	*B*	
16b	?	200	457	82	0.2	0.4	Achieved	1.3	?	*B*	
17	V	29	290	52	0.2	1.8	Achieved	0.6	?	*C*	Ψ
18	V post-grouted	44	224	180	0.8	4.1	Achieved	2.2	92	*A*	Ψ
19a	V	75	327	104	0.3	1.4	Achieved	0.4	?	*C*	ΨΩ
19b	IV/V	>200	429	367	0.9	1.8	Achieved	0.7	?	*C*	ΨΩ
20	V post-grouted	79	351	145	0.4	1.8	Achieved	2.3	96	*A*	Ψ

21a	V	29	188	196	1.0	6.8	Achieved	0.8	?	*C*	Ψ
21b	IV/V	166	256	258	1.0	1.6	Achieved	0.1	?	*C*	Ψ
22	V	51	229	115	0.5	2.3	Achieved	>1.0	?	*B*	Ω
23a	V	36	141	115	0.8	3.2	Achieved	7.7	100	*A*	Ψ
23b	V post-grouted	131	243	144	0.6	1.1	Achieved	7.6	94	*A*	Ψ
24	V	41	286	14	0.1	0.3	Achieved	1.0	100	*A*	Ψ
25	V	20	223	18	0.1	0.9	Achieved	0.9	90	*A*	Ψ
26	V	46	220	83	0.4	1.8	Achieved	1.1	96	*A*	Ψ
27	V	72	290	43	0.2	0.6	Achieved	1.9	99	*A*	Ψ
28	V	108	290	62	0.2	0.6	Achieved	2.3	100	*A*	Ψ

Note: * Data estimated from publication.
Ψ Side shear–displacement data available.
Ω Tested using an Osterberg cell.

The local pile displacements given in Table 1.25 are those deduced at the maximum achieved side shear resistance during each test, at the vertical centre of the saprolite layer. Local displacements were calculated using the measured pile-head displacement and shaft shortening, estimated using extensometers and strain gauges in the pile shaft. The results of side shear resistance are either the maximum achieved during each test or a value obtained by extrapolation to a greater displacement than actually achieved. Side shear resistance was separated from the

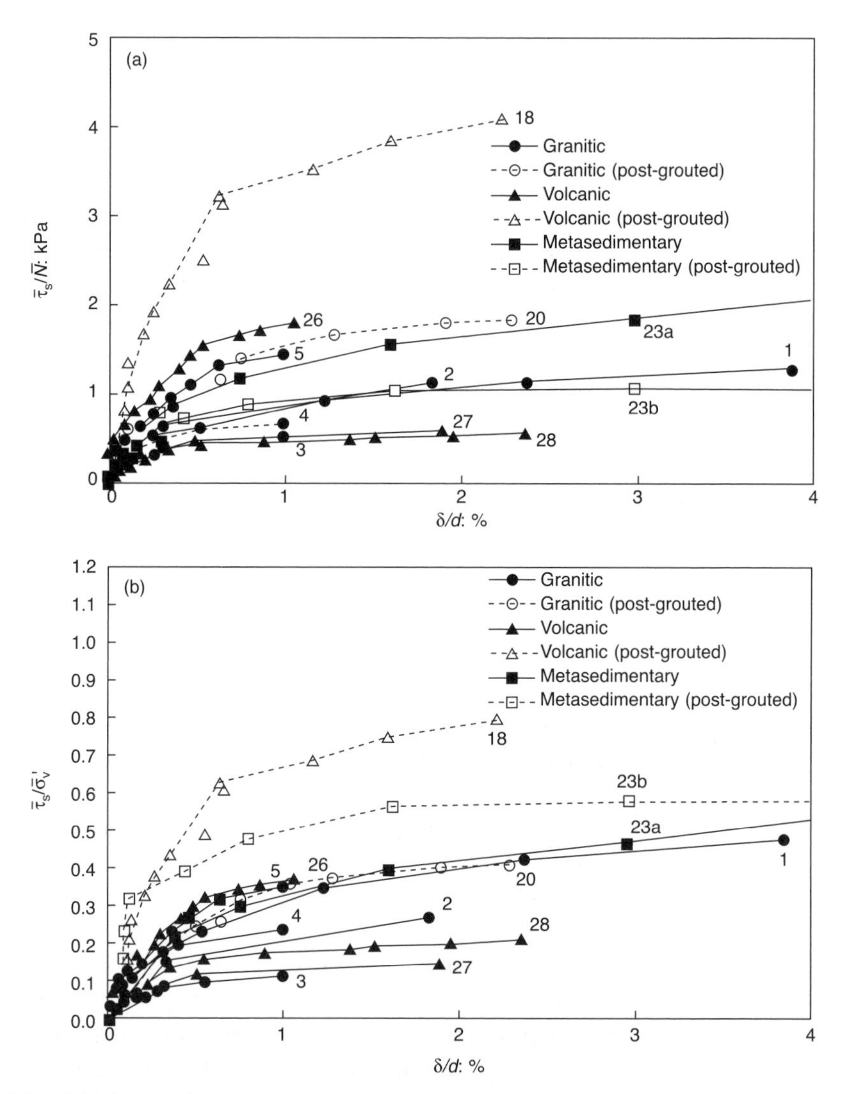

Fig. 1.39 Normalized side shear–displacement curves for piles constructed under water with MR = A or B: (a) normalized by mean STP value $\bar{N}$, (b) normalized by mean effective vertical stress $\bar{\sigma}'_v$ (after Ng et al., *2001b)*

pile toe resistance using calculations combining strain gauge measurements with an estimation of the pile modulus of elasticity.

Degree of mobilization of side shear resistance

Results of normalized average side shear resistance $\bar{\tau}_s$–displacement relationships for the piles constructed under water are shown in Fig. 1.39. These consist of piles excavated by hammer grab within temporary casings including the post-grouted piles and piles excavated by Reverse Circulation Drill (RCD) under water without temporary casings. Only the results with an MR rating equal to *A* or *B* are shown, as the results with an MR equal to *C* add little to the study of the degree of mobilization of side shear resistance. In Fig. 1.39(a) mobilized average side shear resistance is normalised by $\bar{N}$ and in Fig. 1.39(b) by $\bar{\sigma}'_v$. Tabulated values are given in Table 1.25. The local pile displacement δ was calculated using the measured pile head displacement and shaft compression deduced at the vertical centre of each saprolite layer. It should be noted that the average side shear resistance was taken along the depth of each soil stratum (not the entire pile length). With the limited number of strain gauges (i.e. large spacing) installed along the depth of each commercial test pile, the measured data are not sufficient to warrant a detailed investigation of variation of side shear resistance along the pile length or within each soil stratum.

Correlation of side shear resistance with SPT

Piles constructed under water

The average maximum side shear resistance ($\bar{\tau}_{s\,max}$) for the tests on piles constructed under water and having an MR rating equal to *A* or *B* are shown in Fig. 1.40 and other details are given in Table 1.25. Note that no clear difference in performance could be observed between the piles excavated by grab under water within temporary casings and the piles excavated by RCD under water without casings. However, only one case is available in the RCD group with an MR = *A* or *B*. Scatter is evident in both Fig. 1.40(a) and 1.40(b). Using a direct correlation of side shear resistance with $\bar{N}$, the results generally lie between $0.5\bar{N}$ and $2.0\bar{N}$. The mean side shear resistance is $1.2\bar{N}$ (kPa) with a 95% confidence range of 0.8 to $1.6\bar{N}$ (kPa). The figure has a well-defined lower bound of $0.5\bar{N}$ (kPa), which appears to be particularly suitable for the piles in volcanic saprolites. The post-grouted piles on average achieved higher normalized side shear resistance than that of the non-grouted piles, but are significantly scattered and do not appear to be strongly related to $\bar{N}$, having a 95% confidence range of 0.6 to $4.1\bar{N}$ (kPa). The average values, standard deviations and 95% confidence ranges are given in Table 1.26 including those for each saprolite type. Generally it

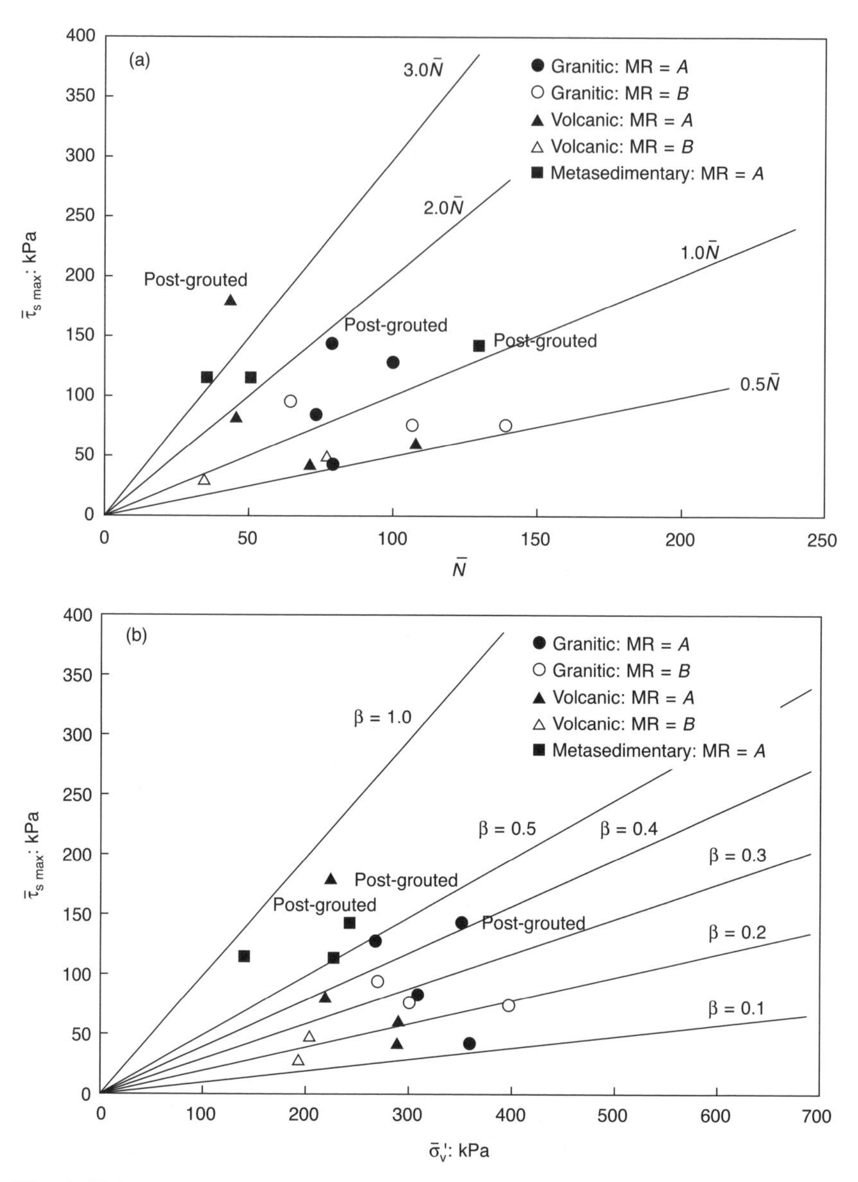

Fig. 1.40 Maximum mobilized side shear resistance $\bar{\tau}_{s\,max}$ of piles constructed under water with MR= A or B*: (a) related to mean SPT value* $\bar{\mathrm{N}}$*, (b) related to mean effective vertical stress $\bar{\sigma}'_v$ (after Ng* et al.*, 2001b)*

is difficult to distinguish the mobilized side shear resistance based on saprolite types. For the metasedimentary saprolites, only two results for non-grouted piles are available and they show about three times higher mobilized side shear resistance than those of piles constructed in the other two saprolites.

Table 1.26 Statistics of mobilized side shear resistance tests with Mobilization Rating MR = A or B *for large-diameter floating bored piles in saprolitic soils in Hong Kong (after Ng* et al., *2001b)*

Method of construction	Saprolite type	$\frac{\bar{\tau}_{s\,max}}{\bar{N}}$ (kPa)					$\bar{\beta}$				
		Sample size	Avg. (kPa)	Standard deviation (kPa)	95% confidence range	Constr. factor	Sample size	Avg. (kPa)	Standard deviation (kPa)	95% confidence range	Constr. factor
Construction by grab/RCD under water with/without temporary casing	Granitic	6	1.0	0.4	0.6–1.3	n/a*	6	0.3	0.1	0.2–0.4	n/a*
	Volcanic	5	0.9	0.5	0.4–1.4	n/a*	5	0.2	0.1	0.2–0.3	n/a*
	Metasedimentary	2	2.7	n/a*	n/a*	n/a*	2	0.7	n/a*	n/a*	n/a*
	All saprolites	13	1.2	0.8	0.8–1.6	1.0	13	0.3	0.2	0.2–0.4	1.0
Post-grouted after construction by grab under water with temporary casing (all saprolites)		3	2.3	1.6	0.6–4.1	1.9	3	0.6	0.2	0.4–0.8	2.0
Construction by RCD under bentonite	Granitic	4	0.7	0.7	0.0–1.3	n/a*	4	0.2	0.1	0.1–0.2	n/a*
	Metasedimentary	2	0.6	n/a*	n/a*	n/a*	2	0.1	n/a*	n/a*	n/a*
	All saprolites	6	0.6	0.6	0.2–1.1	0.5	6	0.1	0.1	0.1–0.2	0.3

Note: * denotes not applicable; Constr. = construction

Piles constructed under bentonite

Figure 1.41 shows the maximum side shear resistance for the piles constructed under bentonite. They were excavated by RCD. All these piles have MR values of *A* or *B*, except Pile 7 having an MR of *C*. Piles with an MR equal to *A* and *B* have an average $\bar{\tau}_{s\,max}/\bar{N}$ value of 0.6 (kPa)

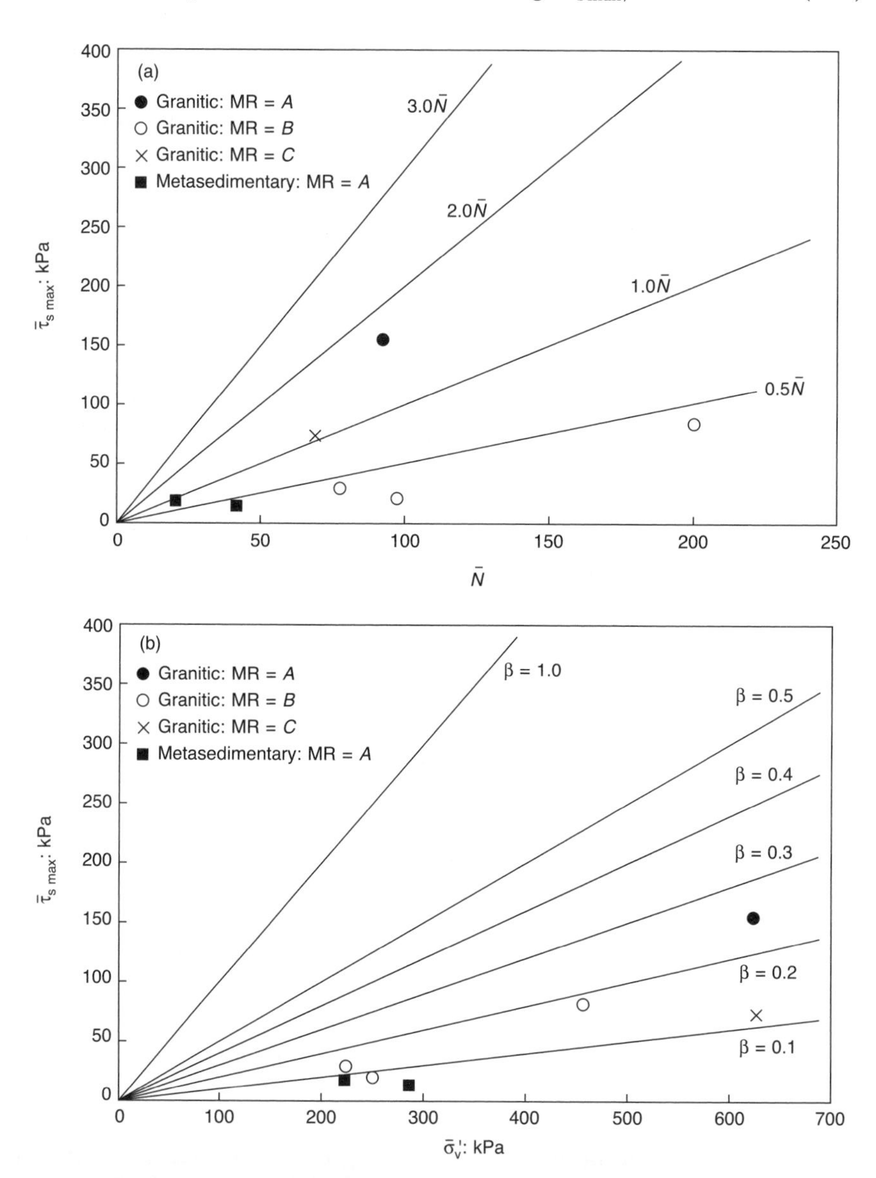

Fig. 1.41 Maximum mobilized side shear resistance $\bar{\tau}_{s\,max}$ of piles constructed under bentonite with MR= A or B: (a) related to mean SPT value $\bar{N}$, (b) related to mean effective vertical stress $\bar{\sigma}'_v$ (after Ng et al., *2001b)*

with a 95% confidence range of 0.2 to 1.1 (kPa) (see Table 1.26) and in Fig. 1.41(b), an average β value of 0.1 with a 95% confidence range of 0.1 to 0.2 (Table 1.26). Therefore, there is less scatter in the effective stress approach than in the $\bar{N}$ approach. Pile 7 with an MR of *C* falls within the 95% confidence ranges for both Fig. 1.41(a) and 1.41(b). The maximum average side shear resistance under bentonite normalized by $\bar{N}$ and $\bar{\sigma}'_v$, is generally lower than for piles constructed under water by about 50% and 30% respectively (see Fig. 1.40).

The results show that the use of bentonite probably reduced the capacity of side shear resistance for the cases reviewed. This is contrary to GEO (1996) and various comparative studies overseas, which report no significant difference in performance with the use of bentonite (Touma and Reese, 1974; Fleming and Sliwinski, 1977). However, GEO (1996) also suggests that strict control of the properties of bentonite is important in preventing the capacity of side shear resistance being affected. Littlechild and Plumbridge (1998) found that longer excavation 'open' times and increased bentonite viscosity decreased the capacity of side shear resistance for piles constructed under bentonite in Bangkok. These factors may contribute to the build-up of a thicker bentonite 'cake' or layer, around the sides of the bore, which leads to a weaker pile–bentonite interface. This is supported by Day *et al.* (1981) who reported that the side shear resistance of the bentonite filter cake is largely a function of the filter cake thickness, thicker cakes producing lower strengths.

Construction factor

It is well recognized that the mobilized side shear resistance at the soil–pile interface depends greatly on the construction methods adopted. For ease of comparison between the influence of different construction methods quantitatively, a construction factor is introduced in the two conventional expressions as follows:

$$\bar{\tau}_{\mathrm{s\,max}} = C_N \bar{A} \bar{N} \qquad \mathrm{kPa} \tag{1.22}$$

and

$$\bar{\tau}_{\mathrm{s\,max}} = C_E \bar{B} (\bar{\sigma}'_{\mathrm{v}}) \qquad \mathrm{kPa} \tag{1.23}$$

where C_N and C_E are construction factors based on the SPT $\bar{N}$ approach and on the effective stress approach, respectively, and $\bar{A}$ and $\bar{B}$ are the corresponding averages of empirical correlation coefficients for side shear resistance normalized by $\bar{N}$ and $\bar{\sigma}'_{\mathrm{v}}$ respectively. In fact, the coefficients $\bar{A}$ and $\bar{B}$ are constants once a reference construction method is chosen.

For comparisons, the average values of the non-grouted piles constructed under water in saprolites with an MR equal to A or B are taken as the references and the construction factor for this group of piles is assumed equal to 1.0. From Table 1.26, the average normalized empirical correlation coefficients (i.e. $\bar{\tau}_{s\,max}/\bar{N}$ and $\bar{\beta}$) for non-grouted piles constructed under water are equal to 1.2 kPa and 0.3 respectively. Comparing with $\bar{\tau}_{s\,max}/\bar{N}$ and $\bar{\beta}$ values (2.3 kPa and 0.6) for post-grouted piles, this implies that the construction factors are $C_N = 1.9$ and $C_E = 2.0$ respectively. On the other hand, the average performance of the non-grouted piles constructed under bentonite is rather poor (i.e. $\bar{\tau}_{s\,max}/\bar{N} = 0.6$ kPa and $\bar{\beta} = 0.1$). The construction factors are $C_N = 0.5$ and $C_E = 0.3$ for this type of pile, respectively (see Table 1.26).

CHAPTER TWELVE

Settlement analysis of piles and pile groups

Overview: methods of analysis

Methods of settlement analysis of piles and pile groups can be broadly classified into three categories:

- Empirical: such as those by Meyerhof (1959) and Focht (1967) for single piles; and by Skempton (1953) and Meyerhof (1959) for pile groups.
- Simplified analytical: using elastic theory to provide design charts.
- Numerical: finite-element and boundary-element methods.

Empirical methods

Poulos and Davis (1980) comment that traditional methods of calculating the settlement of a pile rely on either an arbitrary assumption of the stress distribution along the pile and the use of conventional one-dimensional theory (Terzaghi, 1943), or on empirical correlations. Typical of these correlations are those proposed by Meyerhof (1959) for piles in sand and Focht (1967) for piles in clay. From an analysis of a number of load tests, Meyerhof has suggested that at loads less than about one-third of the ultimate, the settlement, ρ, of a pile could be estimated as follows (provided that no softer layers exist beneath the pile):

$$\rho = \frac{d_b}{30F} \tag{1.24}$$

where d_b = diameter of pile base and F = factor of safety (>3) on ultimate load.

Focht (1967) has examined data from a number of load tests and has related the observed settlement, ρ, at the working load to the computed column deformation ρ_{col} at the working load. Focht has defined a 'movement ratio' as ρ/ρ_{col} and has found that for relatively long highly-stressed piles having $\rho_{col} > 8$ mm, the movement ratio is of the order of 0.5, whereas for relatively rigid piles, having $\rho_{col} < 8$ mm, the movement ratio is larger, of the order of 1.0.

Estimates of the settlement of pile groups have been based either on empirical data or on simplified approaches based on one-dimensional consolidation theory. Among the empirical approaches are those for groups in sand devised by Skempton (1953), who, on the basis of a limited number of field observations, suggested the following relationship between the settlement, ρ_G, of a group and the settlement, ρ_1, of a single pile

$$\frac{\rho_G}{\rho_1} = \frac{(4B+9)^2}{(B+12)^2} \tag{1.25}$$

where B = width of pile group in feet.

For driven piles and displacement caissons in sand, Meyerhof (1959) suggested the following relationship for a square group:

$$\frac{\rho_G}{\rho_1} = \frac{s(5-s/3)}{(1+1/r)^2} \tag{1.26}$$

where s = ratio of spacing to pile-diameter and r = number of rows for square group.

Simplified analytical methods

The analysis of a single vertically loaded pile is described in detail by Randolph (1977) and by Randolph and Wroth (1978a). The relevant steps in the solution will be summarized here. The analysis is based on an elastic soil characterized by a shear modulus G, which may vary with depth, and a Poisson's ratio ν. The soil surrounding the pile is divided into two layers by a line AB drawn at the level of the pile base (see Fig. 1.42). Initially it is assumed that the soil above AB will be deformed solely by the stresses transferred from the pile shaft and that the soil below AB will be deformed solely by the stresses at the pile base. Some modification of this assumption is necessary in order to take account of the interaction between the upper and lower layers of soil; the interaction will serve to limit the deformation of the upper layer of soil, reducing the deformations to negligible size at some radius r_m.

From considerations of vertical equilibrium, it may be shown (Cooke, 1974; Frank, 1974; Baguelin *et al.*, 1975) that the shear stress in the soil around the pile shaft decreases inversely with the radius from the pile. This leads to a logarithmic variation of the deformation w with radius r. The deformation may be written as

$$\begin{aligned} w(r) &= \frac{\tau_0 r_0}{G}\ln\left(\frac{r_m}{r}\right), \qquad r_0 \le r \le r_m \\ w(r) &= 0, \qquad r \ge r_m \end{aligned} \tag{1.27}$$

where τ_0 is the shear stress at the pile shaft, r_0 is the radius of the pile and r_m is the limiting radius of influence of the pile. The deformation of the pile

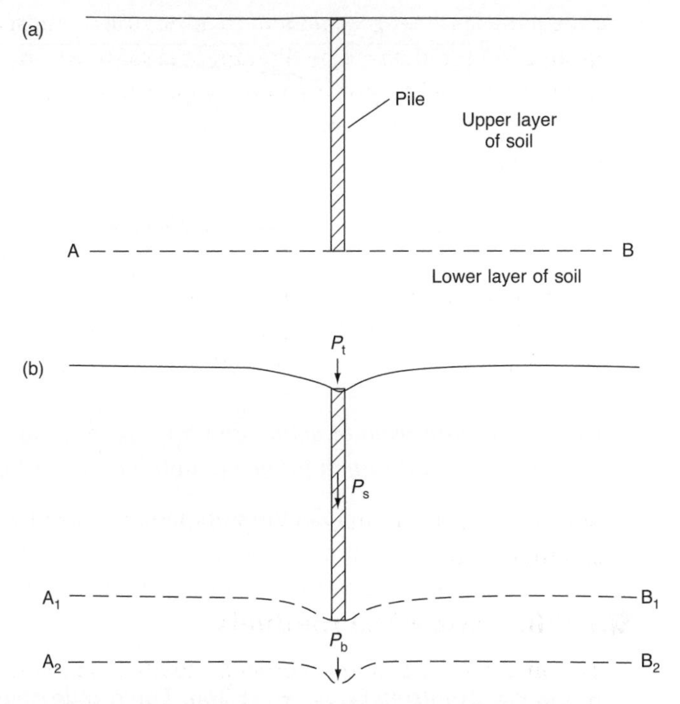

Fig. 1.42 Uncoupling effects due to pile shaft and base: (a) upper and lower soil layers; (b) separate deformation patterns of upper and lower layers (after Randolph and Wroth, 1979)

shaft may be written (Frank, 1974; Baguelin *et al.*, 1975)

$$w_s = \zeta \frac{\tau_0 r_0}{G} \tag{1.28}$$

where $\zeta = \ln(r_m/r_0)$.

The pile base acts as a rigid punch on the surface of the lower layer of the soil. The deformation of the pile base is given by the Boussinesq solution (see Timoshenko and Goodier, 1970),

$$w_b = \frac{P_b(1-v)}{4r_0 G} \tag{1.29}$$

At some distance from the pile base, the loading will appear as a point load. The settlement around a point load decreases inversely with the radius and is given by

$$w(r) = \frac{P(1-v)}{2\pi r G} \tag{1.30}$$

The ratio of the settlements in Equations (1.29) and (1.30) for a given load is

$$\frac{w(r)}{w_b} = \frac{2}{\pi} \frac{r_0}{r} \tag{1.31}$$

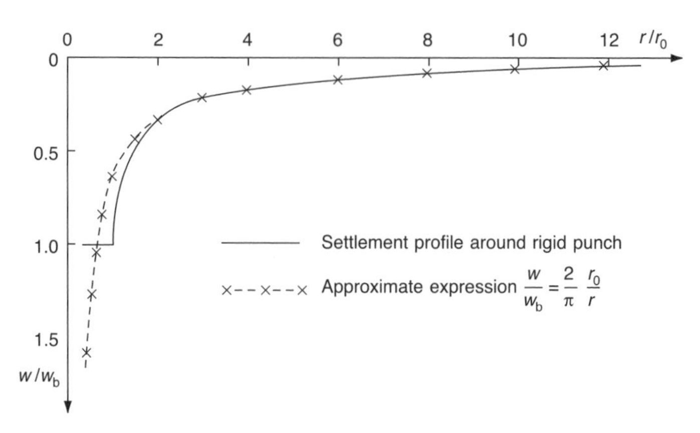

Fig. 1.43 Comparison of actual and approximate surface settlement profiles for a rigid punch (after Randolph and Wroth, 1979)

From St Venant's principle, the settlement caused by the pile base at large radii should equal that due to a point load. Thus the settlement profile at the top of the lower layer of soil in Fig. 1.42 may be approximated by

$$w(r) = w_b c \frac{r_0}{r} \tag{1.32}$$

where $c = 2/\pi$ (from Equation (1.31)). The profile given by this equation is compared with the true settlement profile around a rigid punch in Fig. 1.43. It may be seen that the agreement is very good for $r > 2r_0$ which is the area of interest. The effect of the base loading falls off more rapidly than that of the shaft loading given by Equation (1.27).

For a rigid pile, the shaft settlement is constant down the pile and equal to the settlement of the pile base. Assuming that the shear strain in the soil next to the pile shaft is constant with depth (see Frank, 1974), the shear stress τ_0 will be proportional to the shear modulus at that depth. For soils where the shear modulus varies linearly with depth, it is convenient to introduce a factor ρ, giving the degree of homogeneity, where

$$\rho = \frac{G_{l/2}}{G_l} \tag{1.33}$$

where $G_{l/2}$ and G_l are the values of the shear modulus at pile mid-depth and pile base respectively. The total load transferred to the soil from the pile shaft may now be written (using Equation (1.28))

$$P_s = 2\pi r_0 l (\tau_0)_{av} = 2\pi l \frac{w_s}{\zeta} G_{l/2} = \frac{2\pi}{\zeta} l w_s \rho G_l \tag{1.34}$$

Thus the overall load–settlement ratio for a rigid pile may be written in dimensionless form

$$\frac{P_t}{G_l r_0 w_t} = \frac{P_b}{G_l r_0 w_b} + \frac{P_s}{G_l r_0 w_s} = \frac{4}{1 - \upsilon} + \frac{2\pi\rho}{\zeta} \frac{l}{r_0} \tag{1.35}$$

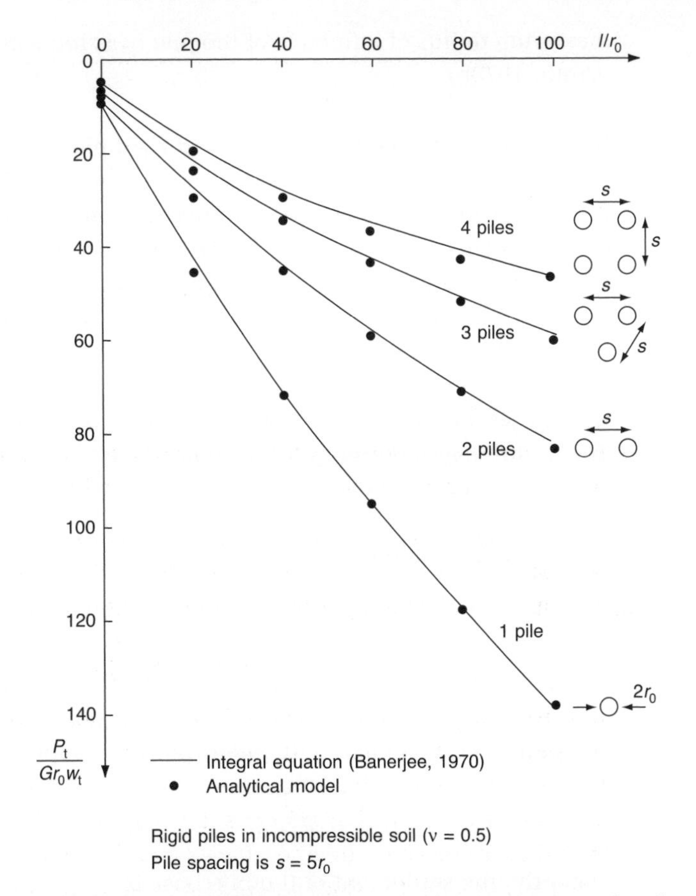

Fig. 1.44 Comparison of load–settlement ratios for symmetric pile groups containing up to four rigid piles (after Randolph and Wroth, 1979)

This equation has been found to give good agreement with the results of finite-element and integral-equation analyses. A suitable value for r_m has been found to be (Randolph, 1977; Randolph and Wroth, 1978a)

$$r_m = 2.5\rho l(1 - v) \tag{1.36}$$

for piles in a deep stratum of soil. Curve 1 in Fig. 1.44 shows a comparison between results from Equation (1.35) and corresponding results (from Banerjee, 1970) obtained by integral-equation analysis for a rigid pile in a homogeneous incompressible soil. Good agreement is obtained for a wide range of l/r_0 – the pile slenderness ratio.

The solution given by Equation (1.35) for a rigid pile may be extended to the case of compressible piles (Randolph and Wroth, 1978a). It is also possible to apply the same solution with minor modifications to under-reamed piles and also to end-bearing piles. In the latter case, the maximum radius of influence of the pile r_m is reduced (see Randolph and Wroth, 1978b).

Interaction between similarly loaded rigid piles

When there are similarly loaded neighbouring piles, the overall displacement of a pile may be obtained by superimposing the individual displacements fields (Cooke, 1974). Due account must be taken of the different amounts of interaction along the pile shaft and at the pile base. It will be assumed that Equation (1.27) gives the settlement profile at the pile mid-depth and Equation (1.32) that at the pile base.

Considering two rigid piles, the overall settlement of one pile may be written as the sum of the settlement due to its own loading plus that due to the neighbouring pile's displacement field. Thus

$$w = w_1 + w_2 \tag{1.37}$$

At the pile mid-depth

$$w_s = w_1 + w_2 = \frac{\tau_0 r_0}{G}\left[\ln\left(\frac{r_m}{r_0}\right) + \ln\left(\frac{r_m}{s}\right)\right] \tag{1.38}$$

where s is the pile spacing, defined as the distance between the centrelines of the piles. Thus s is the average radius of one pile as seen from the other pile. The load–settlement ratio for each pile shaft is now

$$\frac{P_s}{G_l r_0 w_s} = \frac{2\pi\rho}{\zeta + \ln(r_m/s)}\,\frac{l}{r_0} \tag{1.39}$$

Similarly, the settlement of the pile base is

$$w_b = w_1 + w_2 = \frac{P_b(1-\upsilon)}{4r_0 G}\left(1 + \frac{cr_0}{s}\right) \tag{1.40}$$

whence

$$\frac{P_b}{G_l r_0 w_b} = \frac{4}{1-\upsilon}\,\frac{s}{r_0 c + s} \tag{1.41}$$

Thus, the overall load–settlement ratio for each of two similarly loaded piles is

$$\left(\frac{P_t}{G_l r_0 w_t}\right)_2 = \frac{4}{1-\upsilon}\,\frac{s}{r_0 c + s} + \frac{2\pi\rho}{\zeta + \ln(r_m/s)}\,\frac{l}{r_0} \tag{1.42}$$

This result may readily be extended to symmetrical groups of equally loaded piles. The expression for piles in a group of three (at the corners of an equilateral triangle of side s) and a group of four (at the corners of a square of side s) are respectively

$$\left(\frac{P_t}{G_l r_0 w_t}\right)_3 = \frac{4}{1-\upsilon}\,\frac{s}{2r_0 c + s} + \frac{2\pi\rho}{\zeta + 2\ln(r_m/s)}\,\frac{l}{r_0} \tag{1.43}$$

and

$$\left(\frac{P_t}{G_l r_q w_t}\right)_4 = \frac{4}{1-\upsilon}\frac{s}{2.707 r_0 c + s} + \frac{2\pi\rho}{\zeta + \ln(r_m^3/\sqrt{2}s^3)}\frac{l}{r_0} \tag{1.44}$$

In Fig. 1.44, results obtained from equations (1.42) to (1.44) are compared to the results of integral equation analyses obtained by Banerjee (1970), for rigid piles in homogeneous, incompressible soil, at a pile spacing of $s = 5r_0$. As for the case of a single pile, good agreement is obtained.

It is possible to use Equation (1.42) to obtain interaction factors between two piles. Such factors could then be applied in the manner of Poulos (1968) to analyse rigid pile groups. The interaction factor between two particular (similar) piles may be calculated from the inverse of the load–settlement ratio. Thus the increased settlement for a given load may be written as

$$\left(\frac{G_l r_0 w_t}{P_t}\right)_2 = (1 + \alpha_v)\left(\frac{G_l r_0 w_t}{P_t}\right)_1 \tag{1.45}$$

where α_v is the interaction factor. This factor may be calculated from Equations (1.35) and (1.42) for a variety of pile slenderness ratios and different spacings. Figure 1.45(a) shows typical results for piles in homogeneous soil with a Poisson's ratio of 0.4. A comparison with corresponding results from integral equation analysis shows reasonably good agreement, particularly for pile spacings of practical relevance ($5 \le s/r_0 \le 20$). The semi-analytical model tends to under-predict the interaction between piles compared to the integral equation method, especially for piles of $l/r_0 < 40$. However, in a real soil, the non-linear nature of soil deformation will lead to less interaction than predicted from a linear elastic analysis, since the deformation will be confined more to the immediate vicinity of the pile. Thus the method presented here may provide more realistic predictions of pile interaction than a rigorous integral equation analysis.

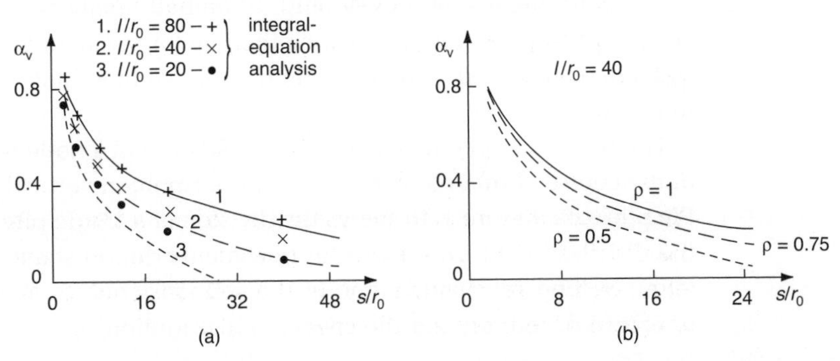

Fig. 1.45 Interaction factors for pairs of rigid piles: (a) homogeneous soil, $\upsilon = 0.4$; (b) non-homogeneous soil $\upsilon = 0.4$ (after Randolph and Wroth, 1979)

Numerical methods

Finite-element analysis

Poulos (1989) observes that finite-element methods (e.g. Desai, 1974; Valliappan *et al.*, 1974; Balaam *et al.*, 1975; Ottaviani, 1975; Jardine *et al.*, 1986) utilize a variety of constitutive soil models where such factors as soil non-homogeneity and anisotropy can be taken into account.

Finite-element methods offer the most powerful analytical approach in that, not only can non-linear soil behaviour be modelled, but the complete history of the pile can be simulated, that is, the processes of installation, reconsolidation of the soil following installation and subsequent loading of the pile (e.g. Withiam and Kulhawy, 1979; Nystrom, 1984). Such analyses are valuable in leading to a better understanding of the details of pile behaviour, but are unlikely to be readily applicable to practical piling problems because of their complexity and the considerable number of geotechnical parameters required.

Boundary-element methods

Boundary-element methods employ either load-transfer functions to represent the interface response (e.g. Coyle and Reese, 1966; Kraft *et al.*, 1981) or elastic continuum theory to represent the soil mass response (e.g. Butterfield and Banerjee, 1971; Banerjee, 1978; Banerjee and Davies, 1978; Poulos and Davis, 1980).

A reasonable compromise between excessive complexity and unacceptable simplicity is provided by boundary-element methods, in which the pile–soil interface is discretized and the characteristics of the soil response are represented in a lumped form by ascribing the behavioural features of the soil to the interface elements. This method has been developed by a number of research workers and is widely used in practice, and attention will therefore be focused on this method. Although a considerable number of formulations exist, most appear to have a common underlying basis. A convenient means of developing a unified analysis is to employ a substructuring technique in which the pile (or piles) and the surrounding soil are considered separately and then compatibility conditions are imposed.

The behaviour of each element is considered at a node which is located at the centre of that element and along a common vertical plane through the pile axis. Figure 1.46 shows the division of a single pile into elements, the distribution of free-field soil movements due to some external cause (e.g. swelling or consolidation of the soil mass due to moisture changes or external loading) and the specified distributions of:

- the limiting pile–soil stresses, f_c for compression loading, and f_t for tensile loading; for shaft elements, the limiting pile–soil resistance

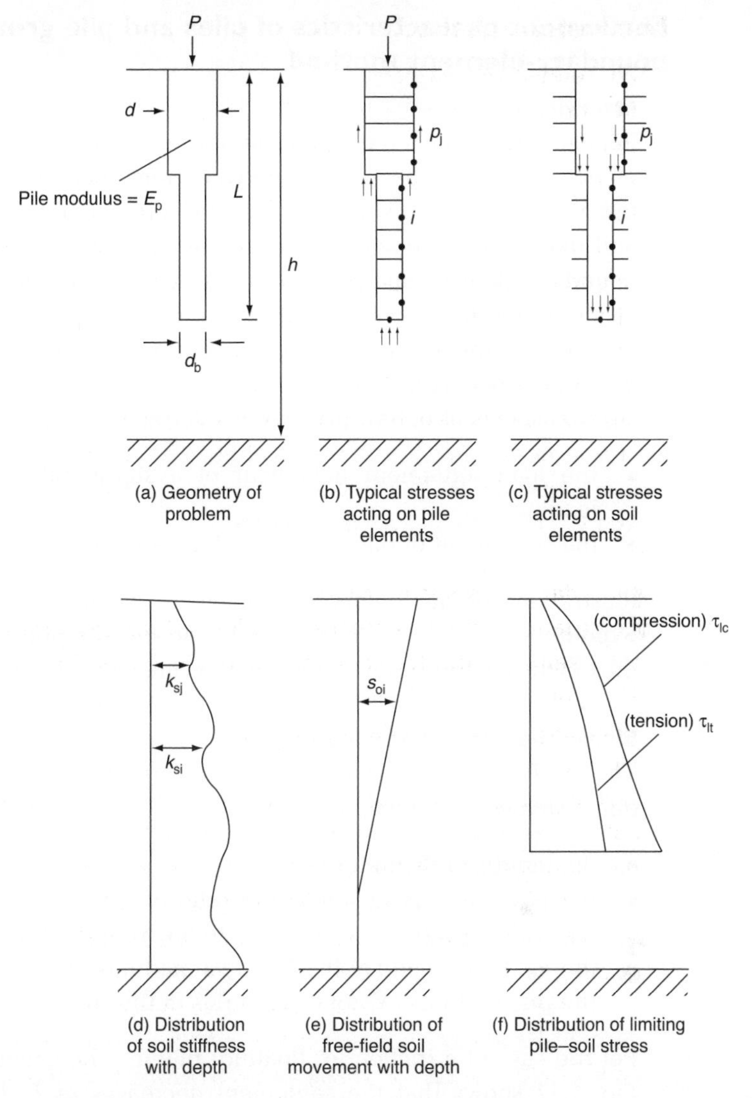

Fig. 1.46 Basic problem of single axially loaded pile (after Poulos, 1989)

will be termed here the side resistance, while for base elements, the term end-bearing resistance will be used; and

- the local stiffness k_s of the soil.

At this stage, no assumptions are made regarding the nature of f_c, f_t and k_s. These may vary with depth, stress (or displacement) level and time.

The responses of the pile and soil elements to an increment of axial load ΔP are analysed in turn. Details of the equation that describes incremental displacements of the pile are given by Poulos (1979).

Settlement characteristics of piles and pile groups using the boundary-element method

Overview

Poulos (1989) points out that most of the more significant characteristics of pile behaviour have been derived from theoretical analyses but in many cases are also supported by measurements made from laboratory and field tests. The solutions described here have been obtained from boundary-element analyses based on elastic continuum theory, with an elastic, or an elastic–plastic, interface model, using simplified distributions of a soil's Young's modulus and side resistance with depth (either constant or linearly increasing).

Two aspects of behaviour are considered:

- the load–settlement behaviour of a single pile under static axial loading
- the settlement of pile groups under static axial loading.

Illustrations of the points will be made frequently with reference to hypothetical problems involving realistic soil and pile parameters so that the practical implications of these points may be more readily appreciated.

Single pile under static loading

The settlement of a single pile is governed largely by the following dimensionless parameters

- the length-to-diameter ratio L/d
- the pile stiffness factor K, the ratio of the Young's modulus of the equivalent solid pile section, E_p, to the Young's modulus of the soil, E_s
- the E_b/E_s, the ratio of the Young's modulus of the bearing stratum at the pile tip to the Young's modulus of the soil.

For the case of a friction (or floating) pile in a homogeneous elastic soil, Fig. 1.47 shows that the settlement decreases as L/d and K increase. Experimental evidence from model tests in a clay presented by Butterfield and Ghosh (1977) demonstrates that the theory can give a realistic prediction of the effects of L/d.

The settlement of a pile is not significantly influenced by the nature of the bearing stratum if the pile is relatively slender and/or compressible. Figure 1.48 compares the settlement of an end-bearing pile relative to a corresponding floating pile, for a typical value of K of 1000. For values of L/d in excess of about 50, the reduction in settlement due to the bearing stratum is less than 40%, even if the bearing stratum is very much stiffer than the overlying soil. Thus, if a reduction in settlement of a long pile is sought, there appears to be little to be gained by founding the pile tip

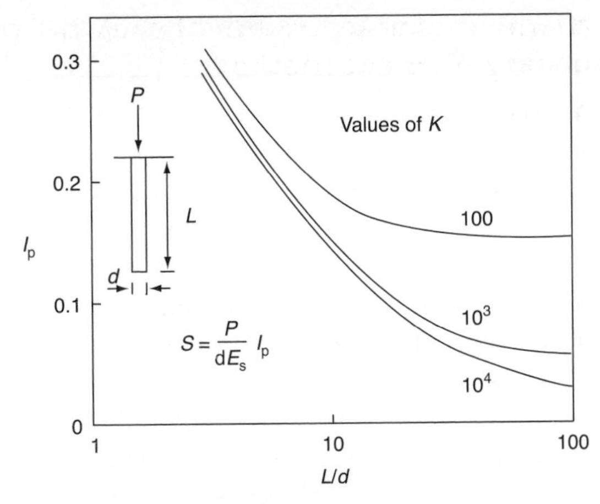

Fig. 1.47 Influence of dimensionless parameters L/d *and* K *on settlement for a single floating pile in homogeneous soil (after Poulos, 1989)*

on a stiffer underlying stratum. Increasing the diameter and/or the stiffness of the pile is likely to be more productive. It should also be noted that there is a critical length for a pile, beyond which further increase in length produces no further reduction in settlement. For a friction pile in a homogeneous soil, this critical length is given by the

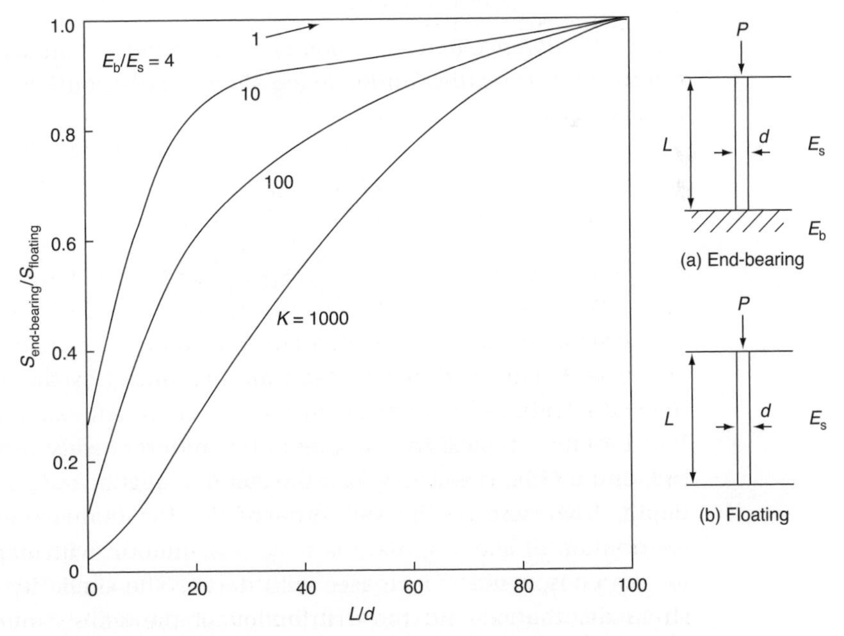

Fig. 1.48 Relative settlement of end-bearing and floating pile for homogeneous soil (after Poulos, 1989)

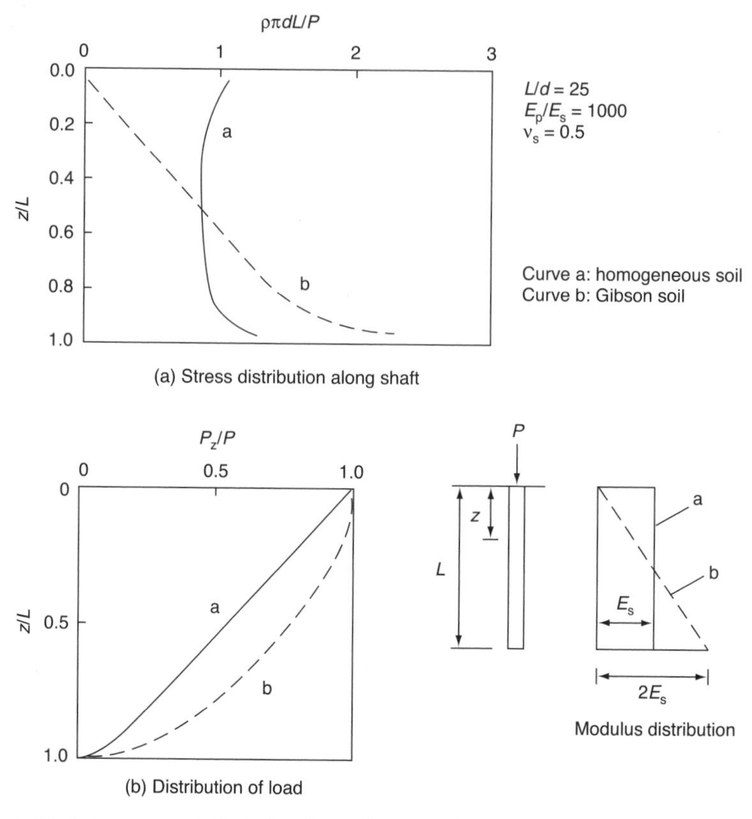

Fig. 1.49 Influence of distribution of soil's Young's modulus on load transfer: (a) stress distribution along shaft; (b) distribution of load (after Poulos, 1989)

approximate expression (Hull, 1987)

$$\frac{L_c}{d} = \sqrt{\frac{\pi E_p A_p}{E_s d^2}} \tag{1.46}$$

where A_p is the area of the pile's cross-section.

The settlement and load transfer are influenced by the distribution of the soil's Young's modulus along the pile shaft. An example is given in Fig. 1.49 for a typical friction pile in two different soils: a homogeneous soil, and a Gibson soil in which the soil modulus increases linearly with depth, from zero at the soil surface. For the homogeneous soil, the distribution of shear stress τ is relatively uniform with depth, whereas for the Gibson soil, τ increases with depth. The similarity between the stress distribution and the distribution of the soil's Young's modulus may explain why load transfer approaches can give reasonable predictions of pile behaviour. For the same average value of Young's modulus along the pile shaft, the pile-head settlements in this case are reasonably similar, and differ by only about 7%. However, the tip settlement for the pile in the Gibson soil is about 18% less than that of the pile in the homogeneous soil.

The major part of the final settlement of a single pile is immediate settlement, and occurs on application of the load because the load is transferred to the soil essentially by shear, with relatively little change in mean stress. For practical values of υ_s' (of the order of 0.3–0.4) and L/d, the theory suggests that the ratio S_i/S_{TF} of immediate to final settlement of a pile in a homogeneous soil is in excess of 0.85, and is almost independent of L/d. This theoretical conclusion is supported by the results of model tests on brass piles in kaolin, reported by Mattes and Poulos (1971) and field maintained loading tests, for example Whitaker and Cooke (1966). A corollary to the above observations is that the *rate* of consolidation settlement of piles in clay is not likely to be an important consideration in design. Time effects stemming from creep at higher load levels are likely to be more important than consolidation time effects (Edil and Mochtar, 1988).

At normal working loads (of the order of 40–50% of the ultimate load), non-linear behaviour of the soil generally does not have a substantial influence on pile settlement. Careful model tests reported by Butterfield and Abdrabbo (1983) support this theoretical conclusion.

For a typical pile in a homogeneous soil, Fig. 1.50 gives some indication of the potential influence of non-linear soil behaviour on settlement The ratio of settlement of the pile in a purely elastic soil S_{elas} to the settlement of the pile in a soil with hyperbolic response S, is plotted as a function of the load level P/P_u where P_u is the ultimate load capacity. The initial tangent Young's modulus of the hyperbolic soil model is assumed equal to the modulus of the purely elastic soil. The ratio S_{elas}/S is generally less than unity (indicating that non-linearity results in an increase in the settlement), the extent of this increase depending on the hyperbolic parameters adopted. For the parameters considered to be most realistic (curves c and d), non-linearity causes an increase of between 11% and 25% in the settlement, as compared with the settlement determined from a purely elastic analysis using the initial tangent Young's modulus of the soil. It should be noted that, in this case, the non-linearity arising from pile–soil slip is only significant at load levels close to failure and therefore that the use of elastic theory with an appropriately reduced secant modulus would give a reasonable prediction of settlement. However, for slender compressible piles, pile–soil slip may have a more dominant influence on the non-linearity of the load–settlement response.

Residual stresses that remain in the pile after installation may influence the pile-head stiffness and hence the calculated pile-head movements.

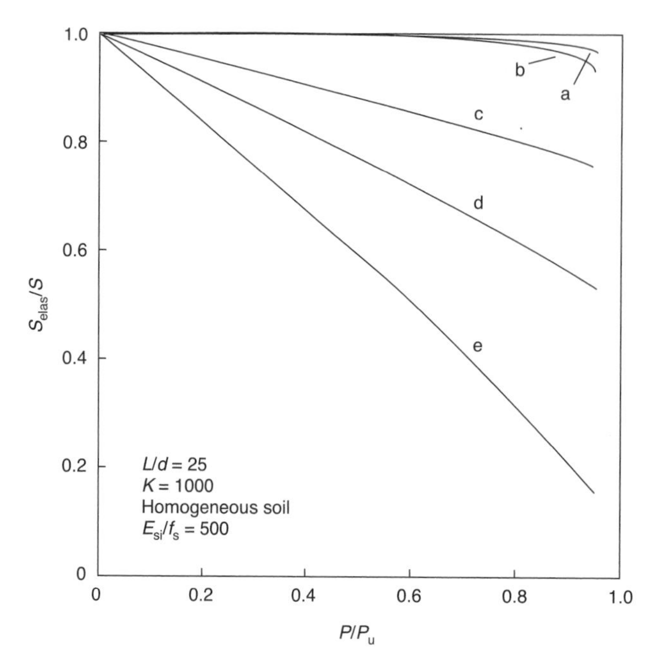

Curve	Hyperbolic parameters	
	R_{fs}	R_{fb}
a	0	0
b	0	0.9
c	0.25	0.9
d	0.5	0.9
e	0.9	0.9

Note: Elastic settlement S_{elas} determined from initial tangent modulus E_{si}

Fig. 1.50 Example of influence of non-linearity on computed pile settlement using a hyperbolic interface model (after Poulos, 1989)

The theoretical analysis of Poulos (1987) has demonstrated that, for cases in which significant residual stresses remain (e.g. for a driven pile in dense sand), the stiffness of the pile head in tension may be smaller than in compression. Figure 1.51 illustrates the point for an elastic–plastic interface. If no account is taken of residual stresses, the pile-head stiffness in tension and compression is the same, but if residual stresses are allowed for, substantially larger movements can occur in tension than in compression. Under zero net load, the residual stresses are such that tensile pile–soil slip occurs over a significant amount of the pile shaft, and a residual compressive load is developed at the tip. Resistance to applied tensile loading comes primarily from the pile tip, the response of which is less stiff than the pile shaft; therefore, the pile-head movements will be greater than if compression is applied.

If the pile–soil interface can strain-soften, both the load–settlement behaviour and the ultimate load capacity will be affected. The ultimate load is no longer statically determinate, but will depend on the relative stiffness of the pile, the ratio of peak to ultimate resistance and the post-peak behaviour. Randolph (1983) presents solutions for a reduction factor to be applied to pile shaft capacities based on peak values of side resistance, in order to allow for the effects of progressive failure along the pile due to strain-softening. These solutions reveal that reductions in

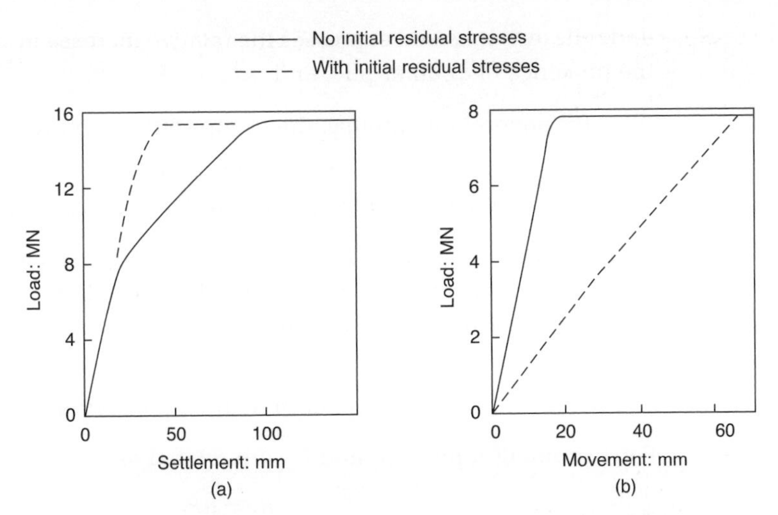

Fig.1.51 Effect of residual stresses on load–deflection behaviour of a pile in sand: (a) compression; (b) tension (after Poulos, 1987)

the peak load capacity are only likely to be significant for relatively long compressible piles.

Pile group under static loading

Before the development of modern analytical techniques, it was commonly believed that no rational relationship existed between the behaviours of single piles and pile groups. In his James Forrest Lecture, Terzaghi (1939) stated

> *Both theoretical considerations and experience leave no doubt that there is no relation whatever between the settlement of an individual pile at a given load and that of a large group of piles having the same load per pile.*

Such statements quite properly encourage caution in dealing with pile groups that contain very large numbers of piles; these are often better considered as a large block foundation. However, for groups that contain relatively few piles, it is possible to link theoretically the settlements of single piles and pile groups. For such groups, under normal working loads, it is convenient to characterize the influence of interaction between piles on the settlement in terms of two dimensionless quantities

(a) for two piles, the interaction factor α is defined as

$$\alpha = \Delta S / S_1 \tag{1.47}$$

where ΔS is the increase in settlement of a pile due to the presence of another equally loaded pile and S_1 is the settlement of a single pile under its own load and expresses the relative increase in settlement due to the presence of another pile, and

(b) for general pile groups, the group settlement ratio R_s, defined as

$$R_s = \frac{\text{settlement of group}}{\text{settlement of single pile at the same average load}} \tag{1.48}$$

An alternative quantity to R_s is the group reduction factor R_G, also termed the efficiency factor by Butterfield and Douglas (1981) and Fleming *et al.* (1985). The factor R_G can be defined as

$$R_G = \frac{\text{stiffness of group}}{\text{sum of individual pile stiffnesses}} \tag{1.49}$$

For a group of n piles, R_s and R_G are related as

$$R_s = nR_G \tag{1.50}$$

A comprehensive review of the load capacity and settlement of pile groups has been made by O'Neill (1983). Here, attention will be focused primarily on group settlement behaviour, as determined from analyses based on elastic continuum theory. Some of the more significant aspects of behaviour are discussed in the following.

Under working load conditions, pile group interaction depends largely on two sets of dimensionless parameters: those related to the soil and pile characteristics, and those related to the geometry of the piles and the pile group. The important soil and pile characteristics are the pile stiffness factor, K, the ratio of E_b/E_s of Young's moduli of bearing stratum to soil, and the distribution of the soil's Young's modulus, E_s, with depth. Figure 1.52 illustrates the influence of these factors on the two-pile interaction factor α. It may be seen that α decreases as K decreases, or as E_b/E_s increases, or as the distribution of the soil's Young's modulus becomes less uniform with depth. Consequently, it should be expected that early published solutions for interaction factors, which are for a rigid pile ($K = \infty$) in a homogeneous mass ($E_b/E_s = 1$ and E_s constant with depth), will overestimate the settlement interaction between piles in more realistic situations. The field measurements shown in Fig. 1.52(c) (O'Neill, 1983) support this contention; however, the theoretical interaction factors for a Gibson soil agree quite well with the measurements.

The primary geometric factors that influence group settlement interaction are the length-to-diameter ratio L/d, the relative spacing between the piles s/d, and the number of piles in the group. The effects of s/d are apparent from Fig. 1.52. For a value of s/d of 4 and $K = 1000$, the variation of settlement ratio R_s with L/d and n is shown in Fig. 1.53 for square groups of friction piles in a Gibson soil of finite thickness $h = 2L$. The ratio of R_s increases as both L/d and n increase. The influence of L/d is

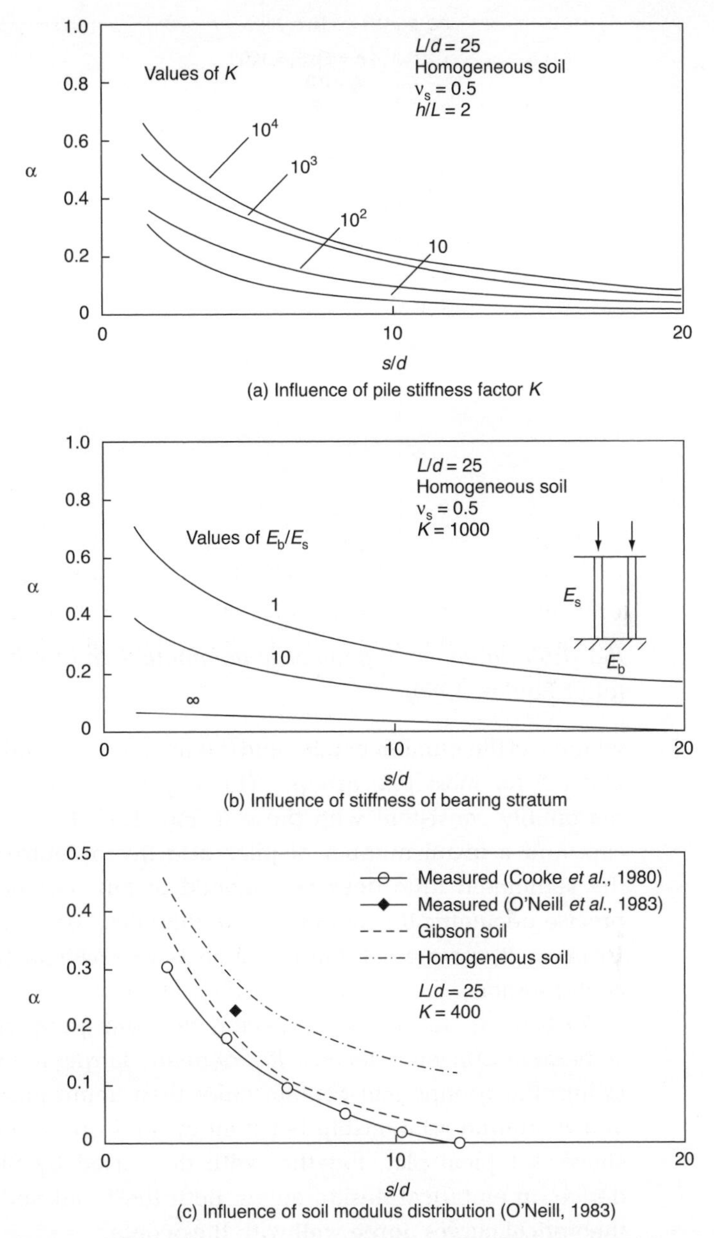

Fig. 1.52 Effect of soil–pile parameters on interaction factors (after O'Neill, 1983; Poulos, 1989)

small for L/d values in excess of 25. Fleming *et al.* (1985) have presented results for larger numbers of piles which suggest that R_s can be approximated as follows:

$$R_s \approx n^{\omega} \tag{1.51}$$

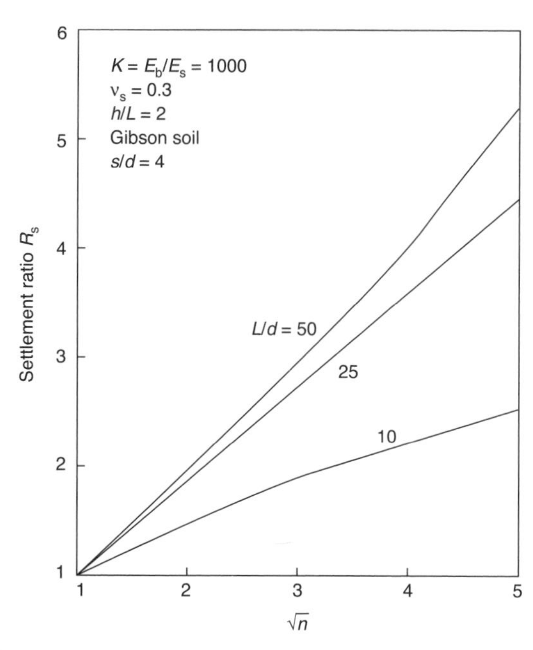

Fig. 1.53 Influence of geometric parameters on group settlement ratio (after Poulos, 1989)

where n is the number of piles and ω is an exponent which lies between 0.4 and 0.6 for most pile groups. This expression gives results which are reasonably consistent with those in Fig. 1.53. For a group with a rigid cap, and a given number of piles at a given centre-to-centre spacing, the settlement ratio does not depend to any significant extent on the precise geometrical configuration of the piles, for example for a group of 16 piles, R_s for a 4×4 configuration is very similar to that for an 8×2 configuration.

For a given set of pile characteristics, the group reduction factor (or settlement efficiency factor), R_G, depends largely on the breadth of the group. For groups that contain more than about nine piles, there is an almost unique relationship between R_G and group breadth. Figure 1.54 shows a typical plot, together with data used by Skempton (1953) to derive an empirical design curve. Both the trend and magnitude of the theoretical curves agree well with these data.

In a pile group with a rigid cap, the distribution of load among the piles is generally non-uniform. In a square or rectangular group, the corner piles carry the greatest proportion of load, while those near the centre carry least. Poulos and Davis (1980), O'Neill *et al.* (1982) and Chow (1986) show that the theoretical trends are supported by field and model test data.

Interaction among piles in a group may be influenced by the stiffness of the soil between the piles. Most of the published theoretical solutions

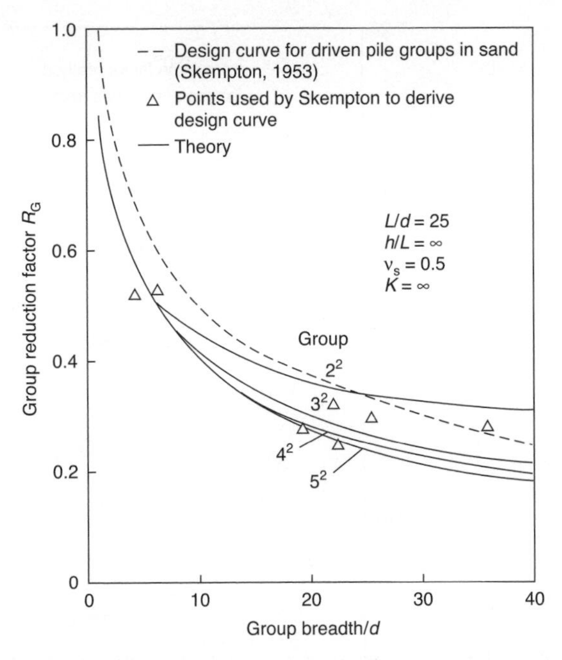

Fig. 1.54 Settlement against group breadth (after Poulos, 1989)

assume a soil to be horizontally homogeneous, with the soil's Young's modulus between the piles the same as the value adjacent to each pile. However, in reality, the soil between the piles undergoes smaller strains and is likely to be stiffer than near the pile–soil interface, and interaction between the piles therefore will be reduced. A simplified analysis of this effect has been made by Poulos (1988a). For groups of piles, the presence of stiffer soil between the piles leads to a smaller settlement ratio R_s and a more uniform distribution among the piles, than is predicted by the conventional analysis.

The relative proportion of immediate settlement S_i to final settlement S_{TF} decreases as the size of the group increases. For groups of piles in either a homogeneous soil, with a typical value of υ_s' of 0.3, the ratio S_i/S_{TF} decreases from 0.93 for a single pile to 0.85 for a 25-pile group. Similar values are found for groups in a Gibson soil. The pile stiffness factor K has little influence on S_i/S_{TF}. For pile groups, the consolidation settlement (and hence the *rate* of consolidation settlement) is more important than for a single pile, but it is still likely to be the minor component of settlement unless soft compressible layers exist beneath the pile tips.

The stiffness of underlying soil layers may have a significant influence on pile group interaction and settlement. Figure 1.55 shows an example of the influence of the relative stiffness of the underlying soil on the

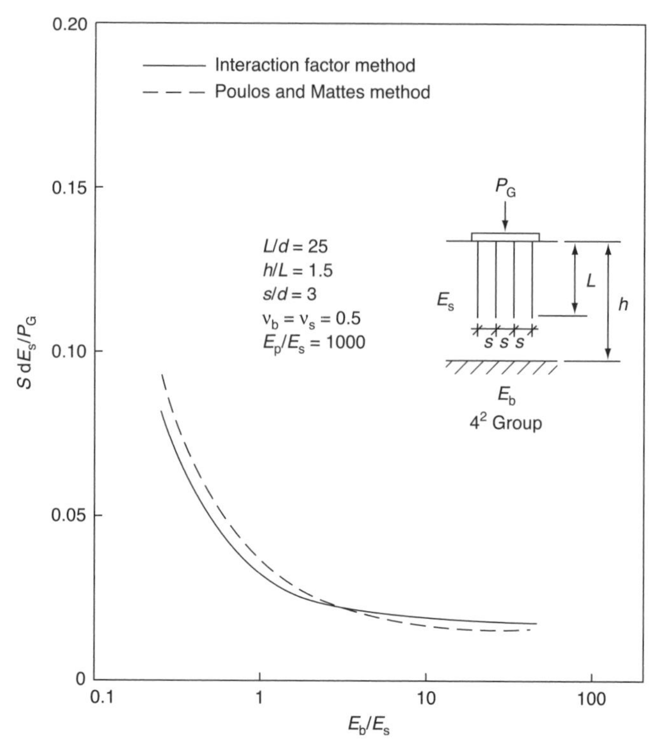

Fig. 1.55 Effect of modulus of underlying stratum on group settlement (after Poulos, 1989)

settlement of a pile group. Clearly, the presence of a softer layer ($E_b/E_s < 1$) may substantially increase the settlement, as compared with the case of a homogeneous soil mass ($E_b/E_s = 1$). Also shown is the settlement computed from the approximate approach described by Poulos and Mattes (1971), in which the group is replaced by an equivalent block. This solution agrees well with the solution computed from the interaction factor method.

The effect on group settlement of the pile cap being in contact with the soil is relatively small unless the pile spacing is large and the group is relatively small. Even for piles at an unusually large centre-to-centre spacing of 10 diameters, the reduction in settlement due to cap contact is only about 5%. Therefore, for most practical purposes, the influence of pile cap contact on settlement at working loads can be ignored.

PART I

Summary

1 A new method of interpretation of the failure load for large-diameter bored piles and barrettes founded on saprolites and weathered rocks is proposed (see Equations (1.7) and (1.8)).

2 Analysis and design procedures of piles are divided into three broad categories, depending on the level of sophistication and rigour adopted (see Tables 1.7 and 1.8).

3 Proposals are made for estimating the geotechnical parameters required for design. These are side resistance, end-bearing resistance, Young's modulus and Poisson's ratio of the soil.

4 For dynamic methods for estimating pile capacity, it is argued that modern methods based on wave mechanics such as CAPWAP and PDA, and on Paikowsky's energy method are roughly twice as cost-effective as standard pile-driving formulae.

5 Research and practice in North America, Australia and Hong Kong with regard to rock-socketed piles are reviewed and suggestions made for the assessment of side resistance.

6 The results of pile loading tests on rock-socketed piles are presented. There are 13 for piles in Hong Kong and 66 for other tests from the literature. The database contains side resistance values recorded at various displacements and degrees of mobilization of side resistance. The term Displacement Index (DI) is introduced which is an approximate measure of the local displacement and degree of mobilization of the side resistance for a given maximum side resistance value. Correlations are given relating side resistance to unconfined compressive strength and to the decomposition grade of the rock.

7 A well-instrumented case study is considered in conjunction with 14 other case studies to provide information on the behaviour of barrettes in Hong Kong. Factors studied include load–displacement behaviour, side resistance, pore pressure response, and changes of lateral pressure at the soil–barrette interface. An interpretation is made of the development of unit side resistance with deformation. The effects of construction on side resistance are assessed.

8 A total of 28 full-scale loading tests on bored piles in weathered materials in Hong Kong are reviewed in order to obtain realistic values of side resistance and settlement and to investigate the influence of various construction factors on the behaviour of the bored piles. Both side resistance and Young's modulus are correlated with the SPT N value for design purposes, for piles constructed both under water and under bentonite.

9 Methods for the analysis of the settlement of piles and pile groups are classified into three categories. These are empirical methods, simplified analytical methods using elastic theory and numerical methods based on finite-element or boundary-element procedures. The applicability of the three approaches to the solution of pile problems in practice is discussed.

PART 2
Multi-propped deep excavations

View across deep excavation construction site at Lion Yard, Cambridge

Overview

The major challenges facing designers of multi-propped deep excavations, particularly in crowded urban areas, are:

1 What stress paths are associated with deep excavations?
2 How can plastic mechanisms of failure be used to evaluate stability?
3 For preliminary design, how do we compute the short-term lateral wall stability and strut loads, the factor of safety against base heave, plus assess the local ground settlement and swelling?
4 How do we assess the lateral pressure of wet concrete acting on diaphragm walls?
5 What are the stress transfer and deformation mechanisms around diaphragm walls?
6 What are the effects and significance of modelling soil non-linearity and wall installation on deep excavations in stiff clay?
7 How can we estimate ground movements caused by diaphragm wall installation.

This Part 2 – effectively a short course on deep excavations – seeks to answer these questions by (1) examining the stress paths relating to deep excavations, followed by (2) an extensive study of the stability of plastic mechanisms of failure. We then present the classic preliminary design methods for (3) above in the form of 'handwritten' Short Course Notes. By using the extended case study of Lion Yard as a theme, we then address (4) to (7). Lion Yard is a deep excavation in stiff clay (Gault Clay, UK) but the concepts and ideas considered can be applied to different soils including soft clays. Finally, we consider some of the lessons learned about deep excavations in the completely decomposed granite soils of Hong Kong.

Sources

In addition to the author's papers, this Part 2 makes verbatim extracts from the following:

- Bolton, M. D. (1993). Mechanisms of ground deformation due to excavation in clay. *Proc. Excavation in Urban Areas, KIG-Forum '93*, Toshihisha Adachi (ed.), Japanese Soc. Soil Mech. & Fdn Engng, pp. 1–33.
- Lings, M. L, Ng, C. W. W. and Nash, D. F. T. (1994). The lateral pressure of wet concrete in diaphragm wall panels cast under bentonite. *Proc. Instn Civ. Engrs Geotech. Engng*, **107**, July, pp. 163–172.
- Malone, A., Ng, C. W. W. and Pappin, J. (1997). Invited paper: Displacements around deep excavations in completely decomposed granite. *Proc. 14th Int. Conf. Soil Mech. & Fdn Engng, Hamburg*, **4**, pp. 2325–2328.

We gratefully acknowledge permissions to make verbatim extracts from these sources from Professor Malcolm Bolton, Mr Martin Lings, Mr David Nash, Dr Andrew Malone and Dr Jack Pappin.

CHAPTER THIRTEEN

Stress paths in relation to a deep excavation

Overview: the stress path dependency of soil parameters

Deep excavations are used increasingly in congested cities to provide underground space. Excavation alters the initial stress states in the ground. Understanding the effective stress changes caused by the excavations is important for designing safe, economical and serviceable excavation supports for foundations and underground structures.

Unlike man-made materials, the behaviour of soils is dependent not only on the current effective-stress states, but also on stress history (Simons and Menzies, 2000). The stress path method presented by Lambe (1967) provides a rational approach to the study of field and laboratory soil behaviour. Because the stability and deformation characteristics of an excavation are influenced by stress history and stress state, anticipated field behaviour should be simulated in the laboratory for appropriate determination of shear strength and stiffness parameters. An understanding of the field stress path is necessary to identify the critical elements that are likely to affect the shear strength. Once the potential changes in the field have been reviewed, appropriate laboratory tests can be assigned to determine the in situ strength and deformation characteristics required for design and analysis.

Idealized stress paths adjacent to deep excavations

Ng (1999) points out that retaining wall construction and excavation causes vertical and horizontal stress relief. The magnitude of stress relief depends on many factors, such as initial stress in the ground, construction method, type of retaining wall and the depth of excavation. The stress changes associated with vertical and horizontal stress relief are very complex. Nevertheless, the states of stress at certain locations around an excavation may be studied qualitatively if simplifying assumptions are made. If an excavation is considered to be long with respect to its width (Fig. 2.1) and if the soil is assumed to be isotropic, then the stress changes for two soil elements (A and P) can be idealized as undergoing plane strain deformations. The soil element A is located at about mid-height behind the

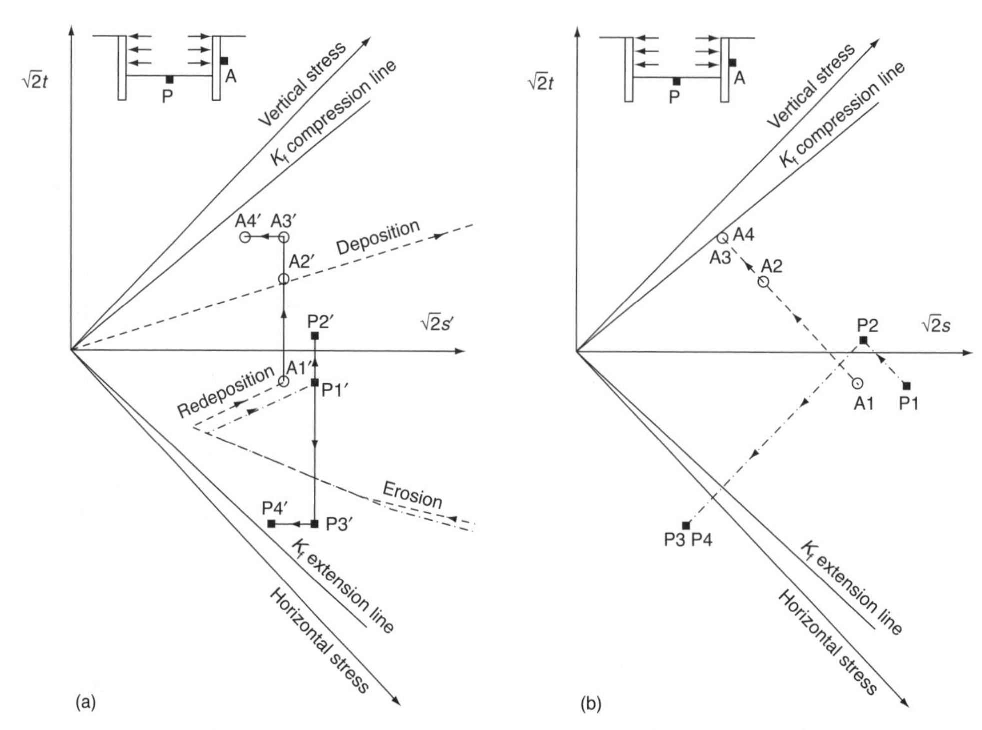

Fig. 2.1 Idealized stress paths associated with stress relief: (a) effective stress path; (b) total stress path (after Ng, 1999)

retaining wall, and the element P is located in the centre of the site below the final excavation level. The stress paths are expressed in terms of mean total stress $s = (\sigma_v + \sigma_h)/2$ or mean effective stress $s' = (\sigma'_v + \sigma'_h)/2$, and the maximum deviatoric shear stress $t = (\sigma_v - \sigma_h)/2 = (\sigma'_v - \sigma'_h)/2$ in the plane of shearing. Axes s and t are oriented at 45° to the (σ_h, σ_v) axes, with a $\sqrt{2}$ scaling factor. The two sets of stress path plots are completely interchangeable.

Before construction, the two soil elements are assumed to have been subjected to slightly different stress histories, which included one-dimensional loading, unloading and reloading associated with natural deposition, erosion, and redeposition processes, respectively. As the result of these natural loading and unloading processes, the initial stress conditions for elements A and P may be, respectively, represented as A1′ and P1′ in the effective-stress space, and A1 and P1 in the total stress space as shown in Fig. 2.1(a) and (b), respectively. The difference between the plotted positions of the two corresponding stress paths in this figure is the magnitude of the pore water pressure.

When a trench is excavated for construction of a diaphragm wall in an over-consolidated stiff clay, a significant reduction in lateral stress takes place in the clay. If changes in vertical stress are ignored and the construction of the diaphragm wall is rapid such that an undrained assumption is valid, the effective stress states for A and P will travel vertically to A2′ and P2′, respectively (Fig. 2.1(a)). On the other hand, the total stress paths A1A2 and P1P2 will move with a gradient of 45° to the horizontal axis, that is $\Delta t/\Delta s = -1$ (Fig. 2.1(b)).

During subsequent excavation stages, the stress changes for element P consist of a reduction in vertical stress, accompanied by a relatively small reduction in horizontal stress. The total stress path P2P3 (Fig. 2.1(b)) will have a slope of 45° to the horizontal ($\Delta t/\Delta s = 1$) if the reduction in horizontal stress can be ignored. The corresponding effective-stress path P2′P3′ (Fig. 2.1(a)) will travel vertically downward assuming that the soil deforms under undrained conditions and passive failure has not been reached. The excess pore pressure generated during excavation is negative. In the long term, the total stress state of P3 will not significantly alter after the construction. On the other hand, the effective stress will reduce as a result of dissipation of negative pore pressure, so that the effective stress path will move toward the extension K_f line. The final position of P4′ may or may not reach the failure envelope depending on soil shear strength and the magnitude of the effective stress changes.

In contrast to element P, the total stress change experienced by element A results from a reduction in horizontal stress. Changes in vertical stress are comparatively small. The total stress path A2A3 will move toward the K_f compression line with a gradient of 45° in $s - t$ space, that is $\Delta t/\Delta s = -1$. If the soil is unloaded rapidly so that undrained conditions are maintained, the corresponding effective stress path A2′A3′ will be vertical. The induced pore water pressure is negative. As the excess pore water pressure dissipates, the effective stress path will shift towards the K_f compression line.

CHAPTER FOURTEEN

Plastic geo-structural analysis

Source

This chapter, 'Plastic geo-structural analysis', is a verbatim extract from 'Mechanisms of ground deformation due to excavation in clay' by M. D. Bolton (1993), published in 'Excavation in urban areas'. We gratefully acknowledge Professor Bolton's kind permission to reproduce this extract.

Overview

In soil mechanics and foundation engineering, engineers have been in the unhappy position of having no simple but robust theoretical models for ground deformation – no equivalent to engineers' beam theory. If quick estimates of soil displacement were called for, engineers have generally had to rely on elasticity theory. It has long been clear, however, that the stress–strain curves of soil elements are highly non-linear, that they depend on the mode of straining and especially on the degree of drainage permitted, and that plastic failure intervenes in a quite complex way. No rational selection of equivalent constant elastic stiffness can emerge, and errors of a factor of 10 in elastic-type calculations are commonplace. Engineers can, of course, ask an expert to run a non-linear finite-element program which is claimed to simulate some more complex features, but there is often a difficulty in communication between engineer and analyst concerning the degree of confidence which may be placed on the values to be used for the special parameters which such programs usually invoke.

Plastic geo-structural mechanism for lateral stress relief on an undrained clay face

Figure 2.2 shows the incremental equilibrium solution for a cut face in soil which is unloaded horizontally by a uniform total stress $\Delta\sigma_{\mathrm{h}}$ while the total vertical stress is constant. Principal stresses are taken to remain vertical and horizontal. The Mohr circle of stress increments shows that an increment in shear stress $\Delta t = \Delta\sigma_{\mathrm{h}}/2$ is induced on all planes at 45° to the face.

Either by reading directly from a stress path test which starts at the predetermined in situ stress state, or by assuming the adequacy of an

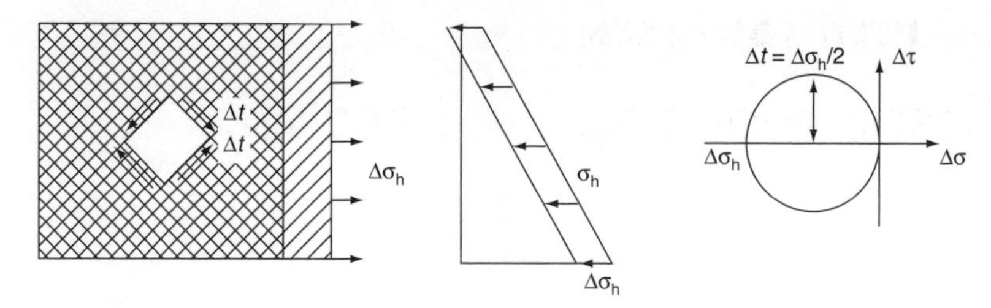

Fig. 2.2 Incremental equilibrium solution for lateral stress relief assuming zero wall friction (after Bolton, 1993)

expression such as (Bolton and Sun, 1991a):

$$\frac{\Delta t}{\Delta t_{\mathrm{p}}} = \left(\frac{\Delta \gamma}{\Delta \gamma_{\mathrm{p}}}\right)^{\mathrm{b}} \tag{2.1}$$

whereby the shear stress–strain changes (Δt, $\Delta\gamma$) are related by a power law, the increment in shear strain $\Delta\gamma$ is then estimated. Figure 2.3(a) shows the typical situation for over-consolidated clay which may be assumed to have commenced a new strain path starting at I with $t_{\mathrm{i}} < 0$, that is $\Delta\sigma_{\mathrm{h}} > \Delta\sigma_{\mathrm{v}}$, and proceeding along the monotonic compression path IM as far as E. Figure 2.3 demonstrates the difficulty for the predictor who has no knowledge of the origin to be used for strain. Here, the clay has been through a vertical loading–unloading cycle IEJ prior to excavation, so that the initial stresses are unaltered, but the apparent origin for strain J is not the real origin. On excavation, the path returns approximately to E where it resumes its original course to M. Experience of cyclic loading suggests that the increment $\Delta\gamma^*$ in Fig. 2.3(b) will probably be less than one-half the increment $\Delta\gamma$ in Fig. 2.3(a). Uncertainty in strain history or initial stress conditions leads to uncertainty in soil strain predictions and, unfortunately, some such uncertainty will generally be unavoidable.

Figure 2.4 shows a plastic deformation field consistent with the earlier assumption that principal directions remain vertical and horizontal. The

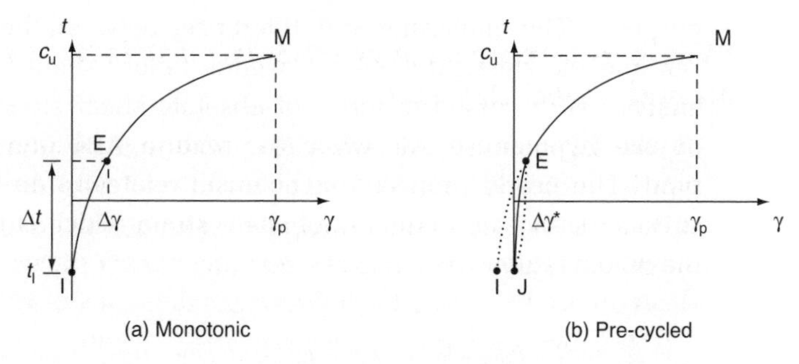

Fig. 2.3 The influence of strain history (after Bolton, 1993)

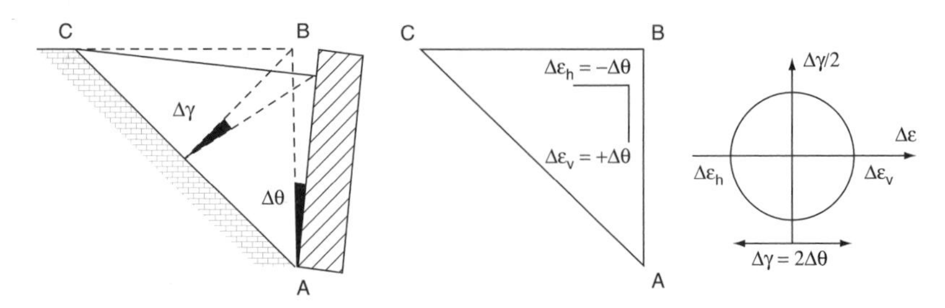

Fig. 2.4 Incremental kinematic solution for wall rotation assuming zero wall adhesion (after Bolton, 1993)

unloaded face is shown rotating outwards by angle $\Delta\theta$, and shearing on planes at 45° produces a similar rotation $\Delta\theta$ in the ground surface behind the wall. Soil beneath the hypotenuse AC remains stationary. The vertical and horizontal strains in the 45° triangle ABC are θ and $-\theta$ respectively (compression positive), and the Mohr circle of strains confirms the geometrical construction that the shear strain increment $\Delta\gamma = 2\Delta\theta$.

Figures 2.2, 2.3 and 2.4 represent the elements of a plastic geo-structural mechanism for lateral stress relief. The engineer can either calculate the face rotation for a given reduction in lateral stress, or can specify a permissible face rotation and ground strain and then deduce the maximum permitted stress relief. If a ground strain of 0.5% is permitted, the face will rotate 0.005 radians, shear strain will be 1%, and the mobilization factor M will need to be in the region 1.5 to 2.0[5].

In reviewing the mechanism, it will be seen that the far-field soil displacements outside the near-field active triangle have been set to zero, whereas the lateral stress reduction is assumed to propagate outwards without limit. The solution would be exact if the far-field stiffness were an order of magnitude higher than the near-field stiffness. The very large stiffness of soil at small strains, and the kinematically unfavourable conditions for far-field displacement if the soil face remains stationary at A, should mean that the proposed mechanism is satisfactory for practical purposes. The similarities and differences between the proposed mechanism and the classical Coulomb wedge failure should be noted. Collapse analysis is discussed in terms of absolute shear stresses approaching c_u on the hypotenuse AC while the wedge ABC remains unrealistically rigid. The newly proposed mechanism relates to an increment of shear stress creating an increment of shear strain which is uniform throughout the whole region ABC.

[5]Bolton (1993) defines a 'mobilization factor M' rather than a 'safety factor F' since it needs to be associated with the permitted degree of soil strain.

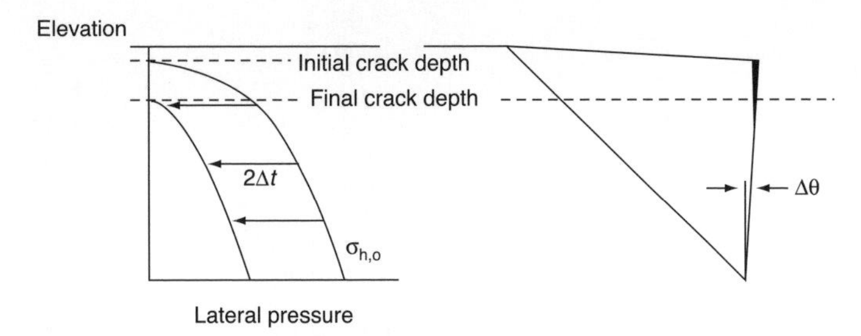

Fig. 2.5 Crack depth as a function of initial lateral pressure and the mobilization of strength (after Bolton, 1993)

The proposed geo-structural mechanism can accommodate tensile cracking in the same approximate way as current active earth pressure calculations. If the crack is taken as dry and open to a depth h_c against the back face of a wall then, as discussed earlier, the horizontal stress at depth h_c will be zero. Below this point the soil can continue to press against the wall with a lateral pressure reduction of $2\Delta t$ consistent with a soil shear strain $\Delta\gamma = 2\Delta\theta$, where $\Delta\theta$ is the wall rotation as before: see Fig. 2.5. It is then clear that the depth of cracking will be such that the pre-existing lateral total stress at that depth equals the calculated amount of lateral stress relief $\Delta\sigma_h = 2\Delta t$. Only in the event that the initial earth pressure coefficient was unity would this also equal the pre-existing vertical stress, and only then would the depth of the crack be equal to $2\Delta t/\rho g$.

Mechanism for undrained excavation against a cantilever wall pinned at its toe

The incremental approach to plastic deformation problems using appropriate stress-path data can be extended to provide solutions to some quite difficult problems. For example, Bolton and Sun (1991b) showed good agreement between an incremental hand calculation and the lateral displacement of model bridge abutments with spread foundations resting on firm clay, observed in centrifuge tests. Figure 2.6(a) shows the kinematics of a stiff in situ wall, socketed into a hard base layer at its toe so that rotation $\Delta\theta$ about its toe is the single degree of freedom permitted to it when an excavation of depth h removes a vertical stress $\Delta\sigma_v = \rho g h$ from the remaining layer of depth d beneath the excavation. Following the compatibility condition sketched in Fig. 2.4, the increment of shear strain must be

$$\Delta\gamma_A = \Delta\gamma_P = 2\Delta\theta \tag{2.2}$$

in both active triangle A and passive triangle P. These shear strains will induce increments of shear stress Δt_A positive and Δt_P negative, so that

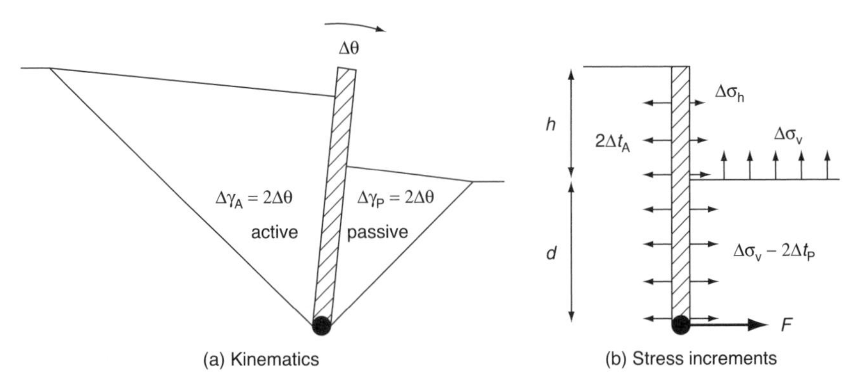

Fig. 2.6 Incremental analysis of a cantilever wall rotating about its toe (after Bolton, 1993)

lateral earth pressure changes $2\Delta t_A$ and $2\Delta t_P$ can be entered as acting to support the wall in the stress increment diagram, Fig. 2.6(b). The effects of stress relief causing the wall to move towards the excavation are the removal of the pre-existing lateral pressure on the face of the wall exposed by the excavation ($\Delta\sigma_h = \sigma_{h,0}$), and a reduction $\Delta\sigma_h = \Delta\sigma_v$ on the buried face of the wall beneath the excavation, each due to the immediate effect of unloading in the absence of wall movement. Any tendency to cavitation has been ignored here, for simplicity, but a crack of depth h_c should be invoked such that the initial lateral pressure will be just eliminated, $\Delta\sigma_h = 2\Delta t_A$.

Taking moments about the toe of the wall produces an equilibrium equation of the form

$$\Delta t_A(h + d - h_c)^2 + \Delta t_P d^2 = (\Delta\sigma_h)_{\text{mean}\, h\geq h_c}(h - h_c)\left(d + \frac{h - h_c}{3}\right) + \frac{\Delta\sigma_v d^2}{2} \quad (2.3)$$

where allowance could additionally be made for the moment effect of any other variations with depth in Δt or, especially, non-triangular profiles of $\Delta\sigma_h$. If both Δt_A and Δt_P are known functions of γ, the stepping of $\Delta\gamma$ will reveal a value which solves for equilibrium and compatibility. The method could allow for any amount of anisotropy, and for $K_0 \neq 1$, as indicated in Fig. 2.7(a). Characteristic stress–strain data, as in Fig. 2.7, are here used directly: there is no need to derive any parameters or even to formalize an equation such as (2.1). The calculation procedure is ideally suited for a spreadsheet.

While mechanisms such as that in Fig. 2.6 permit non-linear analysis based on the typical compression and extension data of Fig. 2.7(a), engineers will often have only data from triaxial compression tests, or no specific

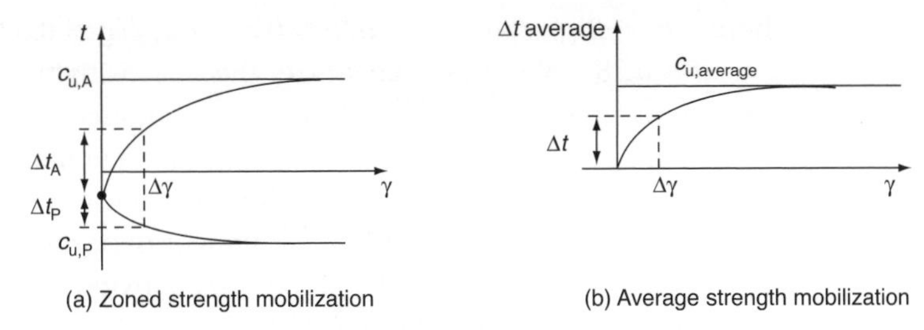

Fig. 2.7 Using stress–strain data (after Bolton, 1993)

data at all, to guide their decisions. If data on anisotropy do not exist, then some 'average' mobilization curve such as that shown in Fig. 2.7(b) might be derived from triaxial compression tests, but the tendency of extension tests to mobilize less strength should not be forgotten. Likewise, if there are no data on the initial lateral earth pressure, the initial magnitudes of shear stress (e.g. t_I in Fig. 2.3) will be unknown and engineers might assume that $t_I \approx 0$. The incremental approach can then be simplified and the final lateral pressures shown in Fig. 2.8 can be used in the equilibrium check. Under such circumstances the calculation of displacements cannot be expected to be perfect, but the geo-structural mechanism can still be shown to offer distinct advantages compared with traditional 'factor of safety' approaches.

The only difference between the simplified final pressure diagram in Fig. 2.8(b) and the incremental diagram in Fig. 2.6(b) is that the mobilized shear strength t_{mob} is assumed to be uniform around the wall, and related to shear strain by the nominal curve in Fig. 2.7(b). Putting $K_0 = 1$ and $\Delta t_A = \Delta t_P = t_{mob}$ in Equation (2.3) we find the equilibrium condition reduces to

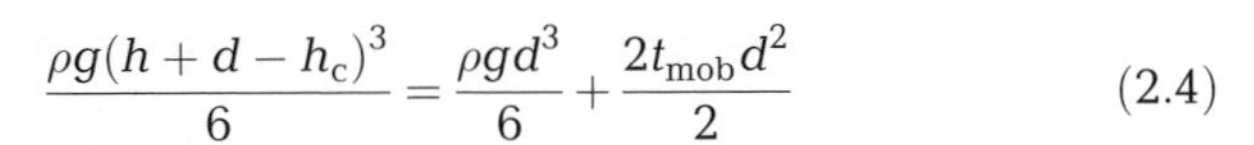

$$\frac{\rho g(h + d - h_c)^3}{6} = \frac{\rho g d^3}{6} + \frac{2t_{mob}d^2}{2} \qquad (2.4)$$

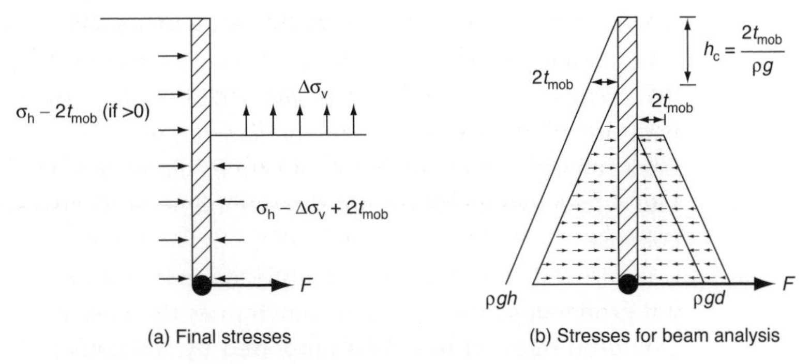

Fig. 2.8 Conventional analysis of equilibrium after excavation with $K_0 = 1$ *(after Bolton, 1993)*

where $h_c = 2t_{mob}/\rho g$. At $d/h = 0.5$, $t_{mob}/\rho g = 0.325$; at $d/h = 2$, $t_{mob}/\rho g = 0.28$. We must recognize the insensitivity of the mobilized strength to embedment ratio. The more significant parameters in the problem are actually those which we have made nominal for the purposes of Fig. 2.8, namely K_0, anisotropy, and variation of strength with depth. For cantilever walls in this range the engineer performing a nominal calculation could safely assume that $t_{mob}/\rho g = 0.33$ giving $t_{mob}/c_u = 0.33\rho gh/c_u$, if an average value c_u can be selected.

Equations (2.2) and (2.4) can then be used to give

$$\Delta\theta = \frac{\Delta\gamma}{2} = 0.5\frac{\gamma_p}{M^{1/b}} \approx \frac{\gamma_p}{2}\left(\frac{\rho gh}{3c_u}\right)^{1/b} \tag{2.5}$$

where the mobilization factor M is given by

$$\Delta\gamma = \frac{\gamma_p}{M^{1/b}} \tag{2.6}$$

Although the mobilization factor M acts numerically in the same way as a safety factor, its relationship to wall rotations is explicit in Equation (2.5). We showed that to reduce $\Delta\theta$ to below 1/200, factor M in clays might lie in the range 1.5 to 2.0, irrespective of wall penetration. However, the improvement in sensitivity which could follow determination of the $\sigma_{h,0}$ and c_u profiles by pressuremeter tests, and anisotropy by appropriate triaxial extension and compression tests, must not be overlooked. The use of finite-element solutions which take account of these complexities (Clough and Hansen, 1981; Bolton *et al.*, 1993) should be facilitated if engineers have access to the more direct sort of calculations referred to above.

Geo-structural mechanism for base heave in undrained clay excavations

Terzaghi (1943) viewed base heave in excavations as a bearing capacity problem. Bjerrum and Eide (1956) took the analogy further by using similar shape factors to get a more accurate indication of the stability of 3D excavations, while Eide *et al.* (1972) made a further allowance for wall adhesion. The collapse mechanisms used in these analyses were based on slip surfaces which could not be mobilized at small deformations, and which are therefore not directly relevant to displacements prior to collapse.

Figure 2.9 shows deformation mechanisms for an excavation which is not propped, so that the sides pinch in as the base heaves. The proposal would be relevant to a face supported by soil nails located within block B. The size of the deforming zone is controlled in Figure 2.9(a) by the width e of the excavation, and in Figure 2.9(b) by the depth d of the soft

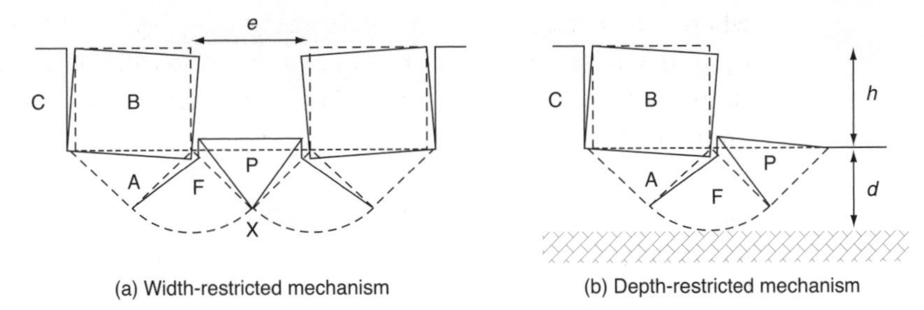

Fig. 2.9 Base heave deformation mechanisms for unsupported excavations (after Bolton, 1993)

material below the base of the excavation. Here, the shape of the foundation bearing mechanism is taken from the statical solution of Prandtl (1920), but the kinematics of the proposed deformation field beneath the level of the excavation are developed from Bykovtsev (1961). If block B offered no resistance to the lateral spreading of active triangle A, and was free to crack away at C from the ground remote from the excavation, the equilibrium analysis of this deformation mechanism would offer the same solution as Prandtl's to the problem of punch indentation in rigid-plastic material – which solution is regarded as formally correct. An infinitely large number of deformation mechanisms can be found which are kinematically admissible, and which offer identical collapse loads.

What is most relevant here, however, is that the particular deformations shown in Fig. 2.9 do not invoke any discontinuous displacements: there are no slip lines, but only diffuse shear strains. These mechanisms can therefore be applied during the strain-hardening phase of soil behaviour prior to peak strength. It is well-known in metal-forming that diffuse shear deformation mechanisms are those observed when strain-hardening is significant (Szczepinski, 1979). Bolton and Sun (1991a) calculated the shear strains in relation to boundary displacements for a simple footing. If the block B in case (a) rotates by angle $\Delta\theta$ towards the excavation, it can be shown that the shear strain increments are: $2\Delta\theta$ in active triangle A, $4\Delta\theta$ in passive triangle P, and variable with a mean value of $4\Delta\theta$ in fan F. A similar result holds in case (b) except that the one-sided mechanism produces only $2\Delta\theta$ of shear strain in triangle P. The mean shear strain increments within the area of the plastic foundation mechanisms can then easily by calculated to be

$$\Delta\gamma \approx 3.3\Delta\theta \quad \text{in case (a)} \qquad \Delta\gamma \approx 2.9\Delta\theta \quad \text{in case (b)} \tag{2.7}$$

It is also easy to see that the maximum vertical heave due to the activation of the near-field plastic mechanism on undrained excavation would be:

$$\Delta v_{\mathrm{n}} = e\Delta\theta \quad \text{in case (a)} \qquad \Delta v_{\mathrm{n}} = \sqrt{2}d\Delta\theta \quad \text{in case (b)} \tag{2.8}$$

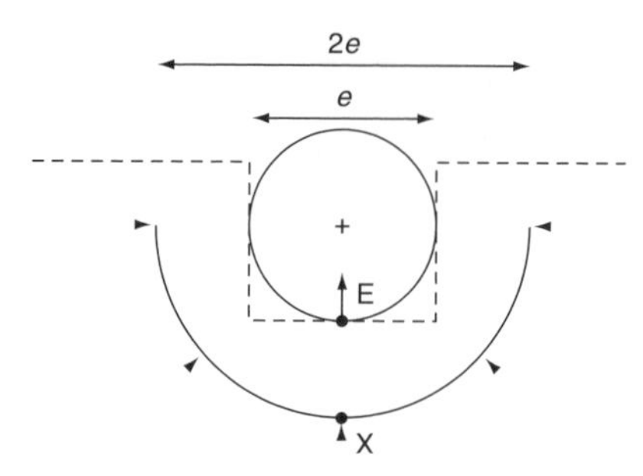

Fig. 2.10 Cylindrical cavity collapse analogy (after Bolton, 1993)

In case (a) there must be some additional heave due to soil strains in the far field, below the apex X of triangle P. The mobilized strength drops rapidly beneath the Prandtl zone, however, so soil non-linearity has the effect of curtailing strains very rapidly. Preliminary investigation using finite-element analysis shows that even in the worst cases in which soil strength does not increase with depth, and the depth of the clay is very large, the extra far-field contribution to undrained heave does not appear to exceed that of the near-field. This approximation can also be appreciated using a cavity collapse analogy. Figure 2.10 shows the excavation idealized as a cylinder of diameter e which contracts by an amount Δv_E at its boundary E, and the far-field zone lying beyond point X as an infinite cylinder with inside diameter $2e$. Continuity of volume in a cylindrical deformation mode demands that

$$\Delta v_E \pi e = \Delta v_X \pi 2e \qquad \text{so} \qquad \Delta v_E = 2\Delta v_X \tag{2.9}$$

Here, the heave due to strains in the far-field beyond X is simply $\Delta v_f = \Delta v_X$, and the total heave $\Delta v = \Delta v_E$, so the heave due to near-field strains must be $\Delta v_n = \Delta v_E - \Delta v_X$. Substituting Equation (2.8) we obtain

$$\Delta v = \Delta v_n + \Delta v_f = 2\Delta v_n \tag{2.10}$$

The cylindrical analogy therefore suggests that the heave due to near-field strains should be doubled in order to estimate the total undrained heave of a deep bed of soil. Applying this rule to Equation (2.8) case (a), and substituting from Equation (2.7) for $\Delta\theta$ in terms of the mean shear strain $\Delta\gamma$ in the near-field, we obtain

$$\Delta\gamma \approx 1.7\frac{\Delta v}{e} \quad \text{in case (a)} \qquad \Delta\gamma \approx 2.0\frac{\Delta v}{d} \quad \text{in case (b)} \tag{2.11}$$

Engineers in practice are free to analyse their works in whatever degree of precision is called for, but some simple serviceability criterion would be desirable during design and decision-making. Taking all aspects of the kinematics into account, a useful rule of thumb appears to be:

$$\Delta\gamma \approx 2\Delta\omega \tag{2.12}$$

where the proportional heave is taken to be the maximum undrained vertical displacement in the excavation divided by the controlling constriction of the mechanism, that is $\Delta\omega = v/e$ in case (a), and $\Delta\omega = v/d$ in case (b).

The plastic equilibrium equation, which engineers have previously employed only at collapse, can be written

$$t_{\text{mob}} = \frac{\rho g h}{N_{\text{c}}} \tag{2.13}$$

where t_{mob} is the mean strength mobilized within the near-field plastic mechanism. Here, Prandtl's $N_{\text{c}} = (\pi + 2)$ would generally offer a safe lower bound neglecting, as indicated above, the strength of soil on the interface BC. Careful consideration must be given to the possibility of cracking on that interface, however. For example, if water could enter a crack at C and be retained, the vertical component of the thrust from block B on to block A would remain constant, but the resultant would be inclined to the vertical at angle

$$\delta = \tan^{-1}\left(\frac{h\rho_{\text{w}}}{2b\rho}\right) \tag{2.14}$$

This would have a large effect on N_{c}, and would also somewhat alter the deformation mechanism. The case of a rigid block under inclined loads causing deformation of a clay layer was considered by Bolton and Sun (1991b).

Taking an appropriate value of N_{c} from the best available analysis of plastic equilibrium, the kinematics of Equation (2.12) can be connected with the equilibrium of Equation (2.13) using the mean data of stress–strain curves as shown in Fig. 2.7, or by the fitting of a power curve such as Equation (2.1) which then offers the estimate of the proportional heave:

$$\Delta\omega = \frac{\gamma_{\text{p}}}{2}\left(\frac{\rho g h}{N_{\text{c}} c_{\text{u}}}\right)^{1/b} \tag{2.15}$$

where the inverse of the term in brackets would previously have been referred to as the safety factor F and now is referred to as the mobilization factor M. In this regard, the similarity between Equation (2.15) and Equation (2.5) can be taken as a general property of these problems of mobilizing plastic strength along a mean power curve.

Figure 2.11 shows analogous mechanisms for excavations which are sufficiently well propped to force the retained soil in block B to sink vertically behind the face. Block B behaves somewhat like material emptying from a rough hopper, where the outer interface C is the hopper wall, and the inner sheeted face might either act as the other wall of the hopper (if

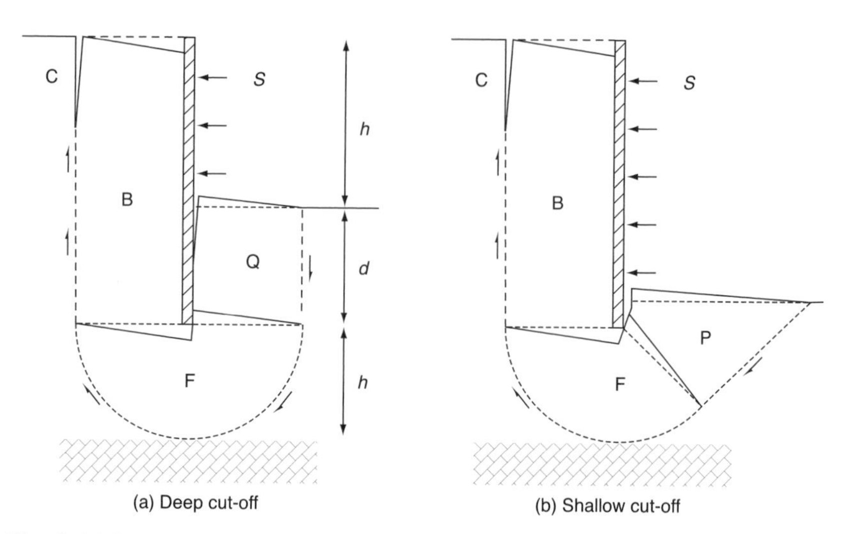

Fig. 2.11 Base heave deformation mechanisms for strongly supported excavations (after Bolton, 1993)

significant friction can be developed against sheeting which is prevented from sinking with the soil) or as the hopper centre-line (if vertical friction at the sheets should be discounted). In what follows, friction against the sheet is ignored, and the subsidence of the surface of block B is assumed to increase linearly up to the face. Other assumptions are possible.

Neither the kinematics nor the equilibrium equations appropriate to strongly supported excavations are currently as well-understood as in the case of unsupported excavations, partly due to uncertainty regarding cracking which was discussed earlier. The tendency of the top left corner of block B to crack away from C could be suppressed if the propping system at S were sufficiently pre-compressed. On the other hand, we have shown that the attempt to mobilize undrained strength at zero total stress leads to unbelievable pore suctions, and the likelihood of degassing into opening fissures. Figure 2.12 shows a detail of the region BC which is consistent with tensile cracking taking place at 45° within a

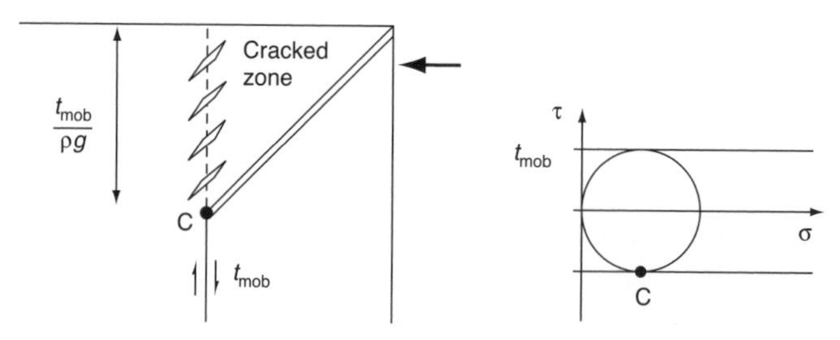

Fig. 2.12 Possible zone of cracking behind a propped face (after Bolton, 1993)

block which subsides vertically. The cracked zone succeeds in resting its full weight on the top of the mechanism, but shear strength is lost over a height $h_c = t_{mob}/\rho g$.

The best approach to equilibrium which can currently be achieved here is to perform a virtual work calculation assuming that the deformations are as indicated in Fig. 2.11. Suppose that the ground rotation is $\Delta\theta$ as shown, mobilizing an average soil shear strength t_{mob}. Now consider an infinitesimal additional displacement $\delta\theta$. The additional loss of potential energy is given by the weight of soil in the triangular depression effectively falling a height h to form the equivalent triangular zone of heave:

$$\delta(\text{P.E.}) = 0.5\rho g r^2 \delta\theta h \tag{2.16}$$

The additional plastic work is due to shear strains $\Delta\gamma = \delta\theta$ in blocks B, Q, P and $\delta\gamma = 2\delta\theta$ in the circular fans F. These work against the shear strength to give dissipation $t_{mob}\delta\gamma$ per unit area. The additional plastic dissipation for case (a) in Fig. 2.11 is therefore:

$$\delta(\text{P.D.}) = (h + 2d - h_c) r \delta\theta t_{mob} + \frac{\pi r^2}{2} 2\delta\theta t_{mob} \tag{2.17}$$

Equating Equations (2.16) and (2.17) and reorganizing, we obtain

$$t_{mob} = \frac{\rho g h}{N_c} = \frac{\rho g h}{2\left[\pi + \dfrac{(h + 2d - h_c)}{r}\right]} \tag{2.18}$$

Likewise the mean mobilized shear strain can be obtained from an average over the whole area of the mechanism

$$\Delta\gamma = \Delta\theta \frac{[(h + 2d - h_c)r + \pi r^2]}{[(h + 2d - h_c)r + 0.5\pi r^2]} \approx 1.5\Delta\theta \tag{2.19}$$

The mean relation between t_{mob} and $\Delta\gamma$ can be used as before to permit an estimation of $\Delta\theta$ for a given geometry of excavation. It must be recalled that assumptions have been made about the shape of the deformed soil mass, the cracking of the retained soil, the sheets generating negligible friction, and the supports remaining undeformed. The calculation has rather the spirit of a 'limit equilibrium' calculation, while being targeted at working displacements.

Figure 2.10 concerned the problem of a relatively stiff sheet structure which was not fixed at bed-rock, and which therefore permitted base heave and external subsidence as clay flowed beneath it. O'Rourke (1992) considered the case of a flexible sheet penetrating the deforming clay, adding the strain energy of the sheet into the work equation. He demonstrated that the consequential small reduction in soil strength required for equilibrium could lead to a significant reduction in soil and wall displacements, drawing on the non-linear stress–strain response of the soil.

The ground deformation mechanisms shown in Figs 2.6, 2.9 and 2.11 are based mainly on theory. Their advantage is that they predict much the same collapse loads as current methods based on slip surfaces, but they invoke distributed shear strains which permit an estimation of displacements under working loads. At the least, they permit a rational choice of the necessary strength reduction factor, taking strains to failure and soil non-linearity into account. They must be used with caution in predicting displacement distributions until further evidence is available to confirm or refine them. They are not a substitute for careful numerical analysis, where that can be justified.

They do, however, suggest an alternative presentation for charts of ground movement around excavations – most of which are expressed as (settlement divided by excavation depth) plotted against (distance from excavation divided by depth of excavation), following Peck (1969). It has now been shown that the kinematics of soil movements is dictated by some controlling dimension which will be different in different cases. For excavations with walls permitted significant lateral displacement but avoiding deep-seated base heave, as in Fig. 2.6, the dimension will be the height over which wall movement is permitted. This will be the depth of excavation in the case of a flexible face supported by struts, but would be the depth of the wall in the case of a stiff reinforced concrete wall penetrating below formation level. For walls on deep clays which do not gain much in strength or stiffness at depth, deep-seated base heave will control the spread of movements, as in Figs 2.9 and 2.11. Movements should then be related to the smaller of the width of the excavation, and the depth of soft clay beneath the cut-off. It would be valuable in future if an appropriate controlling dimension were used to normalize both ground displacements and proportional separations.

CHAPTER FIFTEEN

Short course notes: preliminary design for deep excavations

There follows a set of short course notes such as may be found in a design manual. It therefore comprises a 'tool kit' of summarized methods, charts and tables suitable for preliminary design.

General design aims

- Predict any possible failure and deformation mechanisms
- Devise average and worst credible design parameters
- Assess any adverse geological conditions or features

Preliminary design for ultimate limit states

- Analyse short-term lateral wall stability and determine toe penetration and strut loads using lateral earth pressure design envelopes for simplicity.
    - Pressure envelope for excavations in sand (Fig. 1(a)).
    - Pressure envelope for excavations in soft and medium stiff clay (Fig. 1(b)).
    - Pressure envelope for excavations in stiff clay (Fig. 1(c)).

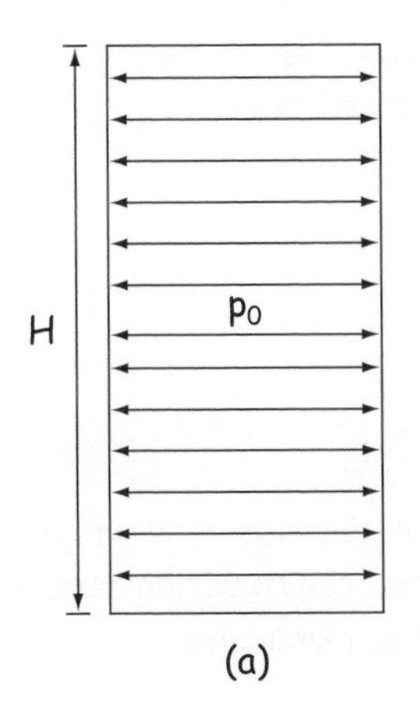

(a)

The pressure p_0 may be expressed as

$$p_0 = 0.65\gamma H K_a$$

where γ = unit weight
H = height of excavation
K_a = Rankine active pressure coefficient $= \tan^2(45 - \phi'/2)$

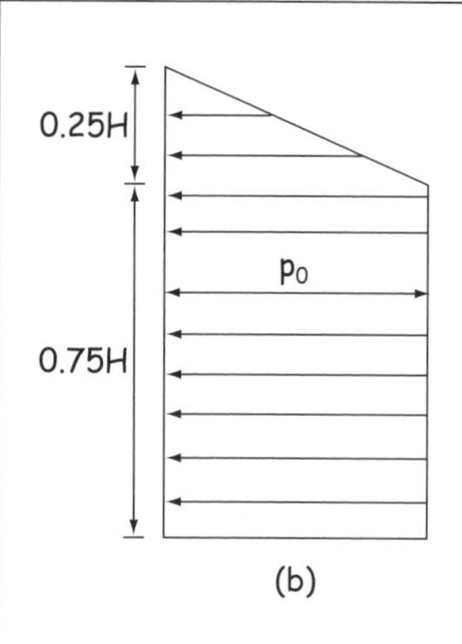

The pressure envelope for soft to medium clay is applicable for the condition

$$\frac{\gamma H}{c_u} > 4$$

where c_n is the undrained shear strength ($\phi = 0$).

The pressure p_0 is the larger of

$$p_0 = \gamma H\left[1 - \frac{4c_u}{\gamma H}\right]$$

or

$$p_0 = 0.3\gamma H$$

where γ = unit weight of clay.

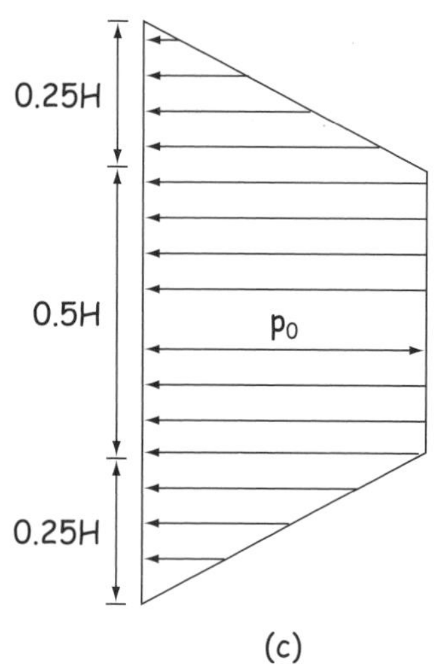

The pressure envelope gives

$$p_0 = 0.2\gamma H \text{ to } 0.4\gamma H$$

with an average of $0.3\gamma H$

and is applicable to the condition

$$\frac{\gamma H}{c_u} \leq 4$$

Fig. 1 Peck's (1969) apparent pressure envelopes for excavations in (a) sand, (b) soft to medium clay, and (c) strutted excavations in stiff clay

- Simplified method for estimating strut loads (Fig. 2).
  - Strut load at A = F_A
  - Strut load at B = $F_{B1} + F_{B2}$
  - Strut load at C = $F_{C1} + F_{C2}$
  - Strut load at D = F_D
- Major limitations.
  - The pressure envelopes are **not** actual pressure distributions which are functions of construction method, support system, construction sequence, relative flexibility of the wall, soil conditions and workmanship.
  - Groundwater is not explicitly considered.

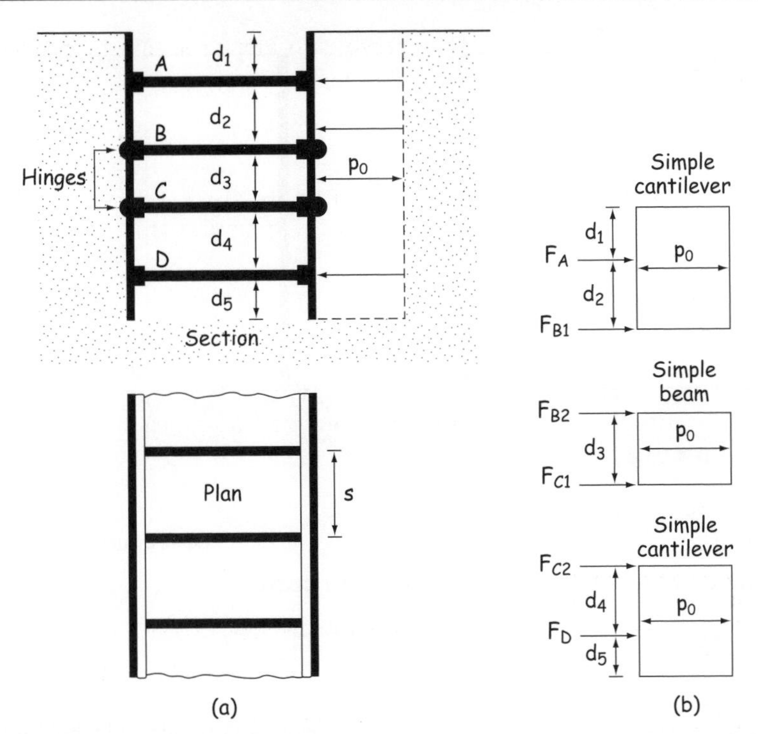

Fig. 2 Determination of strut loads; (a) section and plan of excavation; (b) method for determining strut loads

Design against base heave in clays

An idealized failure mechanism is shown in Fig. 3.

Note: cd and cf are arcs of circles with centres at b and a, respectively.

Assuming that $D \geq 0.7B$, consider the vertical load Q along bd,

$$Q = \gamma HB_1 - c_u H$$

where $B_1 = 0.7B$, c_u is the undrained shear strength, and γ is the unit weight of the soil.

By Terzaghi's net ultimate load bearing capacity relationship:

$$Q_u = c_u N_c B_1 = 5.7 c_u B_1$$

Terzaghi's factor of safety against base heave is

$$F = \frac{Q_u}{Q} = \frac{5.7 c_u B_1}{\gamma HB_1 - c_u H} = \frac{1}{H}\left(\frac{5.7 c_u}{\gamma - \dfrac{c_u}{0.7B}}\right)$$

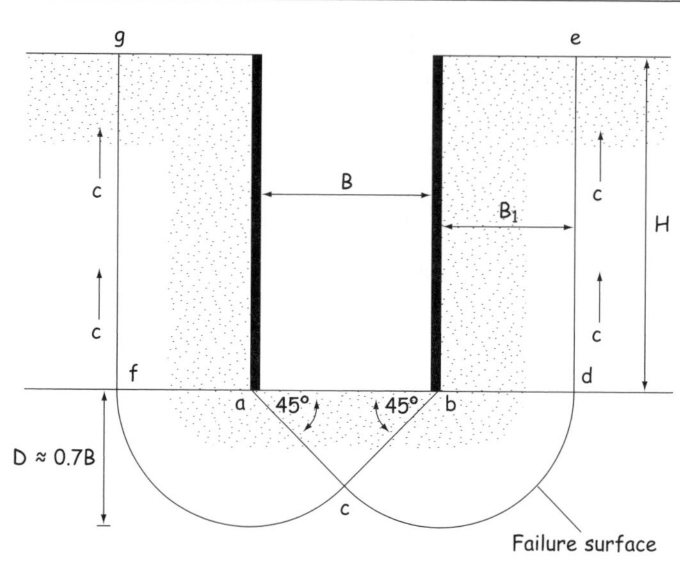

Fig. 3 Factor of safety (FOS) against bottom heave

For $D < 0.7B$

$$F = \frac{Q_u}{Q} = \frac{5.7c_u B_1}{\gamma H B_1 - c_u H} = \frac{1}{H}\left(\frac{5.7c_u}{\gamma - \frac{c_u}{D}}\right)$$

where D is depth to hard layer.

For a relatively long (L) and narrow (i.e. small B) excavation, Bjerrum and Eide (1956) proposed that:

$$F = \frac{c_u N_c}{\gamma H}$$

where N_c is a function of $\frac{L}{B}, \frac{H}{B}$ (see Fig. 4).

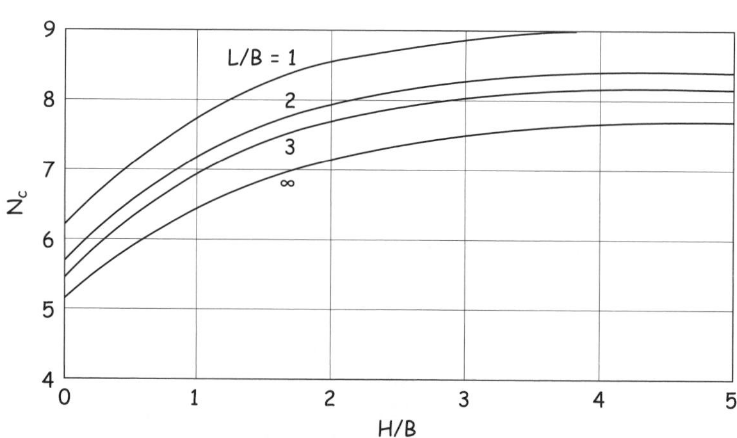

Fig. 4 Variation of N_c with L/B and H/B (based on Bjerrum and Eide's equation)

The variation of normalized maximum horizontal movement $\delta_{H(max)}/H$ with factor of safety F against base heave from field observation (after Mana and Clough, 1981) is given in Fig. 5.

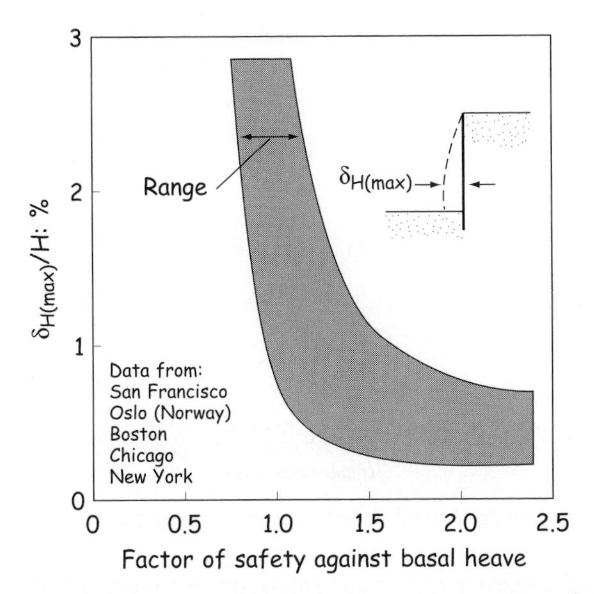

Fig. 5 Variation of $\delta_{H(max)}/H$ with FOS against basal heave from field observation (redrawn after Mana and Clough, 1981)

Design against piping in sands

From the flow net shown in Fig. 6 of a cross-section through a cofferdam, the maximum exit hydraulic gradient can be calculated as

$$i_{max(exit)} = \frac{\left(\frac{h}{N_d}\right)}{a} = \frac{h}{N_d a}$$

where a = length of the flow element at exit, and N_d = number of drops in potential.

It follows that the critical hydraulic gradient is i_{cr} where

$$i_{cr}\gamma_w V = \gamma' V$$

where V is the volume of the flow element, γ_w and γ' are the unit weight of water and effective unit weight of soil, respectively. Therefore:

$$i_{cr} = \frac{\gamma'}{\gamma_w} = \frac{G_s - 1}{1 + e}$$

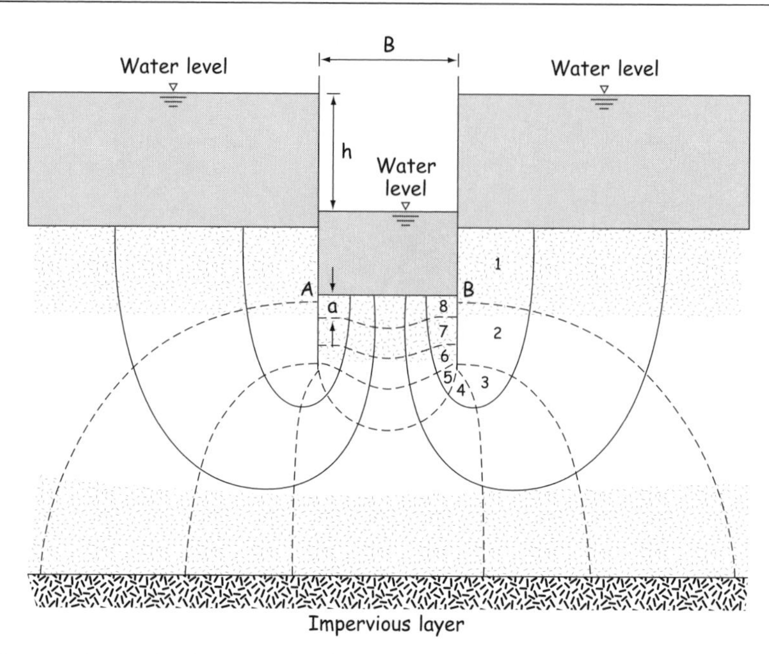

Fig. 6 Determining the factor of safety against piping by drawing flow net

where G_s = specific gravity of the soil solids, and e = voids ratio.

It follows that the factor of safety against piping is

$$F_p = \frac{i_{cr}}{i_{max(exit)}} = \frac{aN_d(G_s - 1)}{h(1 + e)}$$

Preliminary design for serviceability limit states

Simplified deformation models due to unloading (i.e. vertical stress relief) are shown in Fig. 7.

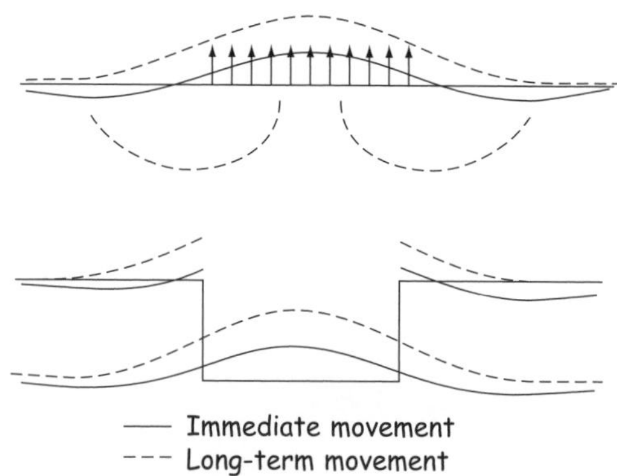

Fig. 7 Effects of vertical stress relief (after Burland et al., 1979)

Lateral wall deformation models due to stress relief are shown in Fig. 8.

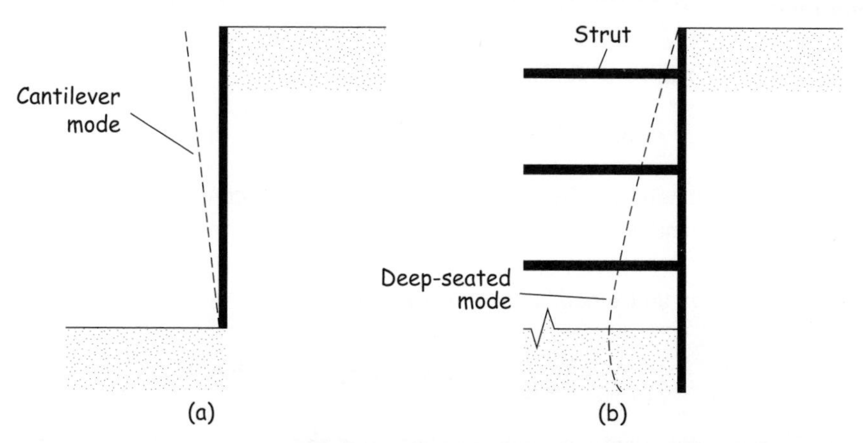

Fig. 8 Nature of yielding of walls: (a) retaining wall; (b) multi-propped excavation

Estimation of ground settlements outside excavations and maximum lateral wall deformations

See Figs 9 and 10.

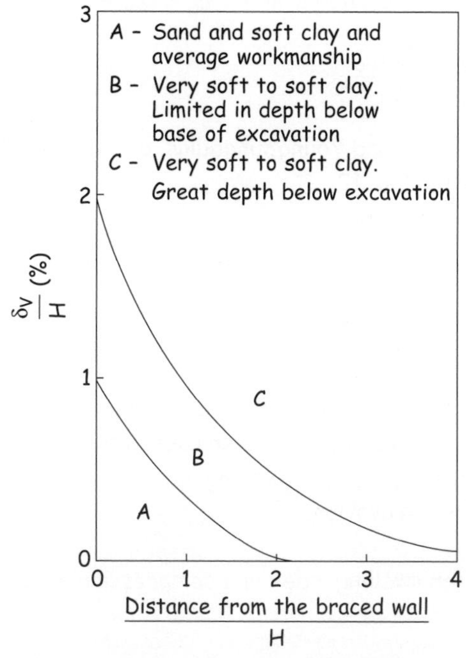

Fig. 9 Variation of ground settlement with distance (after Peck, 1969)

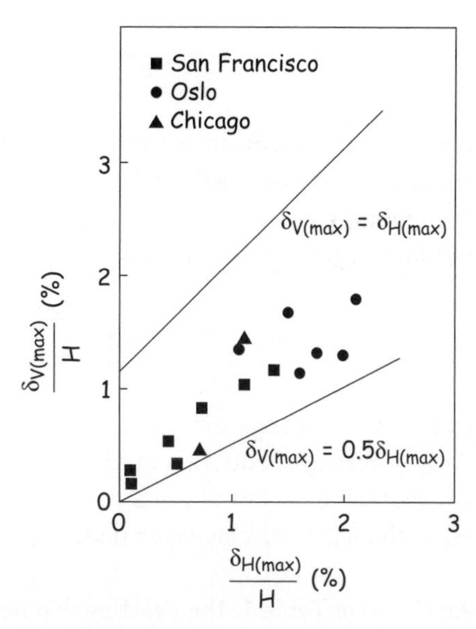

Fig. 10 Variation of maximum lateral wall deflection with maximum ground settlement (after Mana and Clough, 1981)

Estimation of soil swelling inside the excavation in both the short term and long term

Remember that the total vertical soil movement is:

$$\Delta H = \text{elastic heave (undrained)} + \text{swelling}(\Delta\sigma', \Delta u) + \text{creep}(\Delta\sigma' = 0, \Delta u = 0)$$

Upward soil movement as a result of stress relief may be considered as the opposite of downward ground settlement due to footing load.

The estimation of undrained/immediate/elastic heave is:

$$\delta_z = \sum \frac{1}{E'}(\sigma'_z - \upsilon'\sigma'_x - \upsilon'\sigma'_y)\delta h \qquad \text{(Fig. 11)}$$

where E' and υ' are elastic parameters and δh is the thickness of each layer.

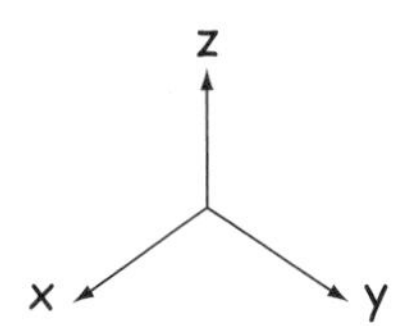

Fig. 11 Reference axes for immediate elastic heave

For the undrained constant volume condition, Poisson's ratio $\upsilon = \upsilon_u = 0.5$ whence

$$\delta_u = \frac{qBI_u}{E_u} = \frac{qB\mu_1\mu_0}{E_u}$$

where I_u is the influence factor, E_u is the undrained Young's modulus, q is the net pressure, B is the width of loading as shown in Fig. 12.

Estimate primary swelling by

$$s = \frac{C_s H_c}{1 + e_0} \log\left(\frac{\sigma'_0 + \Delta\sigma'_{av}}{\sigma'_0}\right)$$

where σ'_0 = average effective pressure on the clay layer before excavation, $\Delta\sigma'_{av}$ = average reduction in effective pressure on the clay layer caused by the excavation, e_0 = initial voids ratio of the clay layer, C_s = swelling index, H_c = thickness of clay layer underneath the excavation.

As shown in Table 1, the relationship between swelling index and compression index may be taken as

$$C_s \approx \left(\frac{1}{5} \text{ to } \frac{1}{10}\right) C_c$$

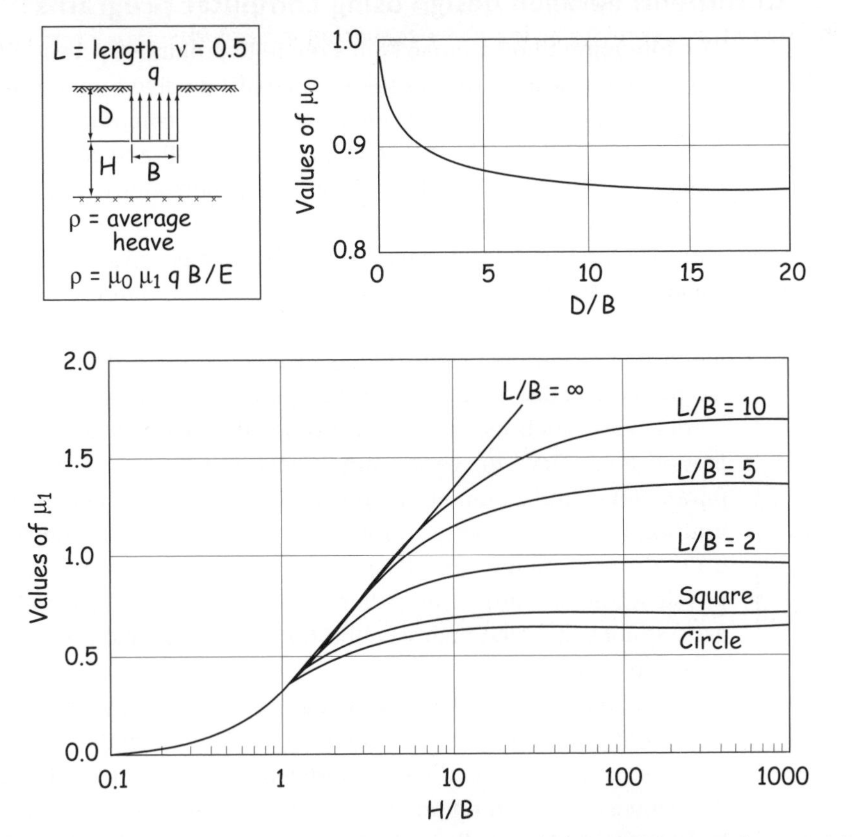

Fig. 12 Diagrams for the factors μ_0 and μ_1 used in the calculation of the immediate average settlement or heave of uniformly loaded flexible areas on homogeneous isotropic saturated clay, after Janbu, Bjerrum and Kjaernsli (1956), as reinterpreted by Christian and Carrier (1978)

Table 1 Compression and swelling indices of natural clays (after Das, 2000)

Soil	Liquid limit	Plastic limit	Compression index C_c	Swelling index C_s
Boston Blue clay	41	20	0.35	0.07
Chicago clay	60	20	0.40	0.07
Ft. Gordon clay, Georgia	51	26	0.12	-
New Orleans clay	80	25	0.30	0.05
Montana clay	60	28	0.21	0.05

Comment: detailed design using computer programs

The foregoing Short Course Notes relate to preliminary design only. For detailed design, due to the complexity involved, many design offices nowadays assist their analyses by using computer programs such as OASYS, FLAC, ABACUS, SAGE-CRISP and SEEP/W – to name just the few that the authors have personal experience of using (there are many others, for example see the Geotechnical and Geo-environmental Software Directory at www.ggsd.com).

For safe and economical designs, it is essential that the programs and parameters are properly calibrated and verified. By this we mean that these idealized numerical models are first checked out as giving sensible results when compared with idealized physical model tests, such as in the centrifuge. It is vital to use physical model tests first to verify any constitutive model, its model parameters and simulation procedures. This is because soil uniformity, homogeneity, properties, boundary conditions and groundwater regime are all specified in the same remit for both physical and numerical modellers. This minimizes the number of assumptions involved in any calibration and verification procedures between them.

After calibrating with physical model tests, programs are then compared with field prototype behaviour. The inevitable lack of fit between numerical predictions and reality, however, will require the exercise of much engineering judgement for these comparisons to be useful. This will be based on a sound knowledge of the modelling assumptions made and an appreciation of the inevitable gap between numerical model predictions and field prototype behaviour. This gap is caused by unforeseeable or unquantifiable factors such as variability (non-uniform and inhomogeneous ground conditions), non-hydrostatic pore water regime, and – perhaps the most indefinable factor of all – workmanship.

For further information and details of calibration procedures or 'benchmarking', reference may be made to 'Finite element analysis in geotechnical engineering' by Potts and Zdravković (2000) and 'Guidelines for the use of advanced numerical analysis' by Potts *et al.* (2002) published by Thomas Telford Ltd (www.thomastelford.com).

CHAPTER SIXTEEN

Instrumentation: a case study of a multi-propped excavation at Lion Yard, Cambridge

Overview: the site

Cambridge is located about 100 km north of Central London. The Lion Yard site is situated in the city centre of Cambridge and is approximately 65 by 45 m on plan as shown in Fig. 2.13. The site is bounded on three sides by Downing Street (DS), St Tibb's Row (STR), and Corn Exchange Street (CE). The remaining side is bounded by the Lion Yard multi-storey car park and shopping precinct development. A five-storey hotel above ground together with a three-storey underground car park was developed on the Lion Yard site. Figure 2.14 shows a cross-section of the underground car park. The structure is supported on large-diameter under-reamed bored piles founded 30 m below ground level in the Gault Clay. The 10 m deep excavation is retained by a 17 m deep perimeter concrete diaphragm wall (0.6 m thick), constructed under bentonite in panels typically 8.5 m in length. Figure 2.15 shows an advanced stage of the excavation. Figure 2.16 shows the extent of the multi-propping involved during construction.

Geology

Folding, uplifting and subsequent erosion at the end of the Jurassic period or in the early part of the Cretaceous period gave rise to an unconformity, so that the earliest Cretaceous formation, the Lower Greensand, rests on Kimmeridge clay in places, and Corallian in others (Worssam and Taylor 1975). The Lower Greensand was deposited in shallow water, and the Gault Clay was subsequently laid down as a result of a widespread marine incursion that spanned the Middle and Upper Albian stages. The conditions were not uniform throughout the period of deposition and so the Gault contains several beds of different lithology. At the end of the Gault periods of time a temporary halt in deposition occurred, accompanied by submarine erosion and reworking of sediments on the seafloor.

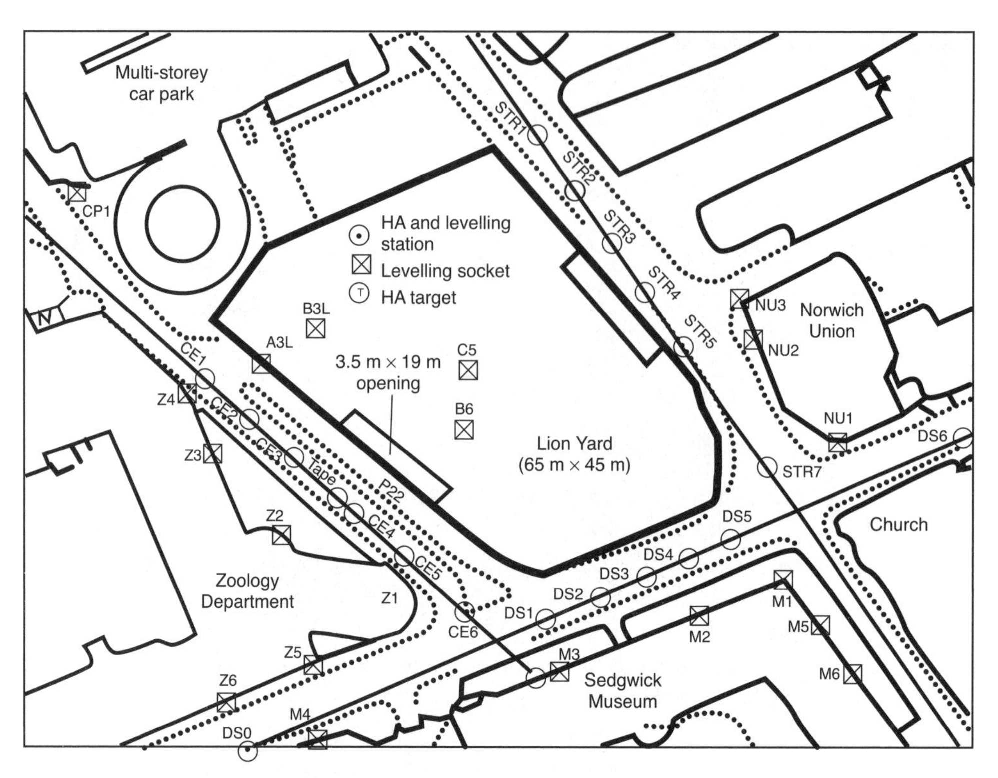

Fig. 2.13 Lion Yard site plan and details of monitoring stations (after Ng, 1992; Ng, 1998)

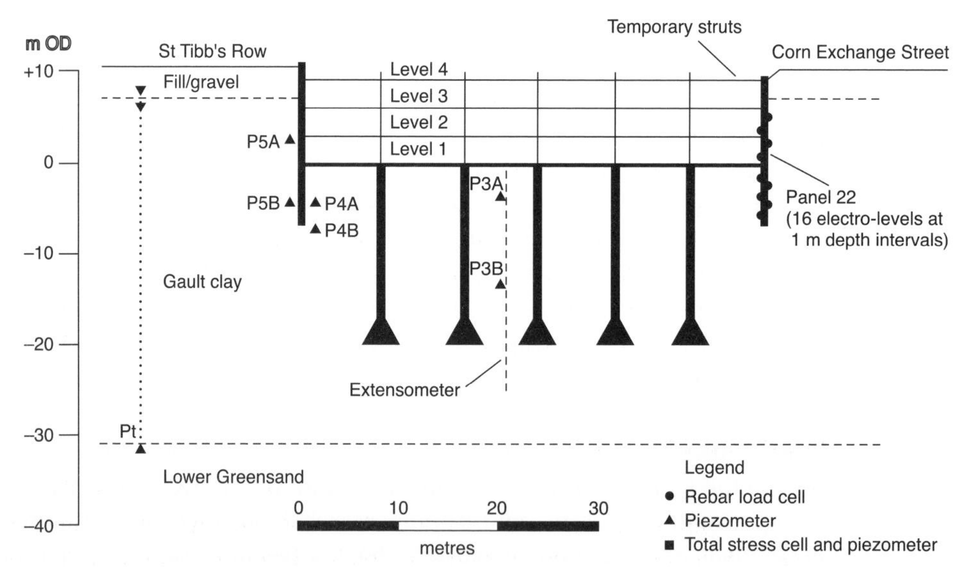

Fig. 2.14 Lion Yard cross-section of site (after Ng, 1992; Ng, 1998)

Fig. 2.15 View across deep excavation construction site at Lion Yard

Following the deposition of the Gault, the chalk was laid down as the sea water cleared and the land areas dwindled, so that less and less terrigenous sediment became available. During the tertiary and quaternary periods, uplift and extensive erosion took place and eventually produced the present landscape. An estimated 200–400 m of chalk has since been eroded (Lings *et al.*, 1991).

In the Cambridge area, the thickness of Gault varies between 27 and 42 m. At Lion Yard, a 38 m thickness of Gault was found during the installation of a standpipe piezometer in the Lower Greensand.

In general, groundwater conditions are approximately hydrostatic below a level of approximately +7 m OD (i.e. 3 to 3.5 m below existing ground level).

Soil profile and properties

At Lion Yard, the existing ground level is at approximately +10 m OD. The ground conditions at the site comprise 3–4 m of made ground and gravel above 38 m of Gault Clay, which overlies the Lower Greensand. The soil profile obtained during site investigation is shown in Fig. 2.17, including

Fig. 2.16 Multi-propping at Lion Yard during construction

self-boring pressuremeter (SBPM) measurements carried out by the Building Research Establishment (BRE).

The Gault Clay in Cambridge is highly calcareous, containing up to 30% by weight of calcium carbonate (Worssam and Taylor, 1975). A similar test result (27.5%) was found in samples taken from Lion Yard (Ng, 1992). The Gault in its natural state is heavily over-consolidated, having natural water contents close to the plastic limit. It consists of stiff to hard silty grey clay of high plasticity. The clay is laminated with closely spaced fissures and joints. The top few metres of clay show signs of weathering, such as cryoturbation. Hard nodules of phosphatized marl are scattered through the clay.

It is extremely difficult to assess initial horizontal stresses in the ground, but these stresses are very important for correctly interpreting the field monitoring data. Reloading on top of a clay deposit has a dominant effect on vertical effective stress but has only a relatively small effect on horizontal effective stress as illustrated by many researchers (Burland and Hancock, 1977; Simpson *et al.*, 1979). These researchers allowed for

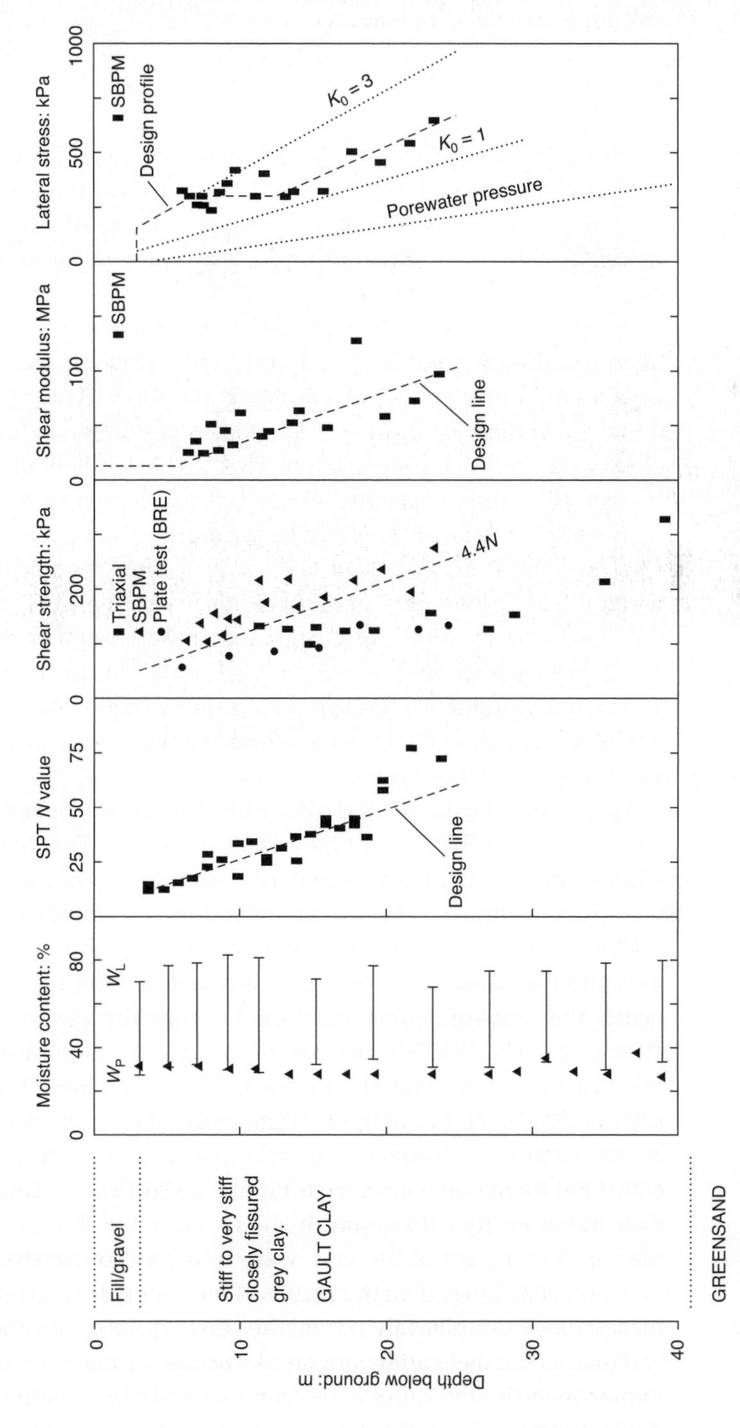

Fig. 2.17 Soil profile and geotechnical parameters at Lion Yard (after Lings et al., 1991)

the effects of reloading by assuming isotropic elastic behaviour and using the equation given as follows:

$$\frac{\Delta\sigma'_h}{\Delta\sigma'_v} = \frac{\mu'}{1-\mu'} \tag{2.20}$$

Simpson *et al.* (1979) show that the ratios of $\Delta\sigma'_h/\Delta\sigma'_v$ due to reloading were 0.18 and 0.33 for isotropic and anisotropic assumptions, respectively, for London Clay. Clearly reloading has a relatively small effect on $\Delta\sigma'_h$ and therefore leads to a reduction of K_0, especially in the top few metres of clay.

At Lion Yard, the geological history is not simple. The Gault Clay has been reloaded by gravel and fill after the erosion of the chalk. A simple calculation indicates that at 1 m below the top surface of Gault Clay, an initial K_0 value of 3.0 prior to redeposition will be reduced to <1 as a result of a 3 m thick redeposition. This is substantially different from the in situ SBPM measurements. The results of these tests suggest that very high initial horizontal stresses are present in the upper 7 m of the Gault Clay at Lion Yard, corresponding to K_0 values ranging from 1.5 to 3.5 when related to the current vertical effective stresses. This upper part of the clay is the most relevant, since the entire basement is located within the 10 m below-ground level. It is thus important to compare the results of the pressuremeter testing with some other tests, and with semi-empirical and theoretical calculations, so that the field monitoring data can be correctly interpreted.

Apart from the Lion Yard site, SBPM tests were conducted in Gault Clay by the BRE at Madingley (Powell, 1990). The Madingley site is situated about 4 km west of Lion Yard and the top of the Gault Clay is at the ground surface, that is, no reloading by gravel occurred. Figure 2.18(a) shows the variation of total horizontal stresses with depth measured in Gault Clay at Lion Yard and Madingley, and estimations using the semi-empirical correlation originally proposed by Schmidt (1966), $K_0 = K_{0nc}\text{OCR}^m$, where $m = 1.2\sin\phi'$. Laboratory test results ($\phi' = 24°$) on reconstituted specimens (Ng, 1992) have been used in the calculation. Since the maximum preconsolidation pressure at Lion Yard is not known, a range of possible preconsolidation pressures (2000–4000 kPa) has been considered. Figure 2.18(b) shows the corresponding distribution of K_0 with depth. It should be noted that prior to reloading, the top few metres of the clay were in a state of passive failure with a K_p value of 3.5, based on the results of laboratory tests ($\phi' = 34°$) on undisturbed block samples taken from the site (Ng, 1992). It can be seen in Fig. 2.18(a) that the measured horizontal stresses at Lion Yard are substantially higher than the measurements using an SBPM at Madingley for the top 7 m of clay, even if the lateral stresses are adjusted to allow for the

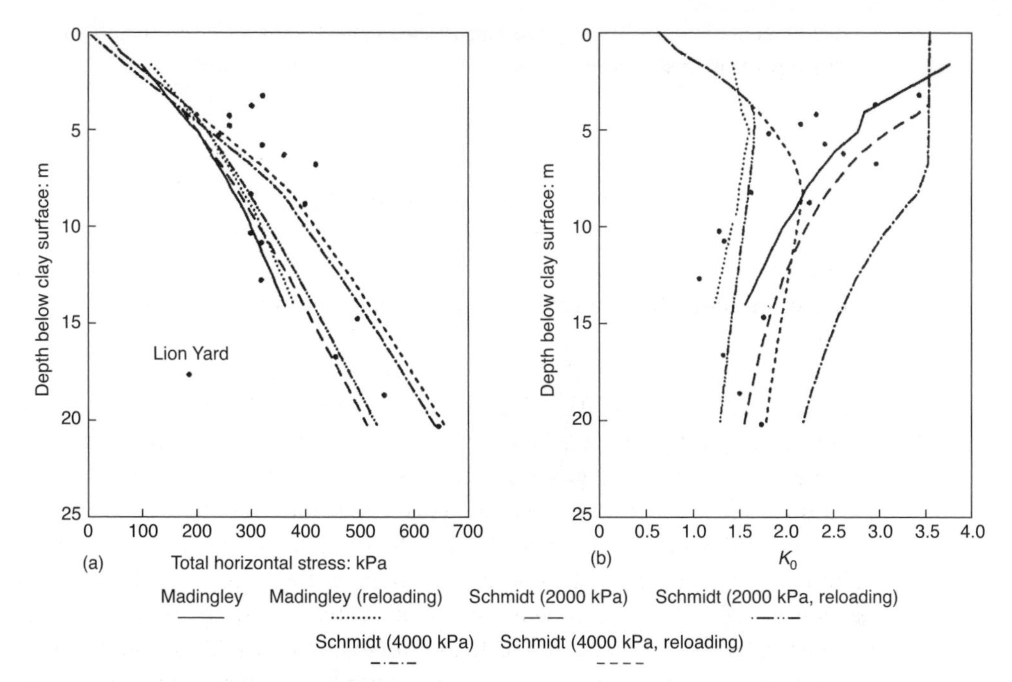

Fig. 2.18 Comparison of measured initial horizontal stresses and K_0 *at Lion Yard and Madingley sites (after Ng, 1998)*

assumed effects of reloading by a 3 m thickness of gravel. Similarly the measured values at Lion Yard are significantly higher than the estimated horizontal stresses using Schmidt's semi-empirical correlation for the possible range of preconsolidation pressures considered.

The geological history of Lion Yard has also been simulated numerically using the Brick model (Simpson, 1992). Details of the numerical simulation procedures are given by Ng (1992) and Ng *et al.* (1998a). Figure 2.19 shows the computed results that agree with the predictions using Schmidt's correlation with the assumed preconsolidation pressure of 2000 kPa and the measurements at Madingley if reloading by gravel is considered. Thus it is believed that the initial horizontal stresses in the ground are likely to have been overestimated by the SBPM tests at Lion Yard. The initial K_0 values are likely to be <2 over the 10 m deep excavation.

Top-down construction

The ground-floor slab (Level 4) was cast as soon as the diaphragm wall and bored piles and associated steel columns had been completed (see Fig. 2.20). Soil was then excavated from beneath the slab by mechanical plant down to the next level (Level 3) and removed through an opening

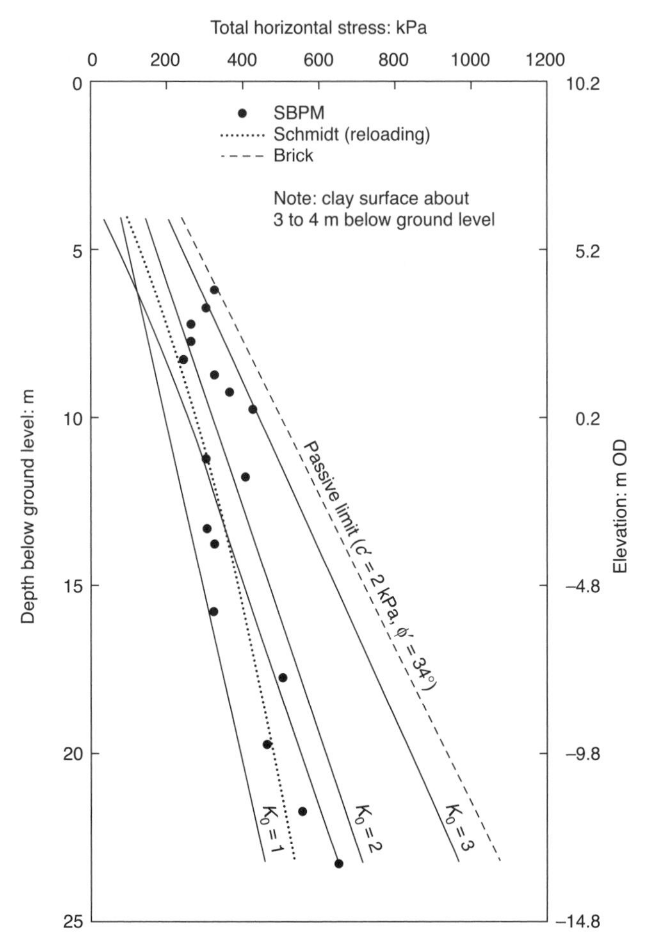

Fig. 2.19 Comparison between results of numerical simulations, measurements and empirical calculations for Lion Yard (after Ng, 1998)

left in the slab. Similar operations were repeated for subsequent stages of construction until the bottom level (Level 1) was reached. Erection of the superstructure was carried out simultaneously. A summary of the major site activities is given in Table 2.1.

On two opposite sides of the site 3.5 m wide and 19 m long rectangular openings were left in the slabs adjacent to the diaphragm wall (see Fig. 2.13). These were at the positions of the sloping vehicle ramps that allow access to the underground car park levels. Temporary steel props were installed across these 3.5 m wide openings at each level to support the diaphragm wall during excavation. The props were not prestressed, but the ends were grouted after installation. More detailed descriptions of the structure and its construction are given by Ng *et al.* (1990) and Ng (1992).

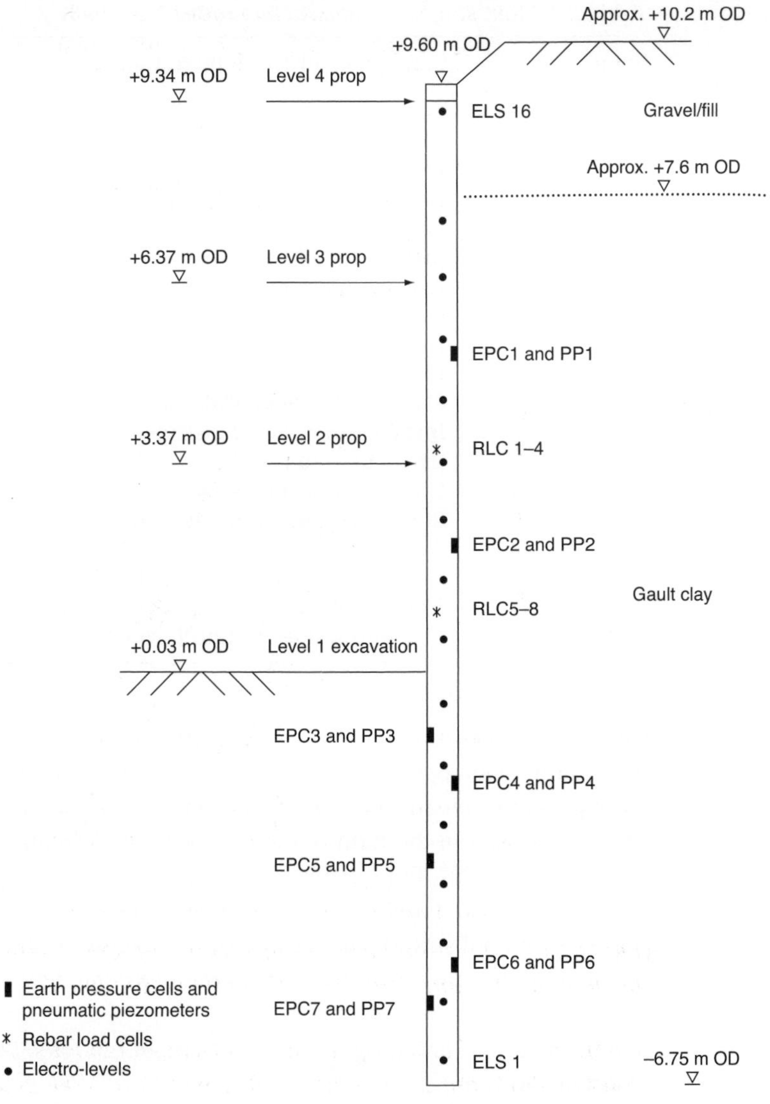

Fig. 2.20 Details of instrumentation of Panel 22 at Lion Yard (after Ng, 1998)

Unusual research opportunities

Overview

The Lion Yard Project provided three unusual features for field monitoring. First, since the basement is located predominantly in the fairly uniform Gault Clay, the interpretation of the field data is not complicated by the presence of other soil deposits. Second, the presence of two temporary vehicle openings enabled the measurements of steel strut loads to

Table 2.1 Main stages of construction (after Ng, 1998)

Stage	Construction operation	Week number
I	Construction of diaphragm wall	1–7
II	Construction of bored piles	7–12
III	Reduced level dig to Level 4	12–13
IV	Casting of Level 4 slab	14–17
V	Installation of Level 4 struts	17–21
VI	Excavation to Level 3	18–20
VII	Casting of Level 3 slab	21–24
VIII	Installation of Level 3 struts	23–23
IX	Excavation to Level 2	23–25
X	Casting of Level 2 slab	26–28
XI	Installation of Level 2 struts	27–28
XII	Excavation to Level 1	29–32
XIII	Casting of Level 1 slab	35–41
XIV	Removal of all temporary struts	45–45
XV	4 months after casting Level 1 slab	56
XVI	9 months after casting Level 1 slab	78
XVII	15 months after casting Level 1 slab	104
XVIII	25 months after casting Level 1 slab	150

supplement measurements of total earth pressures on both sides of a diaphragm wall panel. The success of monitoring the strut loads and earth pressures means that the equilibrium of forces and bending moments can be assessed in the light of the measured wall rotation data (Ng, 1992; Lings *et al.*, 1993). Further insight into the soil–structure interaction can then be gained. Finally, the success of recording both pore pressure changes and subsurface movements in the centre of the site provided an unusual opportunity to study in situ swelling behaviour of the clay due to the excavation.

Instrumentation

To verify the design assumptions and to monitor the performance of the 10 m deep excavation at Lion Yard, various instruments were installed on site. Full details of the instrumentation together with data during the construction were collated in a factual report (Ng *et al.*, 1990).

Monitoring soil–structure interaction at Panel 22

Panel 22 is situated at one of the two temporary ramp openings, the one adjacent to Corn Exchange Street (see Figs 2.13 and 2.14), and during the course of construction it was propped by steel struts positioned across the ramp opening. The struts were removed only after the ramps had been constructed. In Fig. 2.20 detailed locations of the

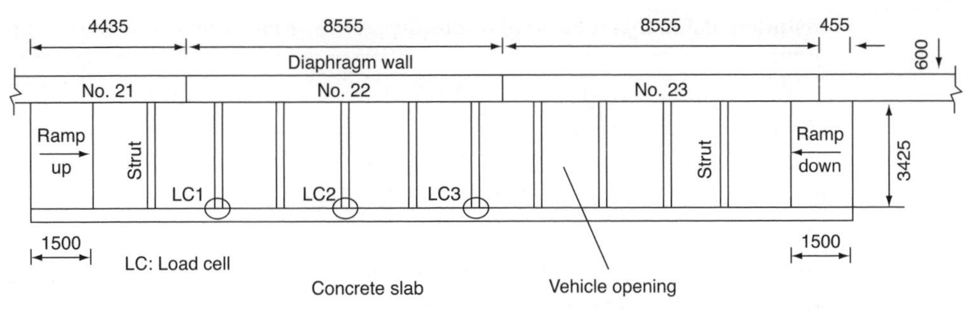

Fig. 2.21 Details of propping system and load cell locations of Panel 22 at Lion Yard (after Ng, 1998)

instrumentation at Panel 22 are illustrated. Altogether seven pneumatic total pressure cells were installed on both faces of the diaphragm wall panel to measure the horizontal total pressures. In addition, seven pneumatic piezometers were positioned 150 mm below the earth pressure cells to measure the pore water pressures at the soil–wall interface. Hence the effective horizontal stresses can be determined at the interface.

Although the car park was constructed top-down, the access ramps on either side of the site were not constructed until after the Level 1 slab had been cast. Thus temporary steel struts were needed across the ramp openings (3.5 by 19 m) to support the diaphragm wall. Figure 2.21 shows the details of the propping system adjacent to Panel 22. Ten temporary steel struts were installed in the ramp openings on each level as the excavation progressed. The struts used were universal columns (UCs) of sections $152 \times 152 \times 23$ mm UCs at Level 4 and $203 \times 203 \times 60$ mm UCs at Levels 2 and 3. Vibrating wire-type load cells were used to measure the strut loads. Of the five struts supporting Panel 22 at each level, three were instrumented with load cells, placed alternately as shown in Fig. 2.21.

Vibrating wire rebar load cells (RLCs), similar to the strut load cells, were also used to measure the reinforcement stresses in Panel 22 (see Fig. 2.20). The load cells were installed at the points at which the maximum bending moments in the wall were anticipated, with the intention of monitoring the rebar stresses for deducing bending moments at all stages of construction, particularly when the excavation reached Level 1. Monitoring data and results of interpretation of the rebar load cells are reported by Ng *et al.* (1992).

To monitor the deflected shape of the diaphragm wall, an electrolevels (ELs) system (Cooke and Price, 1974) was used to measure the rotation of the diaphragm wall. From the measured rotations the relative deflected profile of the wall could be computed. At Panel 22, a train of 16 ELs was installed at 1 m intervals in an inclinometer tube cast into the wall (see Fig. 2.20).

Although the ELs gave the deflected shapes of the diaphragm wall, the absolute displacements of the wall are not known. By measuring the absolute movement at the top of the wall, the absolute displacements of the wall at depth can be determined. To determine the movement at the top of the diaphragm wall (Panel 22), a surface monitoring station was positioned on the theodolite line of sight along Corn Exchange Street offset 10 m from Panel 22. By measuring the distance between a reference point on the top of the wall to the surface station, the absolute movement at the top of the diaphragm wall panel could be determined.

Monitoring soil swelling and pore pressure in the centre of the site

A magnetic extensometer system consisting of 12 arrowhead magnets (AMs) was installed to 35 m below ground level for monitoring short-term heave and long-term swelling of the clay (see Fig. 2.14). The extensometer system was located as far as possible from any internal piles to minimize any restraint from them.

Two pneumatic piezometers (P3A and P3B) were located in a borehole (BH) close to the extensometer system (see Fig. 2.14). In addition, a tensiometer was installed at −1.5 m OD after casting the Level 1 slab. This was to enable the vertical swelling behaviour of the soil to be correlated with the associated pore water pressure changes. To better understand the pore pressure responses around the site due to the excavation, additional pneumatic (P4A, P4B, P5A and P5B) and standpipe piezometers (PIs) were installed (see Fig. 2.14).

Monitoring general ground deformation around site

In addition to the monitoring of the deflected profile of the diaphragm wall and vertical sub-surface movement in the centre of the site, building and road surface settlements and horizontal ground movements were monitored throughout the construction period. The locations of some monitoring stations are shown in Fig. 2.13.

A benchmark levelling socket was installed in a building located 76 m away from the site. At this distance the benchmark is outside the likely zone of influence of the excavation (Burland *et al.*, 1979); the ratio of the distance from the excavation to the maximum depth of excavation is about 7.6. The precise level used in this project was equipped with an external optical micrometer. The micrometer scale could be read to an accuracy of 0.1 mm and by estimation to 0.01 mm. A Wild invar staff graduated at 10 mm intervals was used.

Three theodolite lines of sight along the three sides of the excavation, Corn Exchange Street, Downing Street and St Tibb's Row, were set up to monitor the horizontal ground surface movements normal to the diaphragm walls. Hilti nails were driven into the road for use both as horizontal alignment (HA) or lateral ground surface movement monitoring stations and as levelling stations.

Observed performance

Reductions of lateral stress

During the construction of Panel 22, total earth pressure cells were used to measure the lateral stresses during and after concreting of the diaphragm wall. Subsequently, piling commenced inside the site and 30 m long under-reamed bored piles were constructed with shaft diameters ranging from 0.9 to 1.4 m and bell diameters varying from 2.7 to 4 m. The lateral stresses observed at intervals during these two site operations are summarized in Fig. 2.22. Even though some inaccuracy may exist in the earth pressure measurements using the pressuremeter and the total pressure cells, it can be seen that a marked reduction in total lateral stresses exists at the soil–wall interface due to diaphragm wall construction.

In contrast, the piling operation at Lion Yard did not seem to reduce further the lateral stresses at the soil–wall interface. This was perhaps because the nearest piles were 8 m away from the instrumented Panel 22, and the influence of the piles might not be large enough to cause a noticeable change in the pressure cell reading. More importantly, the effects of the piles might be predominantly in the levels below the toe of

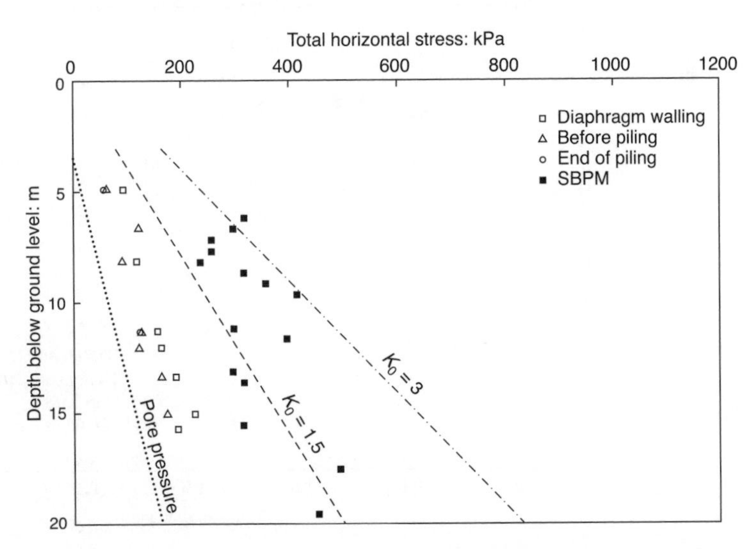

Fig. 2.22 Total lateral stresses at diaphragm walling before and after piling at Lion Yard (after Ng, 1998)

the diaphragm wall as the soil there is thought to have been highly stressed after the installation of the wall (Ng *et al.*, 1995b).

Ground movements

Clough and O'Rourke (1990) summarized the observed settlements during diaphragm wall construction of four slurry walls in granular soil, soft-to-medium clay and stiff-to-very hard clay from various parts of the world. They derived a limiting settlement curve for diaphragm wall construction in a dimensionless form (see Fig. 2.23). Figure 2.23 also shows the design curve proposed by Thompson (1991), who collected settlement data from 13 sites in London Clay associated with the construction of retaining walls. Settlements due to piling were excluded in both cases. It can be seen that the settlements decrease rapidly with distance away from the wall irrespective of the soil types in which the wall was constructed.

On the same diagram, two sets of Lion Yard data are also presented. The solid and open symbols represent the settlements due to diaphragm walling and diaphragm walling plus piling, respectively. Although only a few settlement stations were surveyed before the construction of the diaphragm wall, and the Gault Clay is quite different from the London Clay in various aspects, the comparison can still provide some information on the general pattern of settlements associated with wall installation. It is clear that the observed settlements at Lion Yard fall well within the limiting curve proposed by Clough and O'Rourke (1990) and are also

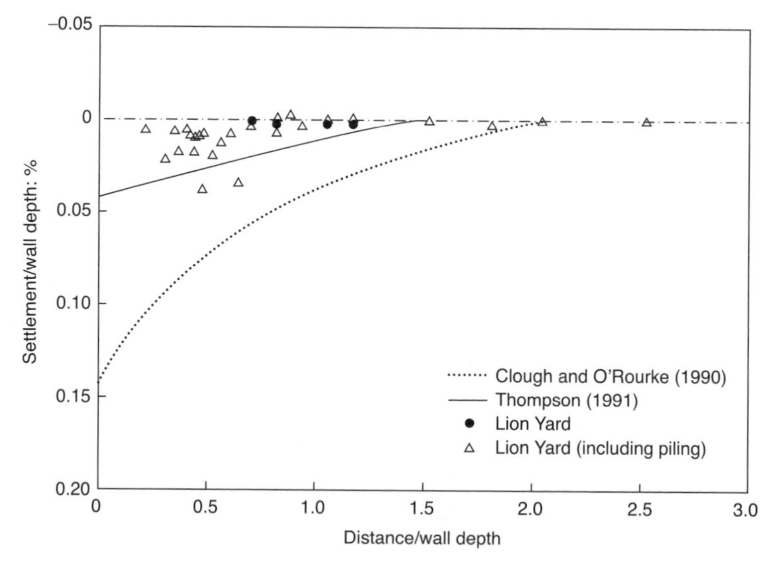

Fig. 2.23 Normalized plot of settlements caused by diaphragm walling at Lion Yard (after Ng, 1998)

considerably smaller than the settlements observed adjacent to similar diaphragm walls constructed in London Clay. Moreover, the observed total settlements due to diaphragm walling and piling follow the same pattern as the two published curves and still fall within Thompson's curve, with only a few exceptions.

Deflection of Panel 22

The interpreted displacement profiles of Panel 22 are shown in Fig. 2.24. The absolute displacement profiles of the wall during the main excavation were deduced from the measured movement of the top of the wall that had been monitored throughout the construction period by using a theodolite and precise taping. At the first stage of the main excavation (L3), the wall was lightly propped and behaved effectively as a cantilever (Stage VI). Because of difficulties in installing the first set of prop load cells, L3 excavation proceeded at Panel 22 before three out of five struts had been grouted up. However, subsequent levels of excavation proceeded after all struts had been properly installed.

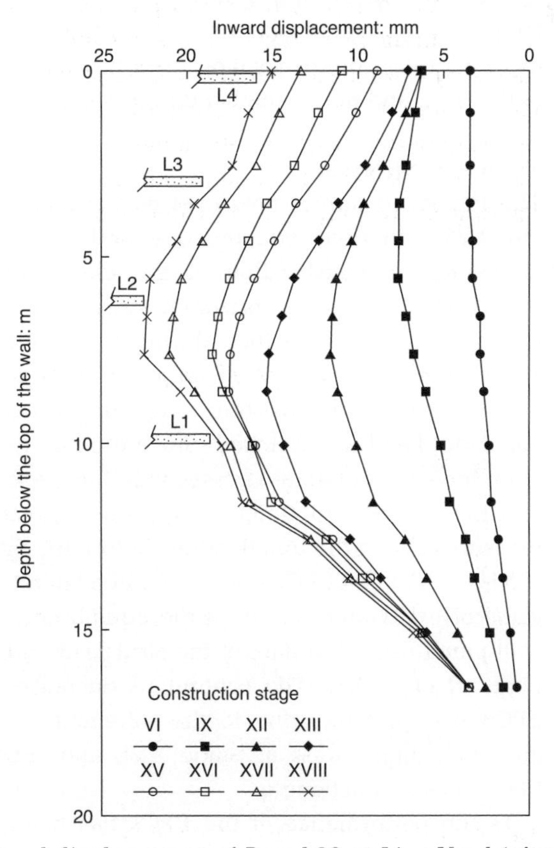

Fig. 2.24 Lateral displacement of Panel 22 at Lion Yard (after Ng, 1998)

At the final stage of the main excavation, a deep-seated deflected shape of the wall was observed, and the maximum wall displacement was only 12 mm. The deduced toe movement was 4 mm at this stage.

Because of the construction of the hotel façade, a new socket had to be inserted into the façade blockwork on the top of Panel 22 for further precise tape measurement. Unreliable readings were subsequently obtained. This may be due to the movement of the façade blockwork or the socket itself. Since the movement at the top of the wall could not be determined reliably after the construction of the façade, the absolute displacement profile of the wall is not known. Two approaches for determining the absolute displacement profiles of the wall were considered: fixity could be assumed either at the top or at the toe of the wall. If the former approach were adopted, the toe of the wall would have moved outward by 6 mm (i.e. away from the centre of the site), which is considered to be unlikely. The latter approach seems to be more sensible. As shown in Fig. 2.24, the wall has continued to creep inward, probably as a result of shrinkage of the concrete slabs. The maximum wall displacement at Stage XVIII is 50% greater than the maximum deflection of the wall at Stage XII. This is consistent with the observed long-term settlement patterns behind the wall. The large movements of the wall above the excavation level have actually reduced the curvature of the wall, and hence the bending moments.

Total earth pressures

The measured lateral pressures on both sides of the wall are given in Fig. 2.25. A marked reduction of lateral stresses at the soil–wall interface following the construction of the diaphragm wall is revealed. The reduction of the total horizontal stress was substantially larger at the wall installation stage than during the whole main excavation period. During the main excavation stages, all except the lowest earth pressure cell (EPC) showed consistent reductions in lateral stress until the lowest basement level was reached. Since then there has been a general trend of an increase in lateral stresses with time, except EPC3. At Stage XVIII, four EPCs (those still functioning properly) recorded an increase in earth pressures above the initial values before the main excavation.

It is evident that EPC3 under-recorded lateral stress by 50 kPa at the final stage of excavation based on the equilibrium analysis of the diaphragm wall panel using the data of the strut load and EPCs and ELs (Ng, 1992; Lings *et al.*, 1993). The continuous decrease of total earth pressure at EPC3 was probably due to the softening of clay. Cell EPC1 recorded abnormal, high stress at Stage XVI, and subsequently both EPCl and EPC2 ceased functioning.

As the performance of the EPCs has been critically assessed in the light of the measured wall rotations and strut loads during the main

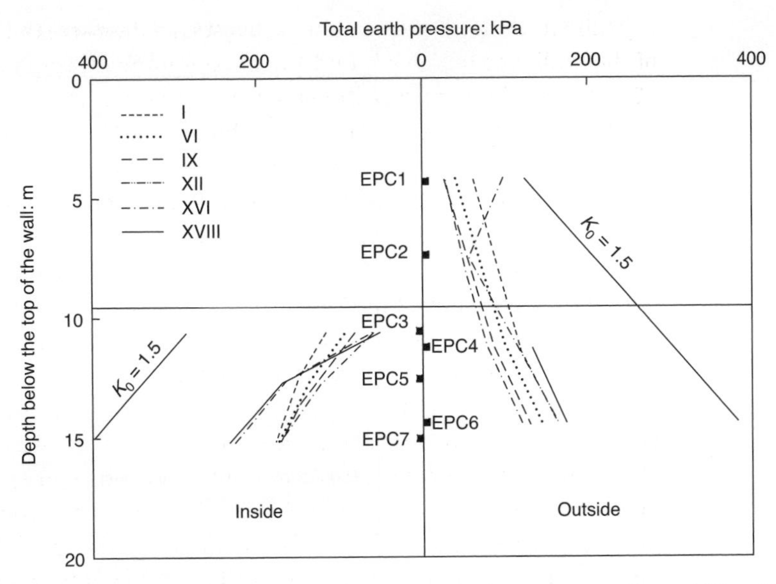

Fig. 2.25 Total lateral pressure during and after construction at Lion Yard (after Ng, 1998)

excavation stages (Ng, 1992; Lings *et al.*, 1993), this gives confidence in the derived effective stress paths that have been discussed in detail by Ng (1992).

Strut loads

The measured strut loads are considered to be some of the most reliable data obtained from the site. These load measurements together with wall rotation data have been used to assess the force and moment equilibrium of the measured total earth pressures at Panel 22 of the diaphragm wall (Ng, 1992; Lings *et al.*, 1993).

It is interesting to compare the highest measured strut loads at each level with the suggested range of design values given by Peck (1969). The maximum strut loads after final excavation have been converted into apparent pressure diagrams and superimposed on Peck's diagram as illustrated in Fig. 2.26. The values measured are close to Peck's suggested lower bound, being equivalent to a trapezium based on $0.27\gamma H$ at Level 2 and $0.22\gamma H$ at Level 3. The Level 4 apparent pressure diagram is somewhat smaller at $0.11\gamma H$, but this was partly due to these struts not becoming effective until the end of the Level 3 excavation. The low strut loads at the end of excavation were attributed to the low lateral stresses in the ground following the construction of the diaphragm wall.

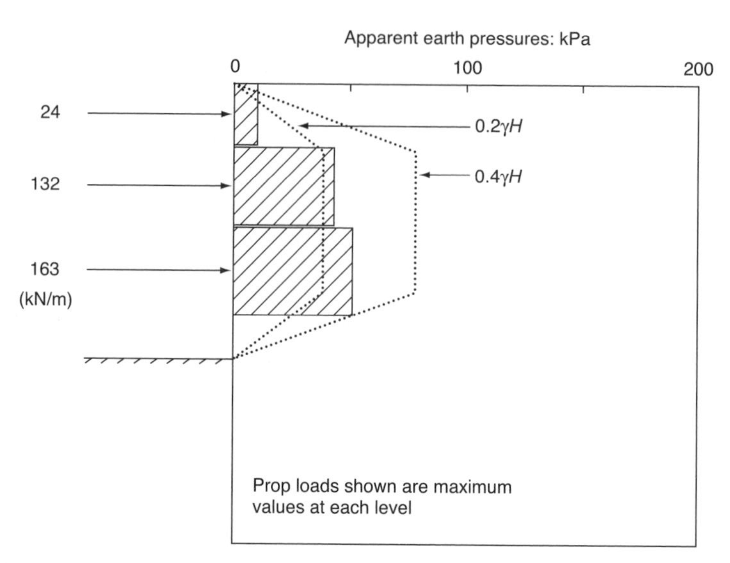

Fig. 2.26 Comparison of measured and Peck's (1969) design earth pressures at Lion Yard (after Ng, 1998)

General ground deformation patterns

Overview

The initial reference data set for the surveying stations were taken at different times and in many cases after the start of diaphragm walling. It is thus rather difficult to make consistent comparisons, and the surveying results taken just before the main excavation have been chosen as a single reference datum for all stations. The general deformation pattern will not be altered by this change since the ground movements during diaphragm walling and piling were small. Thus all the data presented in this section represent only the ground movements caused by the three stages of the main excavation, unless stated otherwise. The accuracies of each individual settlement and HA monitoring station are estimated to be ± 3 mm and ± 2.0 mm, respectively.

Surface movements around the site

Figure 2.27 shows the settlement contours around the site at the end of excavation. The contours were obtained by interpolation of readings from surveying points. The observed settlements along Downing Street are smaller than the measured values along Corn Exchange Street and St Tibb's Row. This was because Downing Street is on one of the short sides of the site and the wall was strongly supported by the Level 4 slab that was used as a working platform for the top-down construction.

Figure 2.28(a) shows a normalized plot of ground movements along the three sides of the site, Corn Exchange Street, Downing Street and St

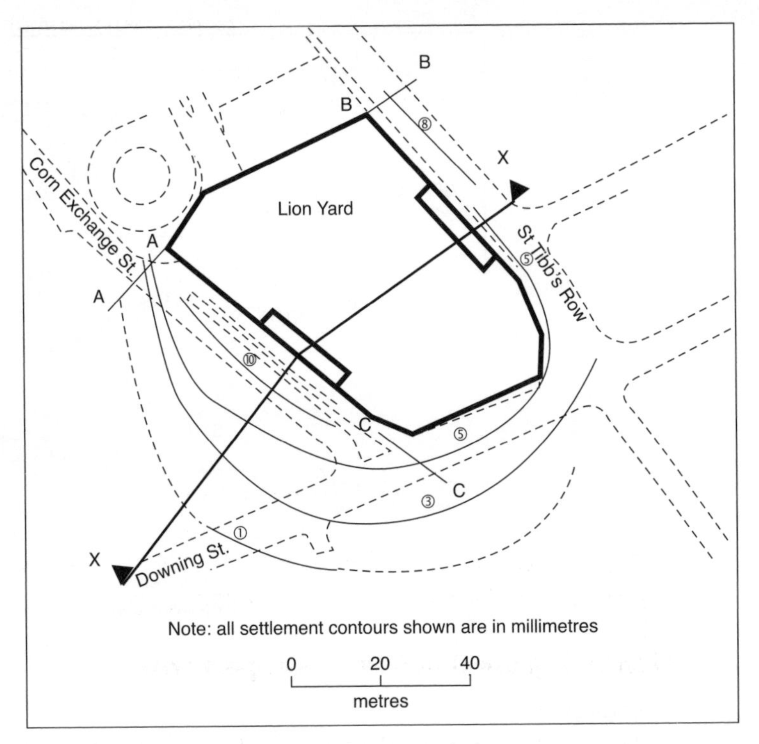

Fig. 2.27 Settlement contours around site at end of construction at Lion Yard (after Ng, 1998)

Tibb's Row, versus offset distance from the wall. The general settlement patterns are very similar along all three sides of the site. A V-shaped settlement profile is apparent on each side of the site. The observed settlements on the two long sides of the site (CE and STR) are greater than on the short side of the site (OS). This is consistent with the settlement contour plot shown in Fig. 2.27. The observed patterns of horizontal surface movements are less clear than the settlement data. The data show a fairly uniform lateral inward movement profile behind the wall along CB but illustrate a decreasing lateral ground movement with increasing distance behind the wall along DS. Lateral movements along STR were excluded because of difficulties in obtaining reliable data.

Figure 2.28(b) shows a normalized plot of the surface movements against distance ratio, which is defined as the distance of a surface station from a reference line on its side of the site over the total length of the diaphragm wall on that side of the site. The reference lines for CE, STR and OS are marked as AA, BB and CC, respectively, in Fig. 2.27. It is easy to deduce that a distance ratio of 0.5 means the middle, while 0 and 1 imply the corners of the site. It would be expected that settlements in the middle would be larger than that at the corners of the site. Figure

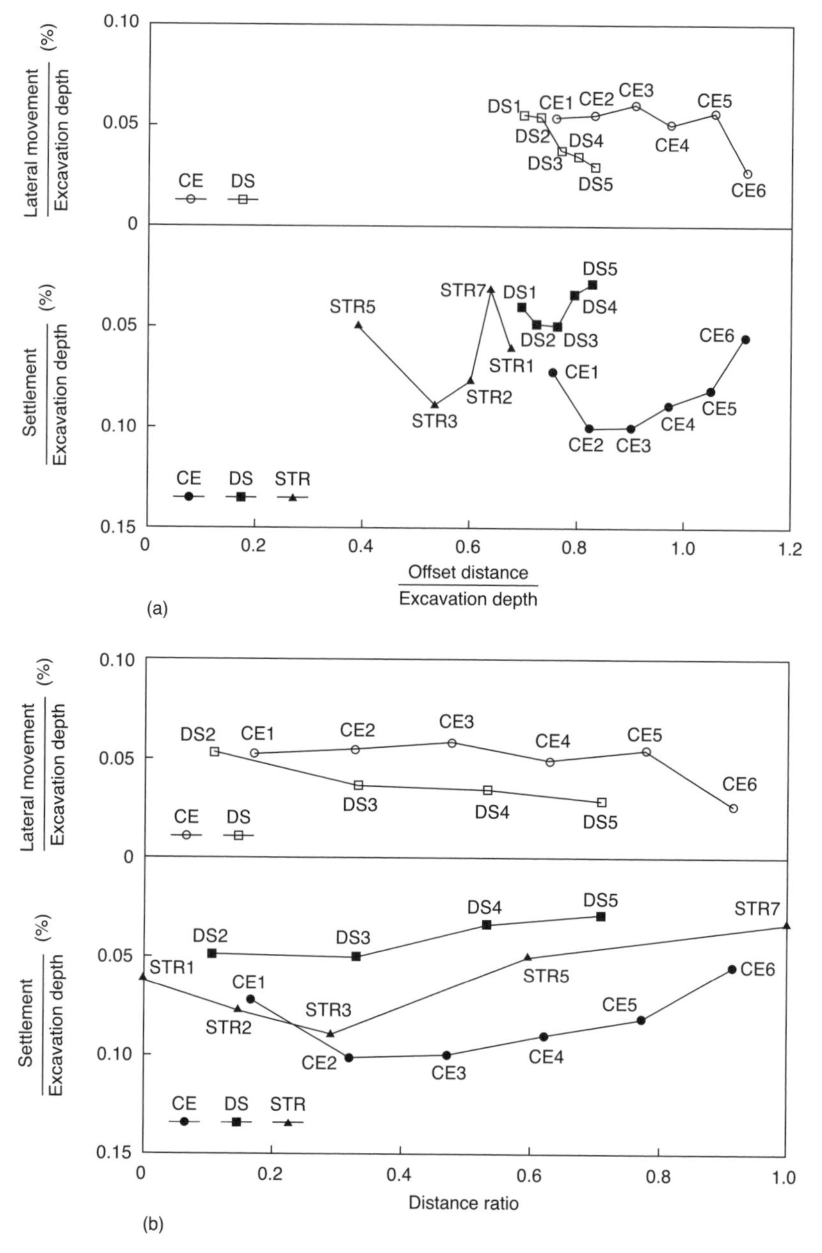

Fig. 2.28 Surface movements around site at Lion Yard: (a) movement versus offset distance; (b) movement versus distance ratio (after Ng, 1998)

2.28(b) shows that the maximum settlements did not seem to occur in the middle of the two long sides of the site (CB and STR) but at a distance ratio between 0.3 and 0.4. This was probably because the depth of excavation increased as the distance ratio decreased; however, the variation of the

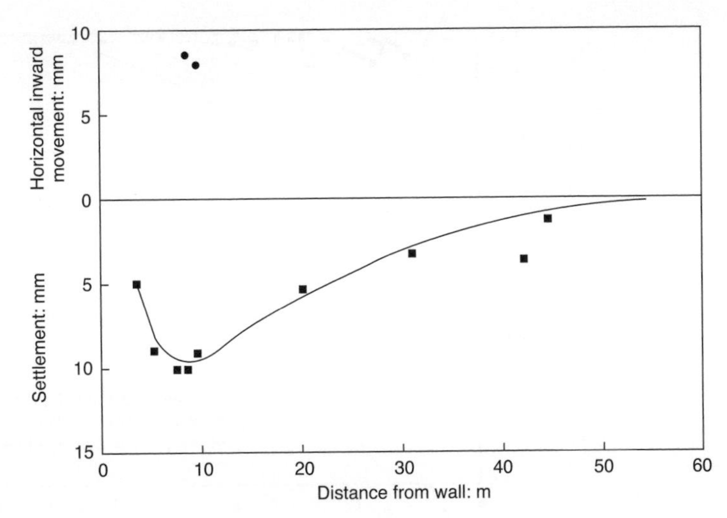

Fig. 2.29 Observed ground movements at various distances from wall to main excavation at Lion Yard (after Ng, 1998)

depth of excavation did not seem to affect the observed lateral inward movements along CB.

Figure 2.29 shows a plot of settlement against distance from the wall. The data points shown in Fig. 2.29 were selected from surveying stations that locate along or close to the X–X section (see Fig. 2.27). A settlement trough can be seen 8 m from the wall, and the settlements extended well beyond a distance of three times the depth of the excavation. Ideally all the data points should have been selected only from one side of the site. Since there are limited data available at Lion Yard, two data points have been used from STR assuming that the settlements are approximately the same in the middle of the two sides of the site. It can be seen from Fig. 2.28(a) and (b) that this is a reasonable assumption.

The measured total settlements (including diaphragm walling and piling) at the end of excavation are superimposed on Peck's (1969) non-dimensional settlement chart in Fig. 2.30. Most of the observed settlements at Lion Yard fall within Zone I for a distance/excavation depth ratio <2.5. However a few data points lie inside Zones II and III at high distance/excavation depth ratios. This seems to suggest that although the observed settlements were small in general, the extent of influence due to construction was quite large. Of course, the measured settlements at these large distance/excavation depth ratios were relatively small.

Characteristic ratios of ground movement

The displacement ratio is defined as the ratio of total horizontal inward movement to settlement, and it is closely related to the mode of wall

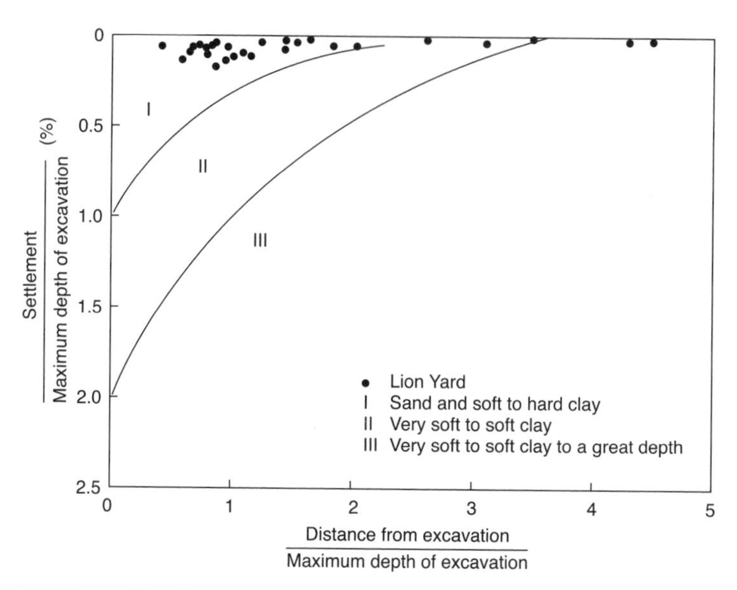

Fig. 2.30 Normalized settlement envelopes adjacent to excavation at Lion Yard (after Peck, 1969; Ng, 1998)

deformation (O'Rourke *et al.*, 1976). If settlement data are available, this ratio can form the basis for estimating the lateral movements of the ground associated with excavation, where lateral surface ground movement data are limited or non-existent.

Figure 2.31 shows a plot of the displacement ratio versus time for the surveying stations in the middle of the two long sides of the site (CB and STR), 9 m away from the wall. However, only data from Corn Exchange Street were considered after the L3 excavation started, to minimize the effects due to different ground levels on each side of the site. A best exponential curve has been fitted through the data points. It can be seen that the displacement ratio was larger than 1 during the early stages of construction, such as diaphragm walling and piling. This was because the ground movements were predominantly caused by the horizontal stress relief. Lateral ground movements were not sufficiently controlled by bentonite pressures. Hence the lateral ground movement was larger than the settlement. Similar observations have been made at Bell Common Tunnel (Tedd *et al.*, 1984).

The displacement ratio gradually decreased as the main excavation proceeded. During this stage the surface inward movements were limited by props or slabs. Lateral movements had to take place at depth, and this resulted in a deep-seated displacement profile associated with large settlements.

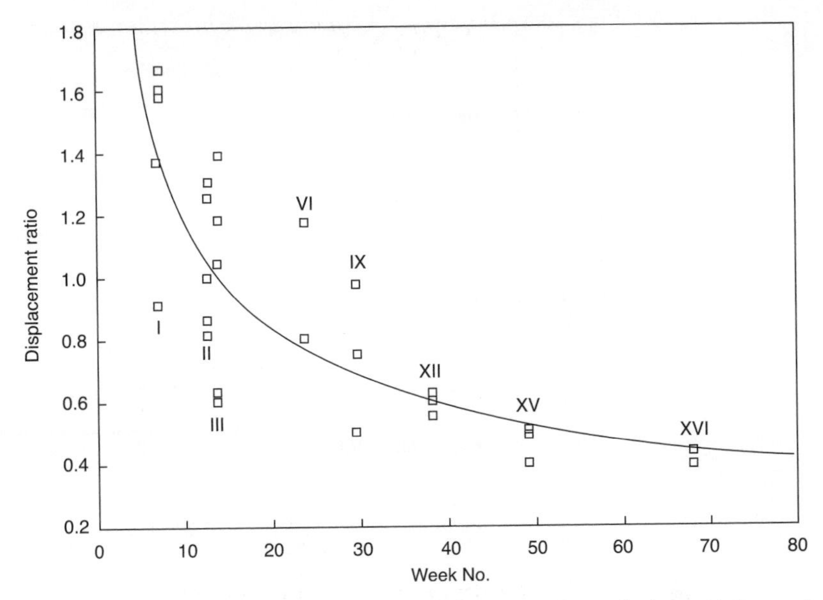

Fig. 2.31 Relationship between displacement ratio and elapsed time of construction at Lion Yard (after Ng, 1998)

O'Rourke *et al.* (1976) defined the coefficient of deformation C_D as follows:

$$C_D = \left(\frac{d/L}{d'/l'}\right) \tag{2.21}$$

The meaning of each symbol is shown in Fig. 2.32 together with the O'Rourke observation curve and the Lion Yard data. It might be coincidental that the Lion Yard data lie close to the curve that was derived from observations made in soft clays in Chicago. On the other hand, it seems to suggest that the deformation characteristics were dominated by the method of constructing the support system rather than by the soil type.

Variations of pore water pressure inside the site

Site investigations prior to construction showed that the initial ground water conditions were approximately hydrostatic below a level of about +7 m OD. The measured piezometric levels inside the site are plotted against time in Fig. 2.33. In Fig. 2.33, piezometers denoted by 'W' are those installed at the soil–wall interface. Piezometers P4A and P4B are located 2.1 m away from the face of the diaphragm wall and positioned at −4.8 m and −7.8 m OD, respectively. All of the piezometers show a consistent pattern of reducing pore pressure with each successive stage of excavation. However, they all rose quickly after each stage of

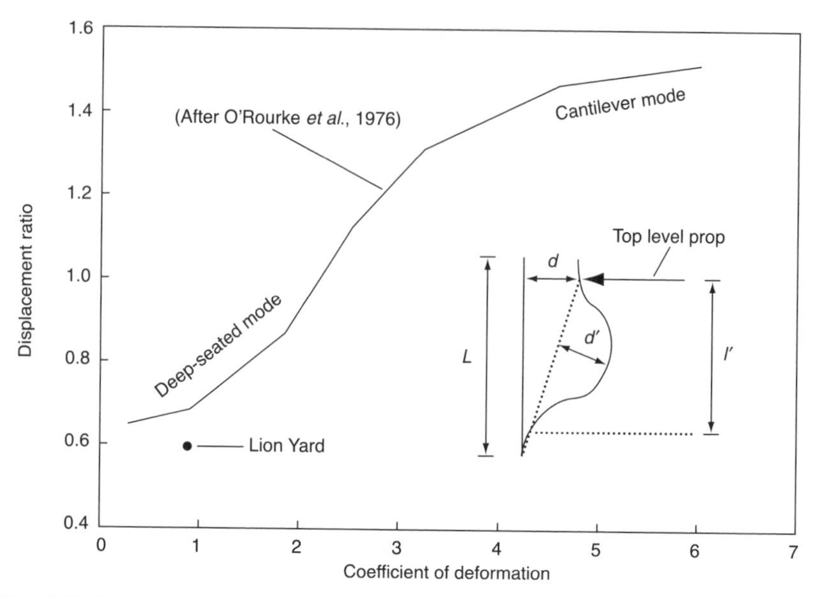

Fig. 2.32 Relationship between displacement ratio and coefficient of deformation at Lion Yard (after Ng, 1998)

excavation, and the rate of rising began to reduce about four weeks after the last stage of excavation. From Stage XV, the rate of pore pressure increases appeared to remain fairly constant for the subsequent 100 weeks.

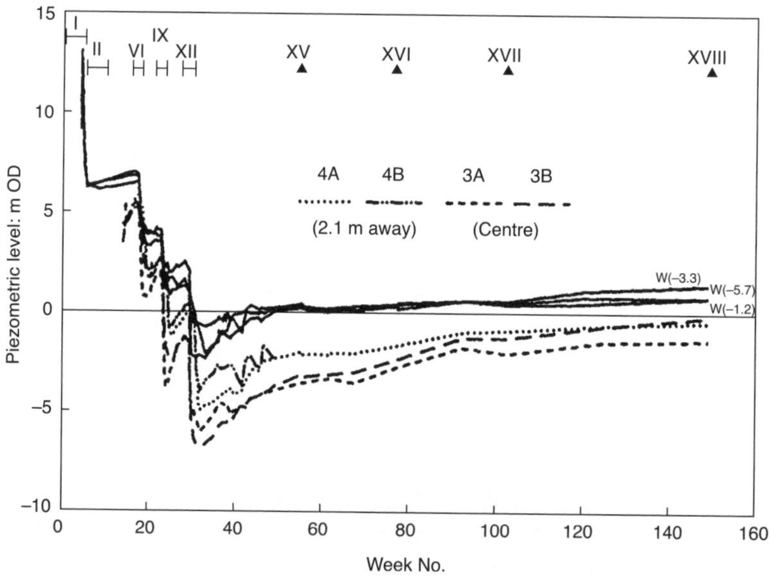

Fig. 2.33 Pore pressure regime inside the site at Lion Yard (after Ng, 1998)

The observed rapid rise of pore pressures inside the site during the construction of the basement can be explained by two separate mechanisms. First, part of the rise of pore pressures can be attributed to undrained loading, since casting of the concrete slabs caused abrupt rises of pore pressure. After careful examination of the piezometric data and a detailed site record, some 40% of the total increase of pore pressures after each stage of excavation can be identified as the result of casting the concrete slabs. Second, vertical stress relief would have caused an increase in soil permeability, which is stress dependent, particularly in this closely fissured Gault Clay. Water could flow in the fissuring system, causing a rapid rise in pore pressure.

An interesting feature revealed by the piezometers in the centre of the site is that during the first two stages of excavation, there was a sign of upward seepage in the centre of the site, because the piezometric level at 3A (installed at −4.48 m OD) was lower than at 3B (installed at −14.48 m OD). However, the seepage pattern seemed to be reversed shortly after the last stage of excavation. This was probably due to the opening up of fissures and surface rain water seeping downward. After casting the lowest slab, an upward seepage condition has been redeveloped with the piezometric level at 3B higher than 3A. The pore water pressures in the centre of the site are gradually approaching their equilibrium states, and the pore water pressures adjacent to the wall are also moving towards their steady conditions.

Tensiometers were used to measure suction inside the site. After the final stage of excavation, suctions measured in the clay 0.5 m to 1 m below the final excavation level were in the range 25–50 kPa. A positive pore pressure of 9 kPa was recorded at Stage XVIII.

In the centre of the site, the calculated values of B against the reduction of vertical stresses inside the site are plotted in Fig. 2.34. The measured values of $\bar{B}$ vary between 0.5 and 0.75 for the early stages, to above 0.9 at the last stage of excavation.

Heave and swelling in the centre of the site

The successful monitoring of both pore pressures and vertical soil movements in the centre of the site has made it possible to study the characteristics of soil swelling in Gault Clay. Figure 2.35 shows the vertical soil movements at three selected locations underneath the lowest basement slab together with the observed pore water pressures in the centre of the site. The arrowhead magnets AM3 and AM7 have been chosen as they are respectively closest to the two piezometers 3A and 3B. Magnet AM9 is at the highest level near the surface of the excavation. At the end of excavation, AM9 measured 26 mm of total heave. This is slightly greater than the 20 mm predicted from an elastic analysis (Ng,

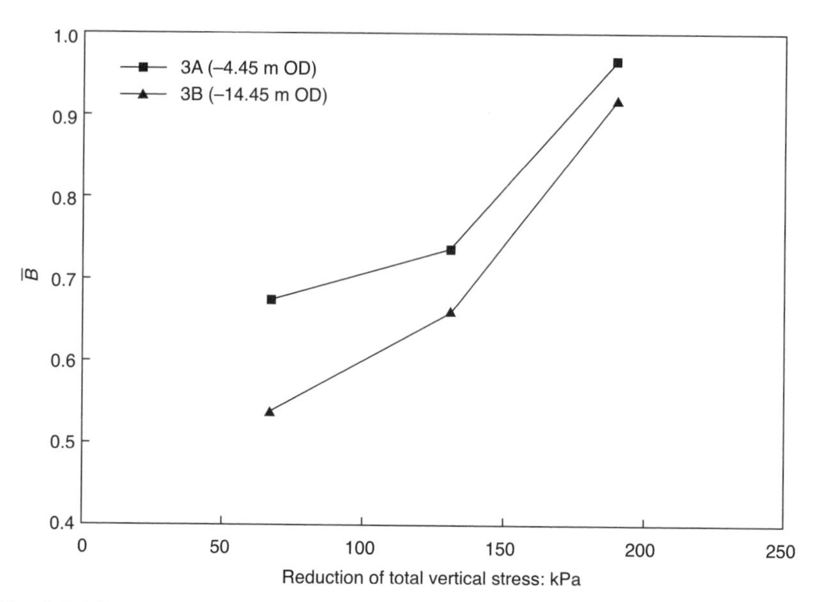

Fig. 2.34 Pore pressure response in the centre of the site at various stages of construction at Lion Yard (after Ng, 1998)

1992). However, by closely examining the extensometer data it has been possible to separate out the movements occurring at times when a change of vertical total stress existed, that is, during excavation, from the total recorded movements. This amounted to 21 mm and, if this is taken to be the elastic response to unloading, it agrees well with the prediction.

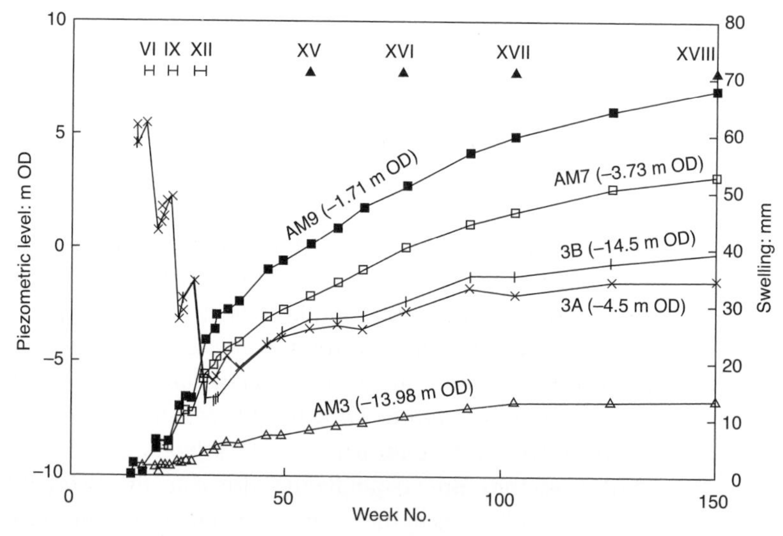

Fig. 2.35 Pore pressures and clay swelling in the centre of the site at Lion Yard (after Ng, 1998)

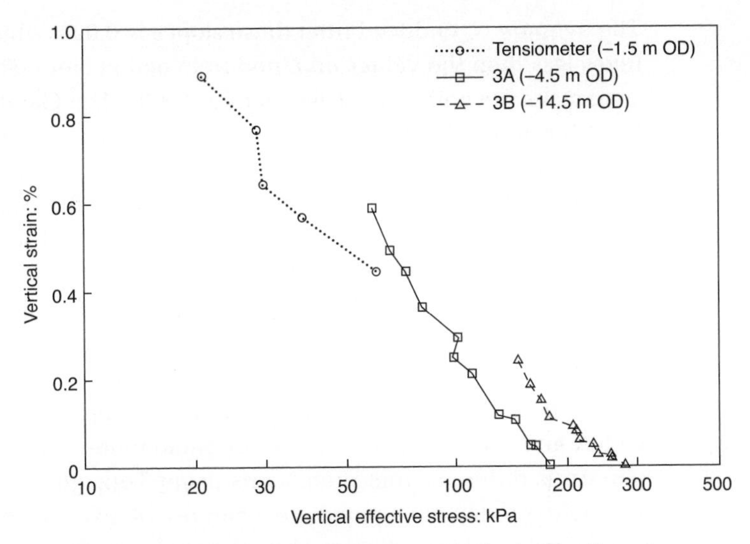

Fig. 2.36 Observed relationship between vertical effective stresses and strains at Lion Yard (after Ng, 1998)

The rate of rise of water pressure decreased with time as did the swelling of the clay. The clay has continued to swell and the highest arrowhead magnet has moved by a total of 68 mm at Stage XVIII relative to the datum magnet. Numerical simulation of the main excavation suggests that the datum magnet has moved −5 mm by the end of excavation (Ng, 1992). Thus the clay surface has probably swelled by more than the 68 mm recorded, although 26 mm of this movement had taken place before the lowest basement slab was cast. The 150 mm void underneath the slab is gradually closing up.

From the differential movements between adjacent magnets, vertical strains can be calculated. The observed strains can provide valuable information for understanding the behaviour of clay. In Fig. 2.36 the deduced vertical strains are plotted against vertical effective stresses. The strains have been calculated at the approximate levels of the two piezometers, and at the tensiometer 1.5 m below the basement slab in the centre of the site. The vertical effective stresses have been calculated assuming the total vertical stresses are equal to the overburden pressures, and the pore pressures recorded by the instrumentation have equalized throughout the surrounding soil. If the observed field behaviour can be assumed to be one-dimensional, the slope of these curves represent the in situ swell index (C_s) as given by the following equation:

$$C_s = (1 + e_0)\frac{\Delta \varepsilon_v}{\Delta \log \sigma'_v} \tag{2.22}$$

The average C_s deduced from these slopes is 0.023, which is two to three times less than the values obtained from oedometer tests on undisturbed specimens at similar stress levels (Ng, 1992). The Gault Clay appears to be stiffer in the field than in the laboratory.[6] This discrepancy could be due to the pore water pressure in the soil lumps lagging behind the fissure system, and thus the vertical effective stress in the soil lumps is higher than in the soil adjacent to fissures. Pile effects and sample disturbance are the other two possible explanations.

Figure 2.37 shows the vertical strain distribution with depth. The maximum observed vertical strain reached about 0.6% at the end of excavation and had further increased to 0.9% at Stage XVIII. The recorded strains at 20 m depth below the final excavated level were 0.05% and 0.1%, respectively, at the same times. The observed swelling pattern is different from predictions using Terzaghi's swelling theory as it would predict no immediate changes of excess pore pressure and hence soil swelling (discussed further below). The diagram also shows two theoretical swelling curves based on the field and laboratory C_s values. For calculating the theoretical strain distributions using Equation (2.22), a pore water pressure distribution was assumed by linearly extrapolating the known pore water pressures at four locations (i.e. at piezometers 3A and 3B, the tensiometer and the Lower Greensand). As expected, the laboratory-based swelling curve is substantially larger than the field observations. It can be seen that the reproduced strain curve based on the observed field C_s value does not line up with the field curve, and the discrepancy between them increases with depth below the final excavated level. This gives a strong indication that the mass permeability of soil indeed varies with depth.

Assessment: the 'two-phase' behaviour of Gault Clay

Classical swelling theory for an intact clay would predict that swelling would proceed from the drainage boundaries inwards. It would take some time for pore water pressure dissipation and associated volume changes to take place in the centre of a clay layer. The pore pressure and soil swelling observations at Lion Yard do not seem to fit with this classical theory, since rapid rise of pore pressures after each stage of excavation and substantial swelling have taken place in the upper part and in the centre of the clay stratum.

[6] This is the case for most soils where stiffness is dependent on strain level. Operational stiffness in the ground is near the maximum small-strain stiffness G_0 found by seismic means (resonant column apparatus, bender elements, surface wave geophysics) and is much higher than the large-strain stiffness found by conventional laboratory apparatus (see Simons and Menzies, 2000; Simons *et al.*, 2002).

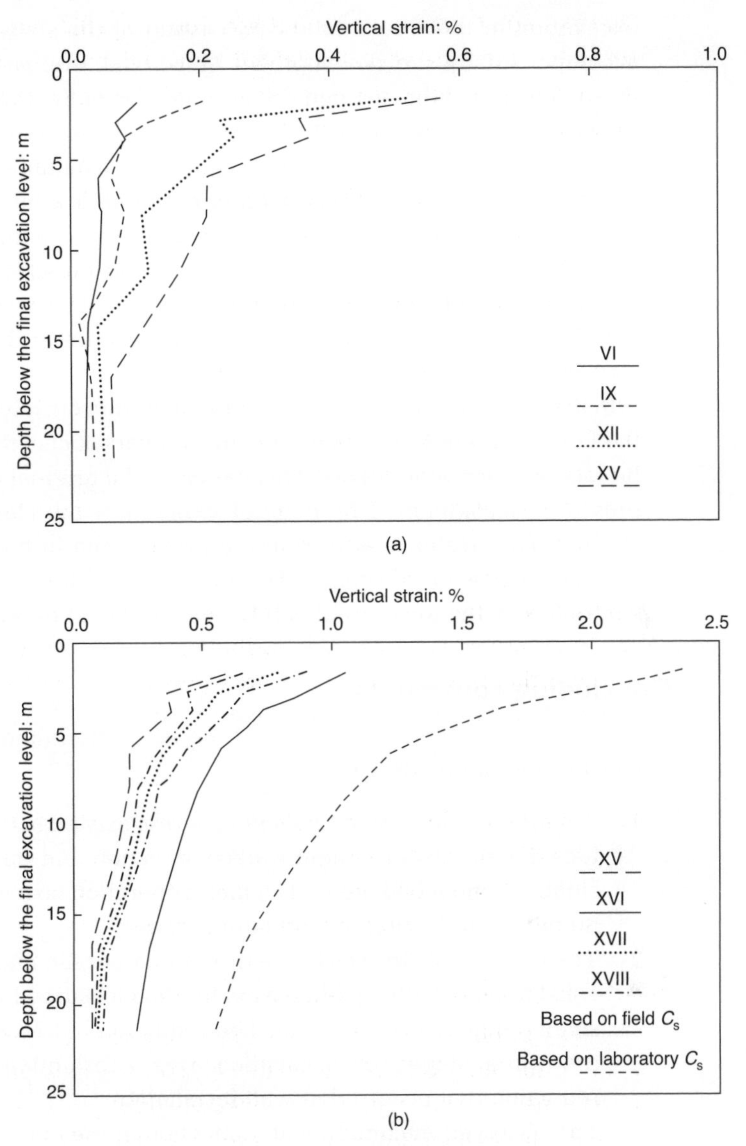

Fig. 2.37 Distributions of vertical strains with depth at Lion Yard: (a) during construction; (b) after construction (after Ng, 1998)

The Gault Clay at Lion Yard seemed to behave like a two-phase material: phase (1) the fissure system; and phase (2) soil lumps. During construction pore pressures rose quickly following each stage of excavation as observed in the fissure system, but the pore pressures in the soil lumps were lagging behind, that is, pore water pressures in these two systems were 'out of phase'. Both piezometers in the centre of the site (3A and 3B) showed a rapid rise of pore pressures after each stage of excavation in the weeks following casting of the slabs, and these were accompanied by the development of vertical strains in the ground at these levels. Water seemed to permeate rapidly through the fissure networks to the upper part and in the centre of the clay stratum but seeped slowly into the soil lumps. Water was available at the final excavated surface as a result of continuous rainfall after the last stage of excavation and from water used to 'flood' the clayboard after casting the L1 slab. Thus the final excavated level could possibly have acted as a drainage boundary for clay swelling. However, there was only a restricted water supply from the Lower Greensand because of the less permeable clay in the lower clay stratum.

In the long term, water has to be drawn in from outside the site and from the Lower Greensand through the lower layers of clay that are thought to be relatively less permeable. This has caused a gradual rise of pore water pressures accompanied by general swelling of the clay throughout the stratum. Pore water pressures in the fissures and in the soil lumps have gradually become 'in phase'. The rate of swell seems to be have been controlled by the lower permeability of the clay at depth.

Concluding comments

Based on the interpreted results of the field observations, the following conclusions can be drawn:

1 A marked reduction in total lateral stress existed at the soil–wall interface due to the diaphragm wall construction, but this only resulted in small ground movements. The piling operation inside the site caused a negligible reduction of total lateral stress.
2 The low lateral stresses in the ground before the main excavation are considered to be the main reason for the measured low strut loads and small deflections of the wall. The reduction of lateral stresses during the three stages of excavation was substantially less than the reduction that occurred at wall installation.
3 The observed maximum wall deflection at the end of excavation was small. However, the wall has continued to creep inward, and the measured maximum deflection 2.5 years after construction is twice as large as the deflection during the main excavation stage. This is consistent with the observed long-term settlement pattern behind the wall. This perhaps suggests that the creep of the wall was due to the shrinkage of slabs. The creeping of the wall resulted in a reduction of wall curvature that is consistent with the observed reductions of tensile reinforcement stress in the wall.
4 The displacement ratio is defined as the ratio of total horizontal to the vertical surface ground movement. For the top-down construction in

Gault Clay, the ratio was >1 during the wall installation and piling; however, it gradually reduced to −0.6 at the end of excavation. The measured settlements at Lion Yard fall within Zone I of the summary chart produced by Peck (1969). It is evident that there was a settlement trough at −9 m behind the wall.

5 The observed rapid rise of water pressures after each stage of excavation was surprising. The pore pressure and clay swelling distributions with depth do not seem to fit Terzaghi's swelling theory, if a single drainage boundary is assumed at the Lower Greensand. The rate of swelling seems to have been controlled by the lower permeability of the clay at depth. The maximum observed vertical strain below the final excavated level reached −0.6% at the end of construction and further increased to 0.9% 2.5 years after construction, whereas the recorded strains at 20 m depth below the final excavated level were 0.05% and 0.1% for the same periods, respectively. The field deduced C_s value is about two and one-half times smaller than the laboratory measured values at similar stress levels. This could be attributed to the difference in effective stresses in the fissures and in the soil lumps, pile effects and sample disturbance.

CHAPTER SEVENTEEN

Lateral pressure of wet concrete in diaphragm walls

Overview

Lings *et al.* (1994) observe that diaphragm wall installation is carried out incrementally by the construction of individual panels according to some planned sequence. For each panel there is first an excavation stage, in which the panel is dug while bentonite slurry provides support to the sides of the trench. This is followed by the insertion of a reinforcement cage into the bentonite slurry-filled trench and the subsequent placing, by way of tremie pipes, of high-slump concrete which displaces the bentonite slurry. Irrespective of the initial horizontal stresses in the ground before construction, the interface stresses at the panel will first be reduced to the bentonite pressure, and then increased to the wet concrete pressure. The horizontal stresses in the ground immediately after concreting a panel are therefore governed by the pressures exerted by the wet concrete. Subsequent curing of the concrete and construction of adjacent panels may cause changes in horizontal stress. Also, depending on soil type, there may be swelling and/or consolidation effects in the surrounding soil following wall installation which may cause further changes in stress. Nevertheless, the wet concrete pressure is the key starting-point that will control the post-installation horizontal stresses.

The process of diaphragm wall installation in stiff clays not only results in significant soil displacements but can also cause substantial reductions in horizontal stress (Ng *et al.*, 1992). It is recognized that these changes can affect the subsequent design predictions for forces in supporting props and bending moments in the wall (Gunn and Clayton, 1992), especially if these predictions have been made using a 'wished-in-place' assumption for the wall installation process. There is now an increasing awareness that the wall installation process may need to be modelled to improve design predictions, and it is important when trying to do this to have an appropriate understanding of the wet concrete pressures.

Clayton and Milititsky (1983) have noted from previously published data on pressures measured during concreting of diaphragm wall panels that wet concrete pressures are approximately hydrostatic at shallow depths, but are less than hydrostatic at larger depths. They state that formwork codes, which give guidance on the likely maximum pressures exerted by wet concrete on timber forms, are inappropriate for trying to predict the magnitude of wet concrete pressures in diaphragm walls, claiming that they relate to very different geometries, much shorter lifts and stiffer mixes.

However, Lings *et al.* (1994) have re-examined the case records referred to above, and have also carried out a detailed study of data obtained at Lion Yard, Cambridge, during the construction of one diaphragm wall panel. This new case record provides a clear picture of the development of wet concrete pressures, and the relevant data are presented later. Lings *et al.* (1994) have also studied some of the recent literature on the pressure of wet concrete on formwork, and have concluded that it does have a useful contribution to make to our understanding of the wet concrete pressures in diaphragm wall panels.

Pressure of wet concrete on formwork

Extensive research has been carried out on the lateral pressure exerted by wet concrete on vertical formwork (Harrison, 1983; Clear and Harrison, 1985). The main objectives of this research have been to understand the mechanisms governing the observed behaviour, and to enable wet concrete pressure to be reliably predicted for use in formwork design. Pressure cells have been used to monitor the concrete pressures acting on formwork at various levels to find out how the pressure at each level varies during the course of concrete placement. These generally show that the concrete pressure at a given level rises to a maximum value, and then remains constant, or may reduce. At greater than a certain depth, referred to here as the critical depth h_{crit} below the surface of the concrete, this maximum pressure has a limiting value P_{max} which is approximately constant with increasing depth. At less than this depth, the maximum concrete pressures are hydrostatic, equalling the full fluid pressure of the concrete. This results in a design envelope of maximum concrete pressure as shown in Fig. 2.38, where the limiting pressure beyond the critical depth is given by

$$P_{max} = \gamma_c h_{crit} \qquad (2.23)$$

The explanation for this behaviour, given by Harrison (1983) under the title 'pore water pressure theory', focuses on the development of effective stresses within the wet concrete due to the twin processes of consolidation within the mix, and hydration of the cement. The behaviour is not primarily dependent on the concrete mix taking an initial set. When the concrete is freshly placed, the solid particles (aggregates and cement grains) are totally suspended in the water, giving a pore water pressure equal to the total

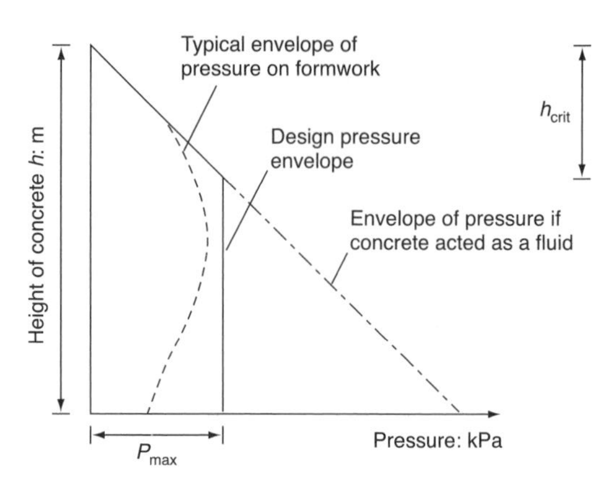

Fig. 2.38 Design pressure envelope (from Clear and Harrison, 1985; after Lings et al., *1994)*

vertical stress. The effective stresses are initially zero, and the wet concrete behaves as a heavy fluid. Because of the substantial excess pore pressures, consolidation will take place, with water flowing vertically through the concrete, and also horizontally towards and through the formwork. At the same time, the process of hydration will itself cause progressive reductions in pore water pressures. These mechanisms result in a transfer of load from the pore water onto the particle structure, causing an increase in effective stress. The horizontal effective stress is only a small proportion λ of the vertical effective stress, where λ is a lateral earth pressure coefficient. Idealized distributions of vertical and horizontal pressures with depth are shown in Fig. 2.39(a) and (b) respectively.

The same pattern of behaviour is observed at a given level on the formwork during filling of the form, as is demonstrated by the example of measured formwork pressures shown in Fig. 2.40, taken from CIRIA Report 108 (Clear and Harrison, 1985). By far the largest proportion of the total horizontal pressure is provided by the pore water, and, consequently, factors which affect the dissipation of pore pressure, such as the permeability of the formwork, will have a marked effect on the resulting pressures. The shape of the pressure diagrams against depth at a particular time (Fig. 2.39) and against time at a particular depth (Fig. 2.40) can be seen to be very similar, and there is a tendency to use them almost interchangeably. However, they will only be the same if the rate of concrete placing is uniform over time.

Figure 2.40 also shows the measured formwork deflection with time. Under the action of only 50 kPa, the formwork deflects nearly 6 mm, which gives an indication of the relative flexibility of the formwork support system in that test. When the total pressures on the form reduce in the second half of the pour, it is noticeable that there is negligible reduction

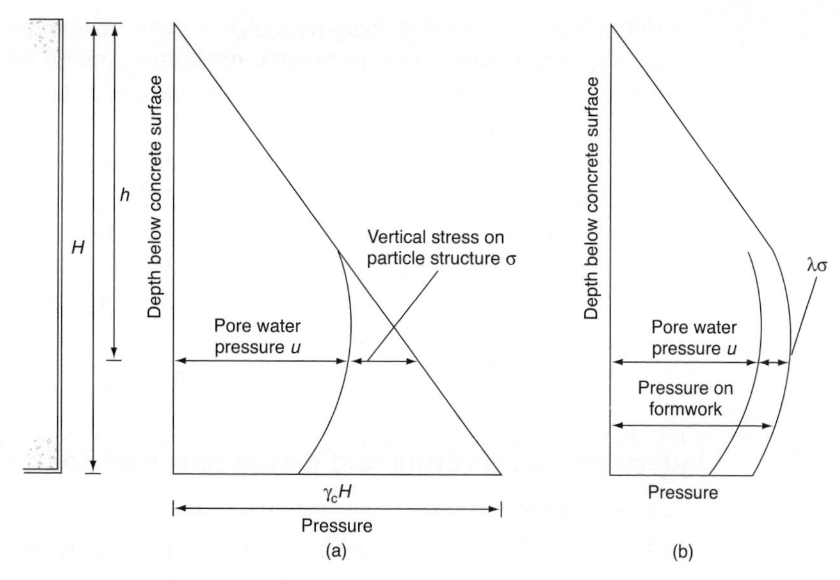

Fig. 2.39 The components of the vertical and horizontal pressures (from Harrison, 1983; after Lings et al., *1994)*

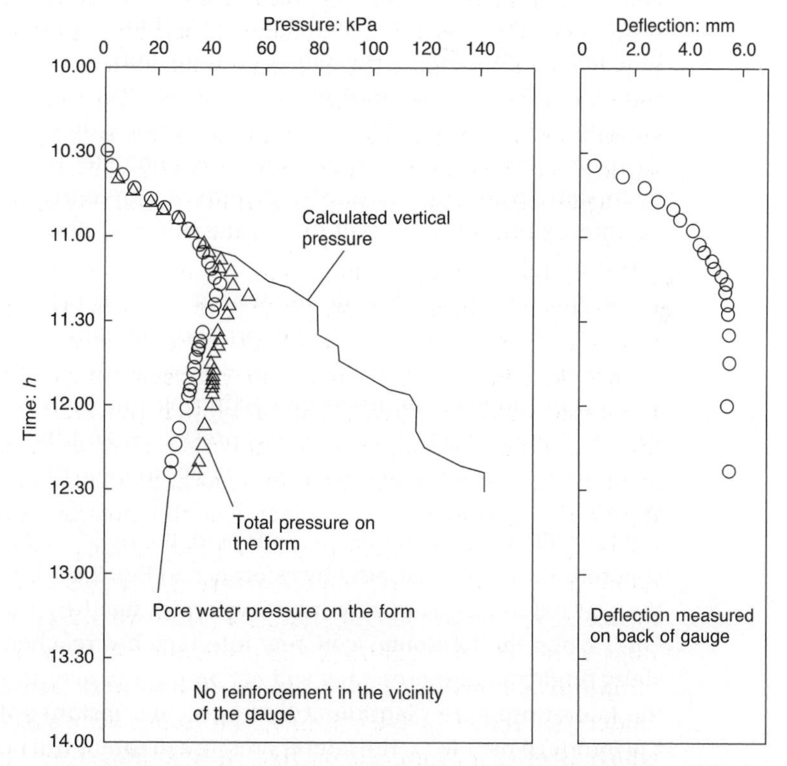

Fig. 2.40 Example of formwork pressures and deflection measurements (from Clear and Harrison, 1985; after Lings et al., *1994)*

in deflection, presumably because there is some hysteresis in the formwork support system. The formwork pressures remain locked in until the forms are struck. There is evidence from other measurements that the stiffness of the formwork support system has a significant effect on the maximum pressures occurring, with stiffer systems giving rise to higher lateral pressures.

The critical depth, h_{crit}, and hence the limiting pressure, P_{max}, on formwork, is governed by a number of factors. The CIRIA Report 108 presents a design equation for calculating the P_{max} value, which depends on rate of placing, cross-sectional shape and height of the form, temperature of concrete at placing, cement type and any additives used.

Underwater concreting and diaphragm wall construction

Compared to the common case of concrete placed by standard methods in formwork (in air), there is very little data on concrete placed by tremie pipe in formwork under water. However, the basic pattern of behaviour is assumed to be the same, and CIRIA Report 108 recommends using exactly the same equation for calculating the critical depth, h_{crit}, as in air. The only difference occurs in the calculation of the limiting pressure, P_{max}, from Equation (2.23), where the effective (buoyant) unit weight of the wet concrete ($\gamma' = \gamma_c - \gamma_w$) is used. This is because it is the increase in pressure, or the net pressure, acting on the form that is of interest to the formwork designer, given by the wet concrete pressure minus the hydrostatic water pressure. The picture is frequently complicated by a varying external water level due to the effect of tides, but this is not of concern in this study.

The placing of concrete in diaphragm wall panels under bentonite is very similar to underwater concreting, the only difference being that the fluid displaced by the rising concrete now has a somewhat larger unit weight. In terms of the pressures acting on the walls of the trench, the values of interest are now the gross pressures, and these can readily be obtained from the net pressures by the addition of hydrostatic bentonite pressures.

If the CIRIA design envelope is adopted, the build-up of pressure during concreting can be illustrated by reference to Fig. 2.41. Initial pressures on the soil before the start of concreting are defined by the bentonite line OB_3. When the bentonite–concrete interface has reached D_2, full hydrostatic fluid concrete pressures will act on the soil over the depth h_{crit} and the lateral pressure diagram at that particular instant will be OB_2P_3. As the interface rises to D_1 the lateral pressure diagram will become $OB_1P_2P_3$. Finally, when the concrete reaches the top of the trench, the lateral pressure diagram will be OP_1P_3. Once the critical depth of concrete has

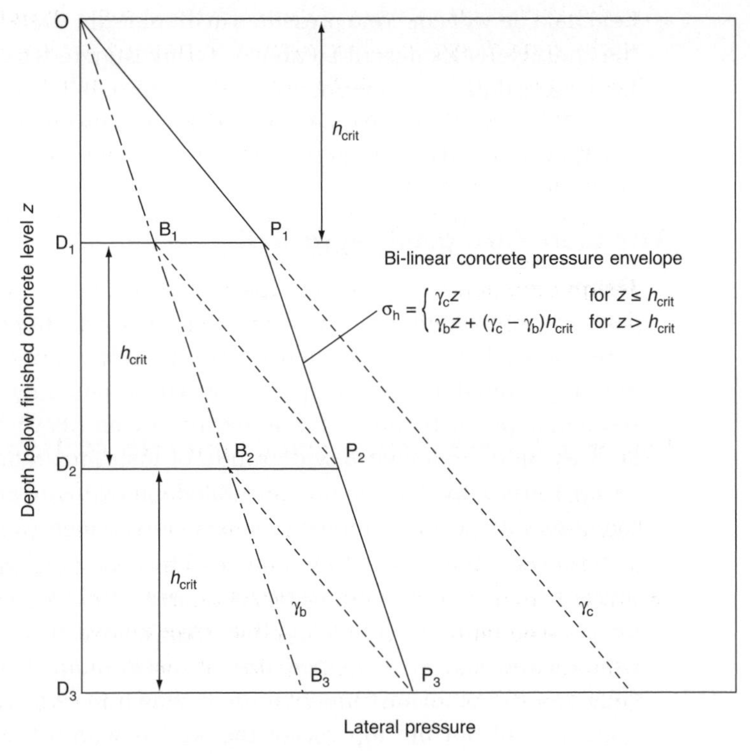

Fig. 2.41 Development of lateral pressures during concreting and the bi-linear concrete pressure envelope (after Ng, 1992)

been reached at a particular level, the pressure remains constant. However, it is important to note that the value of this final pressure varies with elevation.

The resulting theoretical bi-linear concrete pressure envelope (Ng, 1992) after fully concreting a diaphragm wall panel can be expressed algebraically as follows

$$\left.\begin{aligned} \sigma_h &= \gamma_c z && \text{for } z \leq h_{crit} \\ \sigma_h &= \gamma_b z + (\gamma_c - \gamma_b)h_{crit} && \text{for } z > h_{crit} \end{aligned}\right\} \quad (2.24)$$

where σ_h and z are the total lateral pressure and depth below finished concrete level respectively. Within the critical depth, the wet concrete applies the full fluid pressure, as it would if placed in air. At depths greater than the critical depth, the wet concrete pressure increases with depth at a rate given by the unit weight of the bentonite: it is no longer constant with depth as would be the case if placed in air, although the net concrete pressure is, of course, still constant.

Given the similarities between placing concrete under bentonite, and conventional underwater concreting, it seems reasonable to expect any field data on wet concrete pressures in diaphragm walls to display many of the characteristics described above. To investigate whether this is so, the Lion Yard data are examined first in some detail, followed by the case records of the Telefonhuset, Oslo (DiBiaggio and Roti, 1972) and the Seville subway (Uriel and Oteo, 1977).

The Lion Yard diaphragm wall

Overview

The concrete diaphragm wall at Lion Yard in the centre of Cambridge was constructed during the period April to June 1989. The wall is 600 mm thick and approximately 17 m deep, with a typical panel length of 8.5 m. One particular panel (Panel 22) was excavated on 18/19 May and cast on 20 May, and contained a wide range of instrumentation. This included seven 150 mm by 250 mm mercury-filled Glotzl total earth pressure cells, together with seven pneumatic piezometers, which were installed at the concrete–soil interface on both faces of the panel. Although intended to measure earth and pore water pressures during construction of this deep excavation in Gault Clay, they give a clear picture of how the wet concrete pressures developed during concreting. The location of the cells and the other instrumentation is shown in Fig. 2.42, and it will be seen that all but the top 2 m of the wall is within the stiff to very stiff Gault Clay. Full details of the data obtained during field monitoring are given by Ng (1992) and Ng *et al.* (1990) and earlier in this Part 2. A summary of the observed performance of the excavation has been given by Lings *et al.* (1991).

Installation procedure

Each earth pressure cell was attached to a steel backing plate which was fixed, by way of a universal joint, to the end of a hydraulic jack securely mounted within the reinforcement cage for the wall panel. A pneumatic piezometer was also attached to the backing plate below each earth pressure cell, with the tip standing just proud of the active face of the cell. After the cage had been lowered into the bentonite-filled trench, the earth pressure cells were bedded in by jacking them out against the excavated clay surface until cell readings equal to 120% of the bentonite pressure were recorded. No special reaction plates were provided, resistance to the jacking force being obtained through the reinforcement cage and the spacers for achieving correct concrete cover.

Concrete was then introduced by way of two tremie pipes, located approximately at the quarter points of the panel. The concrete used was a nominal C35 mix with Conplast 211 plasticizer, giving an average

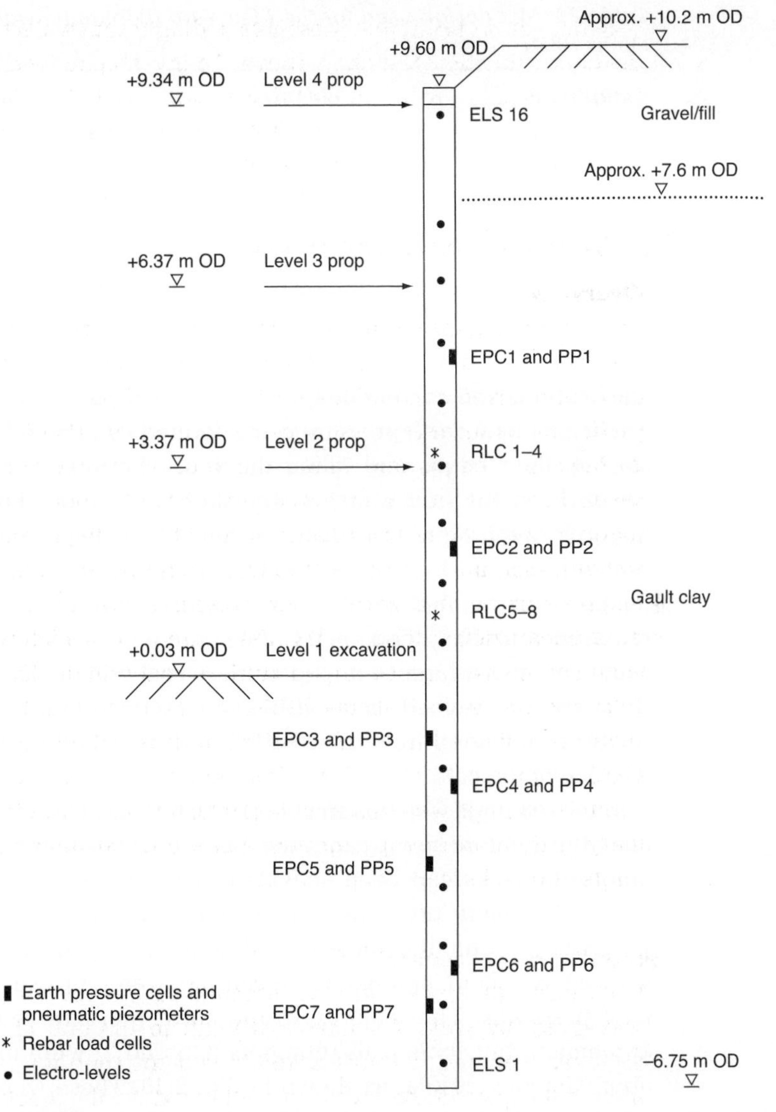

Fig. 2.42 Panel 22 at Lion Yard showing the instrumentation (after Lings et al., *1994)*

slump of 175 mm. The cement comprised 30% PFA and 70% OPC blended at the mixer; full details of the mix design are given in Table 2.2. During the concreting process, the earth pressure cells and jack pressures were regularly monitored, as was the rate of concrete rise in the trench, which averaged 5 m/h. No additional pumping of the jacks was carried out after the initial bedding in, consequently jack pressures fell during concreting, reaching quite low values by the end of concreting.

Table 2.2 Mix design used for the Lion Yard diaphragm wall (after Lings et al.*, 1994)*

Component	Quantities: kg/m^3	Comment
Cement	420	30% PFA and 70% OPC
Coarse aggregate	1030	5–20 mm crushed flint
Fine aggregate	715	Sand
Water	210	w/c ratio = 0.5
Density	2375	2320 measured from cubes

Calibrations and corrections

The Glotzl total earth pressure cell is formed by joining two thin plates of steel at their edges and filling the space between with mercury. The pressure on the active face is transmitted through the mercury to an integral diaphragm transducer, where it is balanced by pneumatic pressure applied by way of two lines from a remote readout unit. With no pressure applied to the active face, a pneumatic pressure is still required to lift and keep open the diaphragm (the cells are read dynamically), and this reference pressure, measured prior to use, is known as the 'P_v value' or pre-load value. It has to be subtracted from all pneumatic pressures obtained from the readout unit to give the pressure acting on the face of the cell.

Initial readings were taken of both earth pressure cells and piezometers after the reinforcement cage had been lowered into the bentonite but before the jacks had been activated. A comparison was made with the known bentonite pressures, calculated from the measured unit weight (10.1 kN/m^3) and the measured depths. The piezometer readings all showed excellent agreement, recording water pressures equal to the total pressures. The majority of the earth pressure cells showed good agreement, but some cells (numbers 3, 5 and 7) were under-reading by approximately 20 kPa, as shown in Fig. 2.43. These were the three cells that had been face-up during attachment to the reinforcing cage, which was then lying flat on the ground, and it is tempting to assume that some form of disturbance occurred. Pre-load values were therefore reduced by appropriate amounts so that the bentonite pressures were then being recorded correctly by all the cells.

Once jacked out onto the soil, earth pressure cells can still be used to measure fluid pressures in the trench, providing the jacking force has been measured. The concrete pressure acting on the backing plate can be inferred from the measured pressure on the active face of the cell, less the average pressure generated by the jacking force. If the cross-sectional area of the jack piston, the height of the pressure gauge above

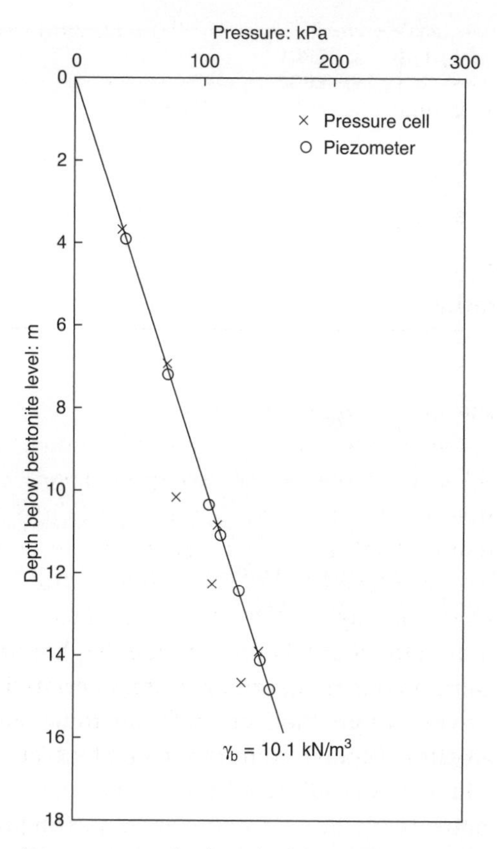

Fig. 2.43 Bentonite pressures before jacking (after Lings et al., *1994)*

each jack, and the density of the hydraulic oil are all known, it is a simple matter to calculate the magnitude of the jack forces from the pressure gauge readings. However, calibration of the pressure gauge itself is vital. At Lion Yard, it quickly became clear that the gauge pressures recorded could not possibly be right, and a subsequent recalibration of the pressure gauge was carried out which enabled appropriate corrections to be applied.

An analysis of the data, corrected as described above, for the stage when the cells had been jacked out onto the soil, but before any concrete had been placed, showed that some of the cells (numbers 4, 6 and 7) recorded the 20% increase in pressure with the application of significantly less than the theoretical amount of jack force. This is believed to be due to non-uniform bedding of the cells against the clay, causing them to over-read. The pre-load values were therefore modified to take account of this, so that when the raw data were processed, the known bentonite pressures in the trench were recovered.

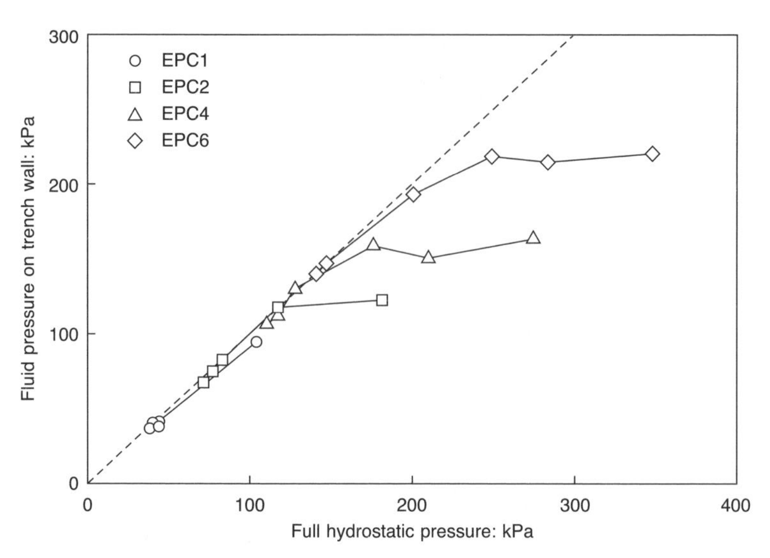

Fig. 2.44 Pressure development during concreting – external face of panel (after Lings et al., *1994)*

During concreting, one cell (number 1) showed a sudden increase in pressure before the bentonite–concrete interface had reached it, but thereafter displayed entirely consistent behaviour. The pre-load value for this cell was adjusted by a suitable amount to remove this unexplained jump.

Pressure development during concreting

The fluid pressures recorded on the wall of the trench during concreting are presented in Figs 2.44 and 2.45 for the external and internal faces of the wall respectively. They are shown plotted against the full hydrostatic pressures for that level, calculated from the known heights of bentonite and wet concrete existing at the time (γ_c taken as $24\,kN/m^3$). In each case, once the concrete has risen past the cell, the recorded fluid pressures increase along the full hydrostatic line, but then reach a plateau, and in some cases show a reduction.

The Lion Yard data have not been plotted here against an actual time axis, but time is implicit in the figure once the concrete level has reached a particular cell. The way the fluid pressure increases at each individual cell fits in well with the general pattern of behaviour observed during concrete pours (see Fig. 2.40).

Comparison with CIRIA Report 108

To make a comparison with the CIRIA design method, it is necessary to take the maximum pressure recorded during concreting at each level. The

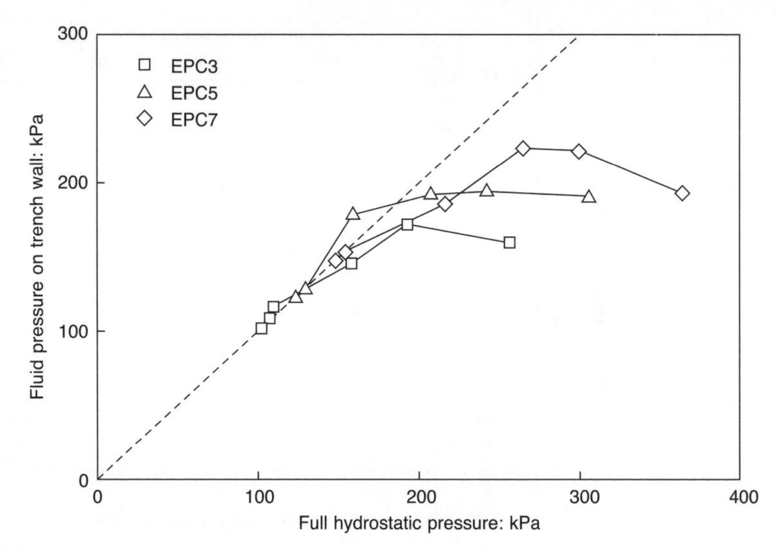

Fig. 2.45 Pressure development during concreting – internal face of panel (after Lings et al.*, 1994)*

appropriate maxima from Figs 2.44 and 2.45 have been plotted against depth below final concrete level, and are presented in Fig. 2.46. Also shown are lines of bentonite and full fluid concrete pressure. A further line at the bentonite slope has been added to form an approximate upper bound to the maximum recorded concrete pressures.

Taking Figs 2.44, 2.45 and 2.46 together, it is clear that the recorded data demonstrate qualitatively all the key features of the theoretical bi-linear model for wet concrete. Individual cells at the various levels show increases in pressure that follow the full fluid concrete pressure until some critical depth is reached, after which they remain approximately constant (points B_3 to P_3 and B_2 to P_2 in Fig. 2.41). When the maximum values are plotted against depth, the data all lie close to an envelope defined by the full fluid concrete pressure up to the critical depth and a line drawn at the bentonite slope thereafter (OP_1P_3 in Fig. 2.41). The data in Fig. 2.46 show that the critical depth at Lion Yard is approximately 5 m.

The Telefonhuset, Oslo

The perimeter diaphragm wall at the Telefonhuset, Oslo (DiBiaggio and Roti, 1972) is 1.0 m thick, with a maximum panel length of 5 m. The 20 m deep instrumented panel was constructed in early December 1969, and contained 15 vibrating wire total earth pressure cells. The cells all recorded the correct fluid pressure when first immersed in the bentonite

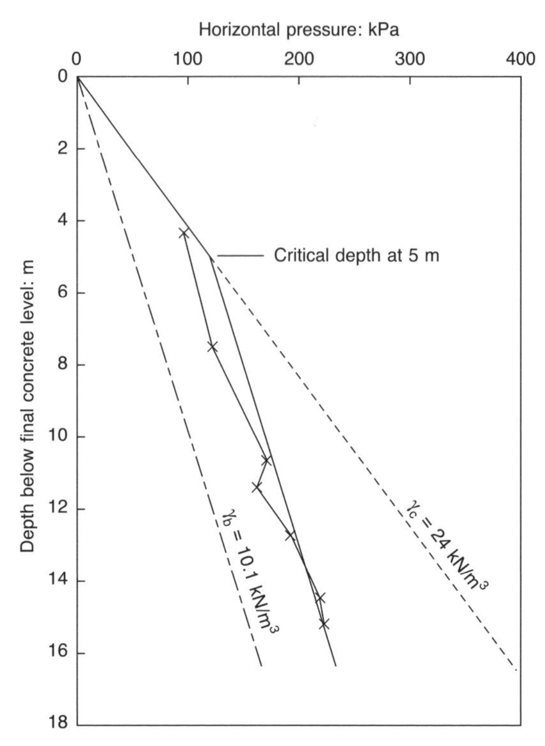

Fig. 2.46 Maximum wet concrete pressures – Lion Yard, Cambridge (after Lings et al.*, 1994)*

slurry, and after the cells had been jacked out onto the surface of the soft clay, constant jacking pressures were maintained until the completion of concreting. The maximum fluid pressures recorded during concreting of the panel, after the effect of the jacking force had been subtracted, are plotted against depth in Fig. 2.47. One cell (number 14) never came into contact with the clay and became embedded in the concrete, but it is encouraging to note that the recorded fluid pressure is very similar to the neighbouring cells.

As before, lines of bentonite and full fluid concrete pressure are also shown in the figure. A further line at the bentonite slope has been added to form an approximate upper bound to the maximum recorded concrete pressures. Again, the data support the view that the envelope of maximum wet concrete pressure is bi-linear with depth, with the critical depth in this case at approximately 6 m. DiBiaggio and Roti (1972) noted the existence of a limiting value of depth, above which pressures were equal to the full hydrostatic pressure of the wet concrete. They estimated this depth to be 5 m, which is the conclusion drawn if a mean line is taken through the data. The CIRIA method is for a maximum design pressure

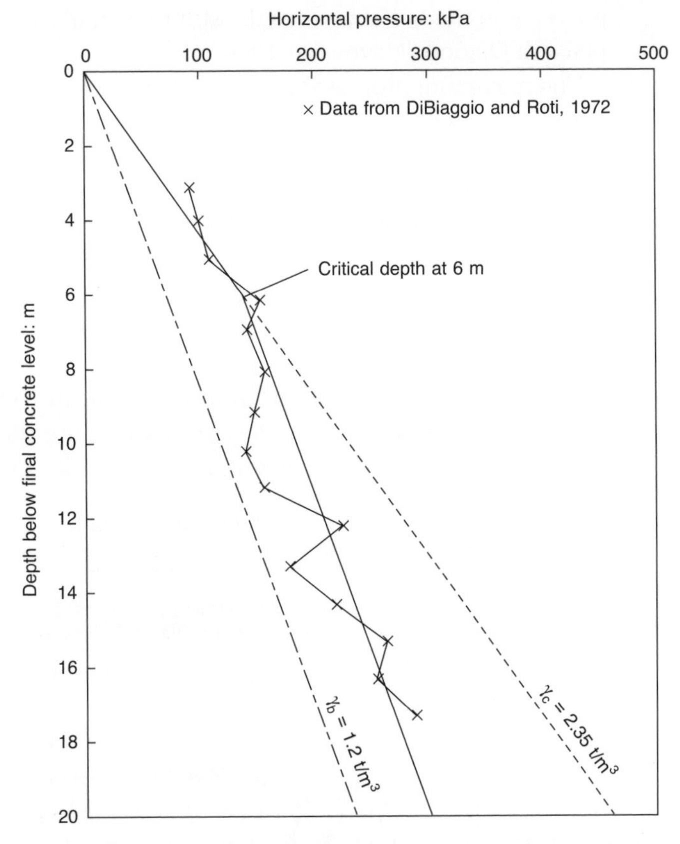

Fig. 2.47 Maximum wet concrete pressures – Telfonhuset, Oslo (after Lings et al., *1994)*

envelope, which is why an approximate upper bound line is shown in this study.

The Seville subway

The diaphragm wall panels forming the circular shaft for the Seville subway (Uriel and Oteo, 1977) are 800 mm thick, 3.4 m long and 34 m deep. Two panels were instrumented with seven 200 mm by 300 mm Glotzl total earth pressure cells each, but data obtained during concreting are only available for one of them (Panel 24). Two different methods of cell installation were used. In the first, cells were jacked out onto the soil; in the second, the cells recorded stresses in the concrete 60 to 80 mm from the soil–wall interface. Unfortunately, those who have described this work (Uriel and Oteo, 1977) do not state which method was used for which cells. The top four cells were in the silt and sand, the fifth in the gravel, and the bottom two in the stiff blue marls. Panel construction took place in October/November 1969.

The maximum pressures recorded during concreting are given, but it is not stated whether any correction (if relevant) has been applied for the jacking force. Pressures with only bentonite in the trench are also given, but it is not clear whether these are before or after activation of the jacks. Several of the cells for this stage record pressures significantly greater than the theoretical bentonite pressure, which could either be due to cell over-read, or the effects of jacking. In both cases it would be appropriate to apply a correction to the data that would bring the points back onto the bentonite line. If the maximum recorded pressures simply show the effects of changing from bentonite to wet concrete, then the same corrections should also be applied to these data. If, however, jack corrections have already been applied, or if the cells were of the

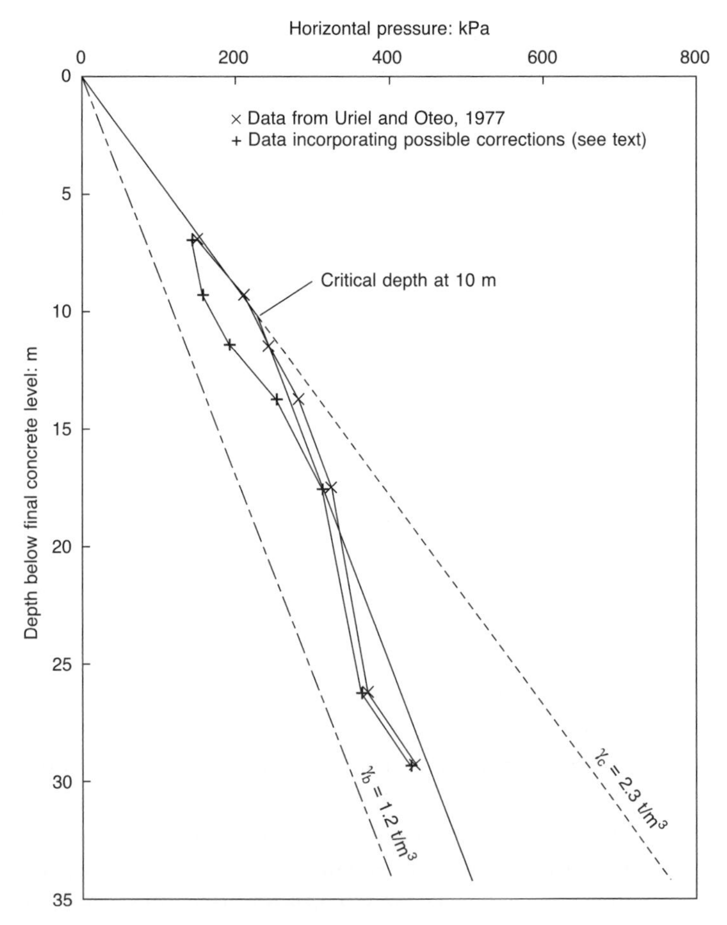

Fig. 2.48 Maximum wet concrete pressures – Seville subway (after Lings et al., *1994)*

embedded type, then corrections may be inappropriate. There is therefore a range of best-estimated maximum concrete pressures, either making the corrections described above, or taking the recorded pressures at face value, and these are presented in Fig. 2.48.

As before, lines of bentonite and full fluid concrete pressure are also shown in the figure. A further line at the bentonite slope has been added to form an approximate upper bound to the maximum recorded concrete pressures. Again, the data support the view that the envelope of maximum wet concrete pressure is bi-linear with depth, with the critical depth in this case at approximately 10 m. This case record provides the least clear data of the three, but with 34 m deep panels it is an important one, being substantially deeper than the other two (17 m and 20 m).

Assessment

Lion Yard data

The case record information presented for the Lion Yard diaphragm wall highlights some of the difficulties that arise when trying to acquire reliable field data. The problems included altered performance (i.e. pre-load value) of earth pressure cells before installation, after jacking and during concreting, and a jack pressure gauge with a large systematic error. The resulting pressures are therefore subject to some uncertainties, but by careful analysis of all the data it has been possible to make corrections for the effects mentioned.

The pressures recorded by cells 1, 2, 4 and 6 (see Fig. 2.44) are much closer to the full hydrostatic line during the early part of concreting than cells 3, 5 and 7 (see Fig. 2.45), a stage when close correlation might be expected. This suggests that the latter may be less reliable than the former, which is possibly linked to the altered pre-load values observed before installation, as noted earlier. Cell number 3 was found to be the least accurate of all the cells when their reliability at the end of the main excavation was assessed by means of an equilibrium analysis (Lings *et al.*, 1992).

The bi-linear pressure envelope

In spite of different bentonite and concrete unit weights, different concrete mixes, rates of placing and placing temperatures, and different wall geometries, there are common features displayed by all three case records. For all of them, a bi-linear maximum pressure envelope forms a boundary to the field data, defined by a full fluid concrete line up to some critical depth, and by a bentonite line thereafter. This behaviour is anticipated by CIRIA Report 108 for underwater concreting, a process very similar to concrete placed under bentonite in diaphragm walls, and is explicitly set out in

Equation 2.24. This way of viewing the data is a significant improvement over the idea of a decreasing 'fluidity ratio' with depth (DiBiaggio and Roti, 1972).

Critical depth prediction

Whereas the bi-linear shape is adequately demonstrated, it is by no means clear whether CIRIA Report 108 can successfully predict the critical depth for diaphragm walls. One difference is that the formwork data are restricted to wall/column heights no greater than around 15 m, whereas diaphragm walls are deeper than this, often by a factor of 2 or more. Vertical drainage path lengths become very great, resulting in slower dissipation of excess pore pressures and consequently increases in wet concrete pressures. Another difference is the greater stiffness of the ground compared to standard methods of formwork support, again tending to increase wet concrete pressures. It therefore seems likely that the critical depth for deep diaphragm walls will exceed the predictions of CIRIA Report 108.

Predicted critical depths are presented in Fig. 2.49 for walls up to 15 m deep over the maximum likely range of placing temperatures of 5–30 °C. Also plotted are the observed critical depths and wall depths from the

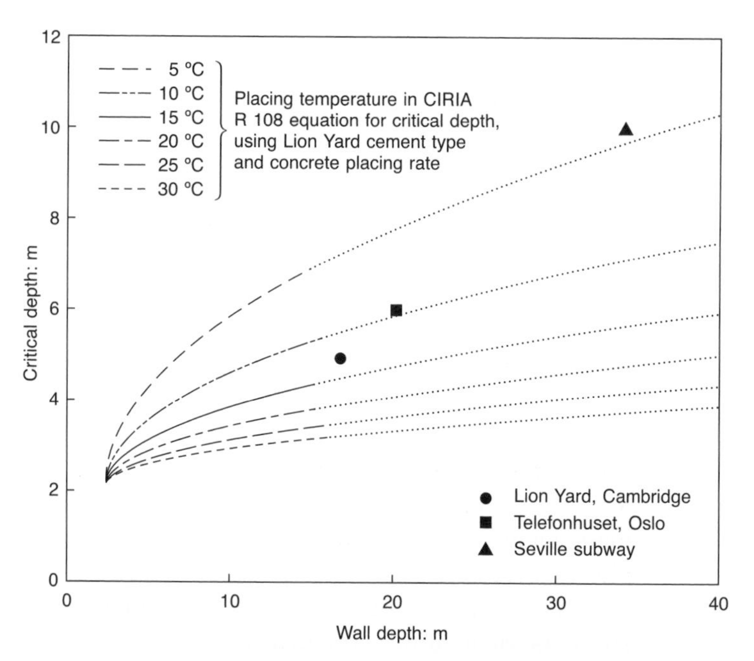

Fig. 2.49 Predicted critical depth from CIRIA Report 108 showing influence of wall depth and concrete placing temperature (after Lings et al., *1994)*

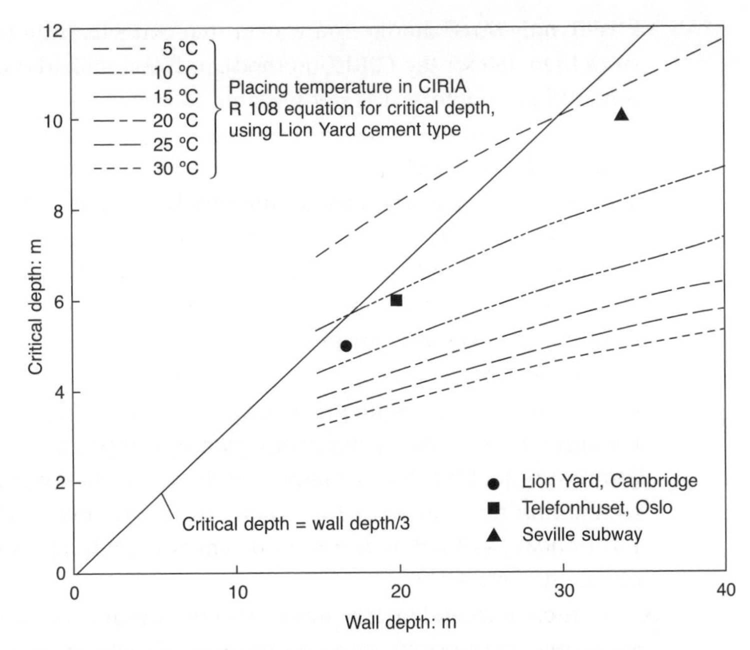

Fig. 2.50 Predicted critical depth from CIRIA Report 108 for diaphragm walls assuming a constant concreting time of 3 h (after Lings et al., *1994)*

three case records, together with extrapolated predictions. The critical depths predicted are based on the particular cement type and rate of concrete placing at Lion Yard, since this information is not given in the other two case records. The plot shows that the predicted critical depth is strongly influenced by the temperature of concrete at placing. Unfortunately, this was not recorded at any of the three sites.

It is likely that concrete placing rates will be faster for deeper walls, since, in practice, stop ends have to be removed at the end of concreting before they bond to the concrete. No allowance was made for this in Fig. 2.49, but the predicted critical depths are replotted in Fig. 2.50 over a realistic range of diaphragm wall depths, on the assumption of a fixed concreting time of 3 h. A more rapid rate of rise is therefore assumed for deeper walls, leading to larger predicted critical depths. Again, the cement type is based on Lion Yard, and the three case records are plotted.

There is nothing in the Cambridge and Oslo critical depths to suggest any significant departure from the CIRIA recommendations. However, the Seville critical depth still appears larger than any realistic placing temperature would suggest, confirming that the CIRIA method probably under-predicts as walls become deeper.

With only three diaphragm wall case records to draw on (compared with more than 350 for the CIRIA method), and insufficient data on the relevant concrete parameters, it is not possible to revise the CIRIA-type approach. However, in all cases the critical depth is less than one-third of the full depth of the wall, as is clear from the line shown in Fig. 2.50. No theoretical reasons can be discerned to explain this observation, but it may, nevertheless, serve as an acceptable method for determining the critical depth when carrying out future analyses involving wet concrete pressures.

Case record requirements

There is an urgent need for more case records to supplement the information presented here. When measuring wet concrete pressures in the field, it is important to check the calibration of the pressure cells in bentonite before any jacking takes place, and it is vital to measure and allow for the jacking force used. A full record should be made of bentonite levels and density, as well as concrete mix details, rate of placing and placing temperature.

Higher concrete placing temperatures increase the speed of the cement hydration reactions, thereby increasing the rate of water adsorption and pore pressure reduction. According to the CIRIA design method, it is one of the key parameters in terms of its effect on resulting wet concrete pressures, yet it was not measured in any of the case records. It is clearly important that concrete temperatures are measured during any subsequent field monitoring. In the light of this marked temperature sensitivity, it may turn out that the effects of wall installation are somewhat different in summer and winter, the stress reductions being greater in the former than the latter.

Behaviour at end of concreting

All the concrete pressures presented cover the period up to the end of concreting. DiBiaggio and Roti (1972) report figures that imply that the average wall–soil contact pressure decreased by 5.5 kPa in the 30 days after concreting at the Telefonhuset. Uriel and Oteo (1977) show reductions in pressure of up to 110 kPa in the 24 h after concreting at the Seville subway. At Lion Yard, pressures reduced by an average of 40 kPa in the first ten days after concreting. These wide variations in observed behaviour may be due to differences in ground conditions or installation techniques at the three sites. Alternatively, they may be due to instrumentation effects, and be more apparent than real.

It is a known feature of mercury-filled Glotzl cells fully embedded in concrete that they expand due to the heat of hydration, deforming the fresh concrete. When they subsequently cool down, they tend to lose contact with the concrete, the coefficient of thermal expansion of the mercury being 10 to 20 times greater than that of concrete. If the cells are intended to measure concrete stresses, the problem is normally overcome by forcing further mercury into the cell by means of a re-pressurizing tube. This may explain the large pressure reductions observed in Seville, where it is known that some of the cells were embedded in the concrete. Even at Lion Yard where the cells were at the wall–soil interface, it is possible that a similar effect took place, albeit in a much smaller way, given the very high stiffness deduced for the Gault Clay, particularly under conditions of stress reversal (Ng, 1992).

In contrast, vibrating wire pressure cells are relatively insensitive to changes in temperature, and this may be the key factor in the small pressure changes observed in Oslo. On balance, it seems likely that the majority of the observed pressure reductions at the other two sites are due to instrumentation effects.

Concluding comments

1. The lateral pressures exerted by concrete placed under bentonite in diaphragm walls should display the same characteristics as other underwater tremied concrete. An extension to the method in CIRIA Report 108 (Clear and Harrison, 1985) leads to the prediction of a critical depth in these circumstances, above which the full fluid concrete pressures apply, and below which pressures increase with depth following the slope of the bentonite line.
2. Data have been presented on the build-up with time and distribution with depth of wet concrete pressures for a diaphragm wall panel at Lion Yard, Cambridge. Two previous diaphragm wall case records have also been reviewed. All three sites clearly demonstrate the existence of a critical depth, and the data have been found to fit within the predicted bi-linear envelope of maximum concrete pressure.
3. It seems likely that the CIRIA method gives a reasonable prediction of the critical depth for walls up to 20 m deep, but probably underpredicts the critical depth for deeper walls. At all three sites, the critical depth was found to be less than one-third of the wall depth. When using wet concrete pressures in future analyses of wall installation, the bi-linear pressure diagram should be used, with a critical depth at one-third of the wall depth.
4. CIRIA Report 108 shows that temperature has a major influence on the resulting wet concrete pressures; consequently, the effects of wall installation on horizontal stresses in the ground may vary according to the time of year when construction took place. It is important to measure the concrete temperature at placing in future monitoring work.

CHAPTER EIGHTEEN

Stress transfer and deformation mechanisms around diaphragm walls

Overview

Embedded retaining walls such as secant bored pile walls and diaphragm walls are becoming widely used in the construction of deep sections of retained cuttings and cut-and-cover tunnels in road schemes and excavations in urban cities. For walls of this nature, the performance of the retaining system is likely to be governed by the relative stiffness of the wall and the surrounding soil, the type of support, the initial stress state in the ground and the stress changes caused by constructing the wall (Clough and O'Rourke, 1990; Ng, 1992). For urban areas developed in heavily over-consolidated soils, where the initial in situ horizontal stresses substantially exceed the vertical stresses, the construction procedures adopted for installing deep retaining walls have been observed to exert considerable influence on the stress regime and deformation in the ground adjacent to the structure. Tedd *et al.* (1984) reported that during the construction at Bell Common Tunnel, where a secant pile wall was being installed in stiff London Clay, about 30% of the total surface ground movement occurred. They also found that the wall installation caused significant reductions in total horizontal stress, which was measured by spade cells. Lings *et al.* (1991) reported that during the construction of a diaphragm wall in stiff fissured Gault Clay at Lion Yard, Cambridge, a marked reduction in lateral stresses was found. Symons and Carder (1993) reported that large reductions in lateral stress occurred during construction of contiguous bored piles and diaphragm walls in three other London Clay sites. If the retaining wall systems are not designed and constructed properly, the construction-induced deformations adjacent to buildings and services could result in serious litigation associated with damages caused by the deformations. For instance, the construction of a diaphragm wall in marine deposits and decomposed granite for the Chater Station caused substantial cracking in the nearby Courts of Justice building in Hong Kong (Humpheson *et al.*, 1987).

Several research projects, both in the field and in the laboratory, have investigated the influence of diaphragm wall construction on stress changes and deformations in the ground. Farmer and Attewell (1973) measured both the horizontal and vertical movements of a diaphragm panel excavation supported by bentonite slurry in about 3 m of fill overlying London Clay. The trench excavation was 6.1 m long, 0.8 m wide and 15 m deep. They showed that the maximum horizontal deformation occurred at about one-third of the panel depth from the ground level. At a distance of about two-fifths of panel depth (or about the panel length) from the trench, the horizontal movements were insignificant. Clough and O'Rourke (1990) examined a number of case histories in various ground conditions and separated movements caused by wall installation and other construction activities. Empirical design charts for assessing ground movements were provided. Powrie and Kantartzi (1996) used the technique of centrifuge modelling to investigate the effects of diaphragm wall installation. The modelled panels, which were constructed solely in an over-consolidated kaolin clay, were 1 m wide and 18.5 m deep (prototype) with varying panel lengths. Based on the centrifuge test results, they reported that a plane-strain excavation resulted in significantly larger deformation than a trench excavation with finite length. The results showed that the vertical settlement at a distance greater than one panel depth away from the panel was negligible.

In modelling diaphragm wall installation effects, there are generally two approaches commonly adopted. The first one uses a 'wished-in-place' wall approach, together with a reduced lateral earth pressure coefficient for the soil above the toe of the wall, as reported by Powrie and Li (1991). They indirectly modelled wall installation effects by reducing K_0 to 1 in the soil above the toe of the wall. It has been shown that this approach does not correctly model the stress changes both above and below the toe of the wall (Ng, 1992; Ng *et al.*, 1995b). In the second approach, a full construction sequence from excavation to bentonite slurry, followed by concreting and hardening, is modelled by using two-dimensional models. Gunn *et al.* (1993) used the finite-element method with a plane strain assumption in the vertical plane to demonstrate the effects of wall installation. Fully hydrostatic concrete pressure was used in the analyses rather than the more appropriate bi-linear concrete pressure suggested by Ng (1992) and Lings *et al.* (1994). Full hydrostatic concrete pressure would not have developed along the entire depth of the panel because the concrete sets during pouring. Reduction in lateral stress associated with wall installation was found to be significant in a proposed retaining wall. However, no results of ground deformation were reported in their studies. Rather large and unrealistic ground deformation would be expected in their plane-strain analyses because the horizontal arching mechanism was ignored. Based on a series of centrifuge model tests, Powrie and Kantartzi (1996) reported that a plane-strain excavation (infinitely long panel) resulted in significantly larger deformation than a trench excavation with a finite length.

An improved numerical methodology for investigating the three-dimensional (3D) installation effects was developed by using two simple perpendicular plane sections (Ng *et al.*, 1995b) to model stress transfer mechanisms and ground deformations. In the study, a horizontal plane analysis with plane-stress assumption was first used to compute the lateral stresses behind the wall for modelling the horizontal arching effects and predicting horizontal displacements. Average displacements were then used as boundary conditions for a subsequent vertical section analysis. The average lateral stresses behind the wall were predicted and compared with field measurements. These analyses appeared to be able to explain the complex stress transfer mechanisms that resulted in large lateral stress reductions, but with small ground deformations in the stiff clay. However, the approximate analyses have certain limitations. First, stresses cannot be redistributed simultaneously in both vertical and horizontal directions. Second, a wall displacement shape with depth had to be assumed during the vertical section analysis, with the plane-strain assumption. Hence, deformations along the length of the panel could not be computed. In this chapter, a more realistic and truly 3D modelling of the construction sequence of a diaphragm wall panel is presented. The objective of the modelling is to gain some fundamental understanding of and new insight into the stress transfer mechanisms and ground deformations during a diaphragm wall panel installation. Computed results are compared qualitatively and quantitatively with some published field measurements and recent experimental data of centrifuge model tests in stiff clays.

Numerical models and parameters

A typical diaphragm wall panel, 8 m long (L), 0.6 m wide (W), and 15 m deep (D), constructed in Gault Clay at Lion Yard Cambridge (United Kingdom), is adopted for the numerical simulation. To model the 3D installation effects of the diaphragm wall panel construction, the finite-difference program FLAC3D was used. The ground succession at the site is about 3 m gravel and made ground overlying 38 m of Gault Clay, which in turn overlies the Lower Greensand. The initial groundwater table is located at about 3 m below ground. All soils have been modelled as simple linear elastic, perfectly plastic isotropic materials with a Mohr–Coulomb yield surface. Because the 3D soil–structure interaction of diaphragm walling is rather complex, this relatively simple soil model was therefore chosen for ease of interpretation of computed results. Model parameters are summarized in Table 2.3. During the construction of the

Table 2.3 Soil parameters used in analysis (after Ng and Yan, 1998c)

Gravel	Gault Clay
Unit weight (γ_s) (18 kN/m^3)	Unit weight (20 kN/m^3)
Coefficient of earth pressure (K_0) (0.35)	Coefficient of earth pressure (1.5 (assumed))
Drained angle of friction (ϕ') (40°)	Undrained angle of friction (0°)
Effective cohesion (c') (0 kPa)	Undrained shear strength (varying, see text)
Drained Poisson's ratio (0.2)	Undrained Poisson's ratio (0.495)
Drained Young's modulus (E') (50 000) (kPa)	Undrained Young's modulus (varying, see text)

diaphragm wall panel, gravel and Gault Clay are assumed to remain drained and undrained, respectively. The undrained shear strength (c_u) of the clay was found to increase linearly with depth and it can be characterized by the equation $c_u = 60 + 10z'$ kPa where z' = depth below the surface of the clay in metres (Lings *et al.*, 1991; Ng, 1992). The shear modulus measured by a self-boring pressuremeter also increases with depth. After conversion, a relationship between undrained Young's modulus (E_u) and shear strength, $E_u/c_u = 1000$, was found for moderate shear strains (typically between 0.01 and 0.5%). Based on the results of geophysical measurements at the nearby test site at Madingley (Powell and Butcher, 1991; Butcher and Powell, 1995) and triaxial stress path tests on intact Gault Clay specimens with internal strain measurements, it was found that the stiffness of Gault Clay could be as high as $E_u/c_u = 4000$ at very small shear strains (Ng *et al.*, 1995a; Dasari, 1996). All the soil parameters adopted here are identical to the ones used by Ng *et al.* (1995b).

It is widely accepted that the stress–strain behaviour of soil is highly non-linear and the stiffness of soil decreases as shear strain increases (Atkinson and Sällfors, 1991; Tatsuoka and Kohata, 1994). To account for the variation of stiffness with strain within a linear analysis, the soil has been idealized into two regions. The first is adjacent to the diaphragm wall panel and the second is the rest of the soil. For soil elements close to the excavation panel, a relatively low stiffness ratio, $E_u/c_u = 1000$, has been assigned to these elements to allow for lower stiffness at moderate shear strains as a result of the effects of the excavation process. For all other soil elements, a higher stiffness ratio, $E_u/c_u = 4000$, has been adopted.

Finite-difference mesh and modelling procedure

Two planes of symmetry are used in generating a finite-difference mesh for the analysis. Figure 2.51 shows the mesh, together with the two planes of symmetry, x = 0 and y = 0. The planes, x = 0 and x = 50, are

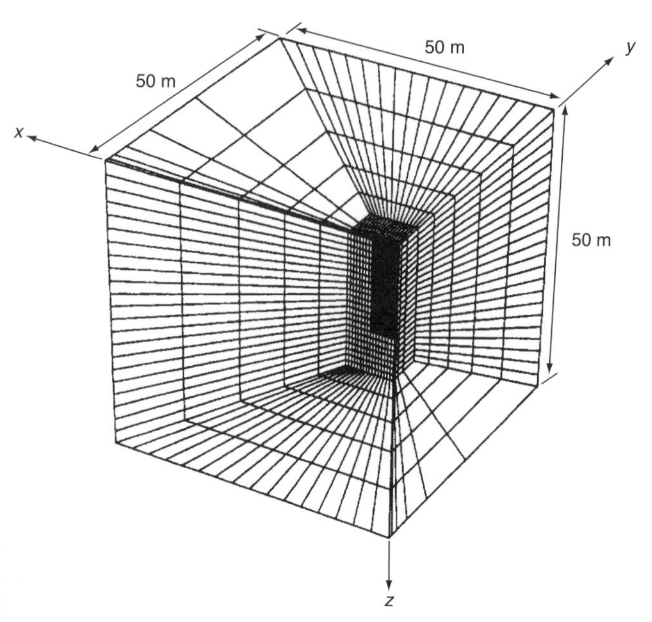

Fig. 2.51 Finite-difference mesh (after Ng and Yan, 1998c)

allowed to move freely in the y- and z-directions, but not in the x-direction. Similarly, the planes, y = 0 and y = 50, are free to move in the x- and z-directions, but not in the y-direction. At the boundary plane, z = 50, all movements are restrained. For modelling the 0.6 m wide by 8 m long diaphragm wall panel, an idealized panel of 0.3 m 'half-width', 4 m 'half-length' (Fig. 2.52), and 15 m deep is adopted in the analyses. Elements

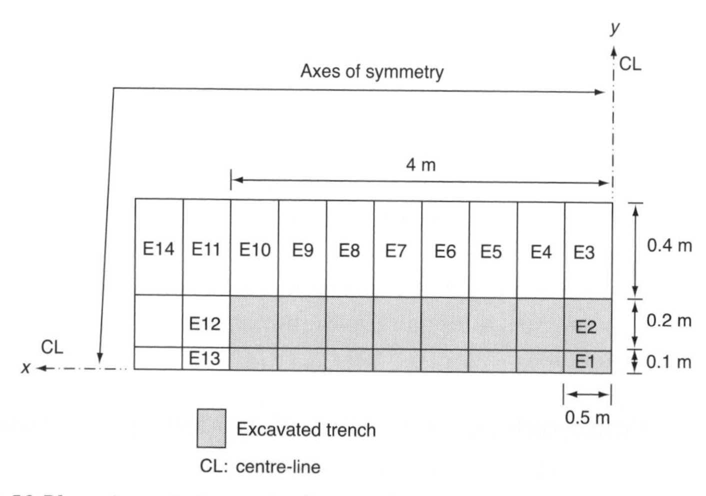

Fig. 2.52 Plan view of elements close to the excavated trench (after Ng and Yan, 1998c)

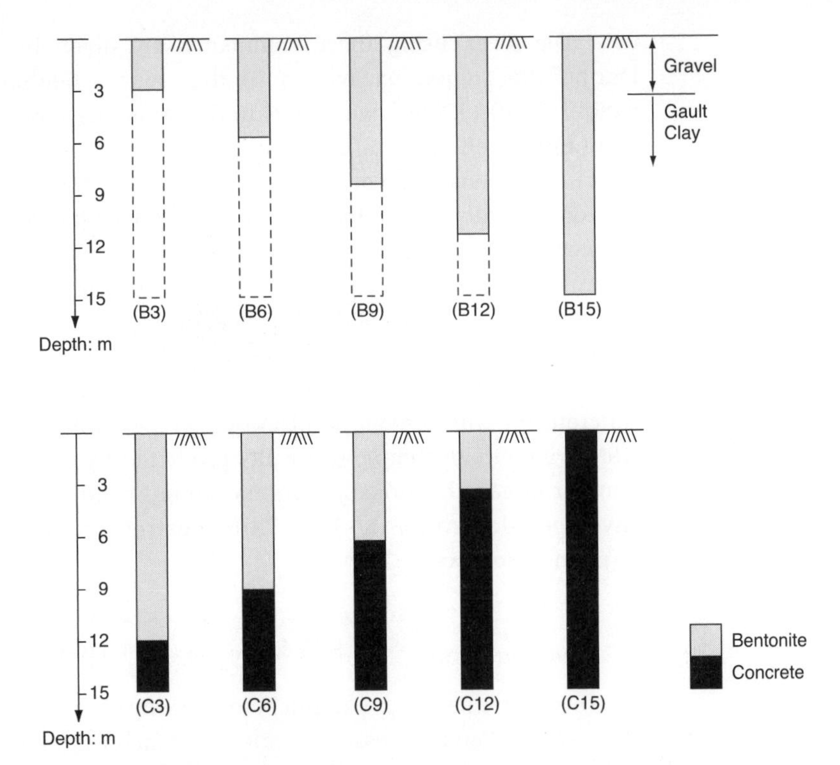

Fig. 2.53 Modelling sequence for constructing the diaphragm wall panel (after Ng and Yan, 1998c)

within a zone of 6 m from the panel are assumed to have a relatively low stiffness ratio of $E_u/c_u = 1000$, whereas all the other elements are given the higher stiffness ratio of $E_u/c_u = 4000$. Quadrilateral elements (or zones) are adopted by the finite-difference program.

Internally, each quadrilateral element is subdivided into four constant stress triangular 'sub-elements'. Stresses acting at the four centroids of the subelements are averaged to represent the stress acting at the centre of the quadrilateral element.

Figure 2.53 shows schematic diagrams illustrating the construction sequence adopted in the finite-difference analyses. The modelling procedures are summarized as follows:

1 Excavate the 0.3 m wide by 4 m long panel at 3 m deep intervals using the NULL command and apply normal hydrostatic bentonite pressure (i.e. the normal pressure is a function of unit weight, 10.1 kN/m^3, and depth of bentonite) on the trench faces simultaneously using the APPLY NSTRESS command as shown in Fig. 2.53 (B3–B15). Nulling a zone sets the initial stresses in that zone to zero. When a face is exposed by nulling an adjacent zone, nothing is done to the exposed

face. Details of using these commands are given by Itasca (1996). During the staged excavation of the trench, bentonite of a unit weight of 10.1 kN/m^3 was measured during the construction of the diaphragm wall panel at Lion Yard (Ng, 1992). It is expected that any minor variations (typically less than 10%) in the unit weight will not significantly affect the computed results. Practical construction problems such as the formation of bentonite cake (Ng, 1992; Lings *et al.*, 1994) due to chemical reactions between the bentonite and the adjacent soil are not included in the analysis.

2 Cast the concrete panel by increasing the lateral pressure inside the panel from the bottom (C3–CI5) using a theoretical bi-linear wet concrete pressure envelope proposed by Ng (1992) and Lings *et al.* (1994), who have compared results predicted by the equation with a number of case histories in various ground conditions. The bi-linear envelope adopted for the Lion Yard construction process (Ng, 1992) is given as follows:

$$\left.\begin{aligned} \sigma_{\mathrm{h}} &= 24\mathrm{z} && \text{for } \mathrm{z} \leq 4.6\,\mathrm{m} \\ \sigma_{\mathrm{h}} &= 64 + 10.1\mathrm{z} && \text{for } \mathrm{z} > 4.6\,\mathrm{m} \end{aligned}\right\} \qquad (2.25)$$

where σ_h is in kPa and z = depth below the top of a panel. This equation suggests that lateral pressure exerted by concrete placed under bentonite in a diaphragm wall panel follows a bi-linear pressure envelope, in which full fluid concrete pressures apply only above a critical depth. Below this depth, pressures increase with depth following the slope of the hydrostatic bentonite line. This proposed envelope takes account of the effects of concrete set on lateral pressure during concreting.

3 Cure the concrete panel by replacing the trench with elastic concrete elements and removing the applied stresses on the trench faces simultaneously.

Because the wall installation process is usually rapid, an undrained assumption is made for the Gault Clay in the analyses.

Normal stress distributions behind panel

Figure 2.54(a) shows the notations describing the 3D stresses adopted. The horizontal arching mechanism refers to the redistribution of total horizontal stress laterally via the shear stress component (τ_{xy}). The downward load transfer mechanism denotes the redistribution of lateral earth pressure (i.e. normal stress) adjacent to the panel vertically downward by shearing (τ_{zy}).

The computed distributions of total horizontal stress (σ_{yy}) at different construction stages at two locations, E3 and E10, are shown in Fig. 2.54(b) and (c), respectively. During the staged excavation of the

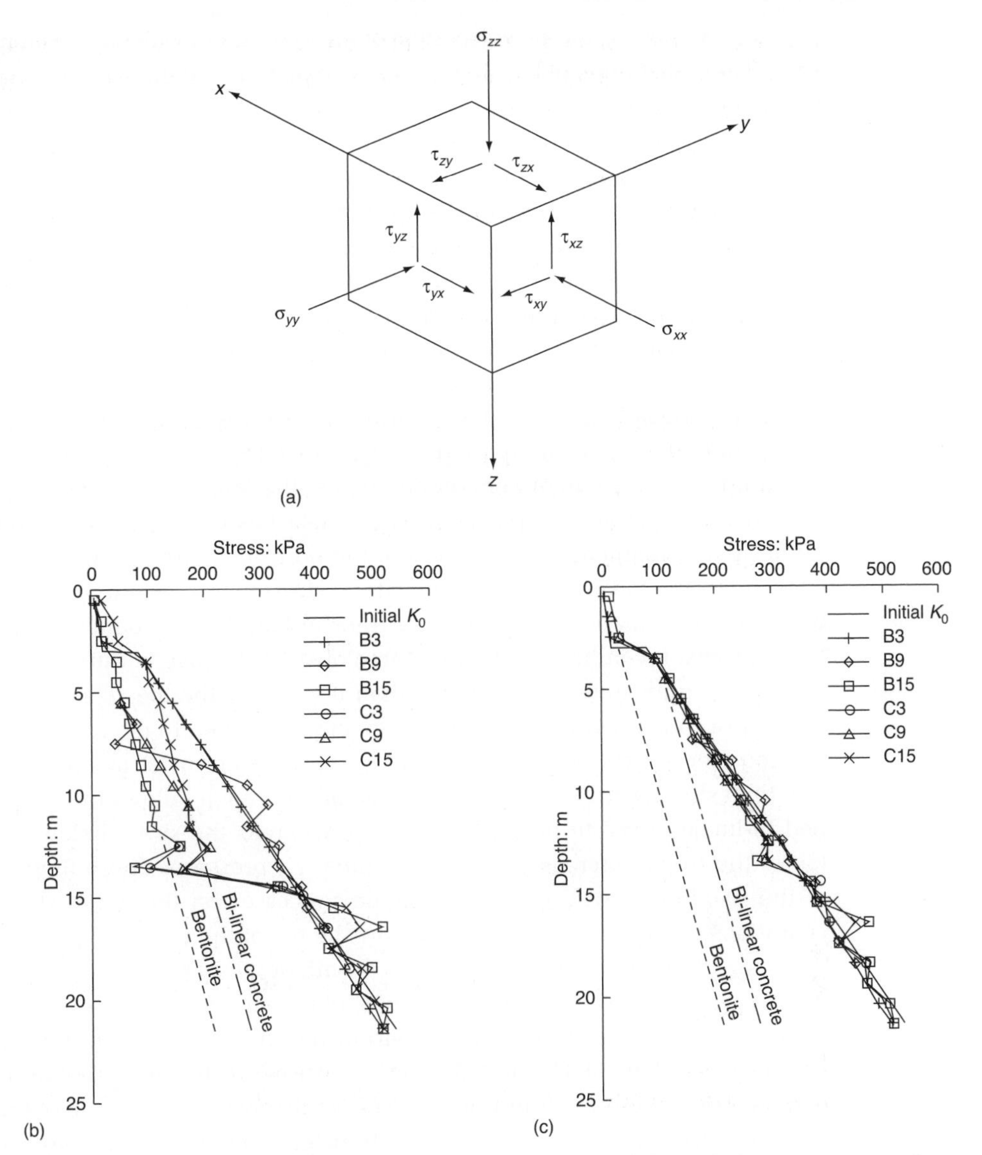

Fig. 2.54 (a) 3D stress state; total horizontal stresses acting at different construction stages (b) on E3; (c) on E10 (after Ng and Yan, 1998c)

trench, bentonite is used to provide temporary support. The initial horizontal stress, σ_{yy}, behind the trench, reduces to the hydrostatic bentonite pressure (see Fig. 2.54(b)) above the toe at each stage of excavation. Beneath the toe, there is an increase in the horizontal stress at each stage of the excavation. The lateral stress reduction causes both the horizontal arching and the downward load transfer mechanisms. The progressive downward load transfer through the shear-stress component

(τ_{zy}) that occurs during the excavation of the diaphragm wall panel can be seen. The distribution of the shear-stress component varying with depth is discussed later.

If the stress reduction were solely attributed to the downward load transfer mechanism, the area of stress reduction above the toe would have been equal to the area of stress increase beneath the toe. However, this is not the case and the computed stress distribution implies that the significant stress reduction that occurs during the bentonite stage cannot be due solely to the downward load transfer mechanism. Some of the horizontal stress reduction must be attributed to the horizontal arching mechanism through the shear-stress component (τ_{xy}), which was not modelled simultaneously during the vertical section analysis in the pseudo-3D approach reported by Ng *et al.* (1995b).

During the subsequent concreting stages, the lateral stress increases substantially in the 3 m gravel zone because the concrete pressure is higher than the initial K_0 pressure. For the horizontal stress in the Gault Clay above the toe of the trench, the lateral stress increases progressively upward as a result of the upward load transfer and the horizontal arching mechanisms. It can be seen that the area of stress 'recovery' during concreting is significantly larger above than it is below the toe. This seems to suggest that stress redistribution is mainly attributed to the horizontal arching mechanism rather than to the upward load transfer mechanism. The final stress distribution above the toe of the trench follows the specified bi-linear concrete pressure envelope, whereas the stress below the toe of the trench increases above the initial K_0 pressure. The influence of the diaphragm wall panel installation processes on the initial stress conditions ceases at a depth of $\frac{1}{3}D$ below the toe of the panel.

The distribution of the horizontal stress with depth behind the trench is qualitatively consistent with the results obtained from the pseudo-3D analysis (Ng *et al.*, 1995b). By comparing the areas of stress reduction and increase above and below the toe of the wall, respectively, it can be deduced that about one-third of the stress reduction above the toe of the panel is attributed to the vertical load transfer mechanism, whereas the rest of the stress is redistributed laterally via the horizontal arching mechanism.

The computed horizontal stress (σ_{yy}) distribution with depth at E10 (see Fig. 2.54(c)) during each stage of construction is very different from the distribution at E3. There is only a small reduction of the horizontal stress just above the toe of the trench and a slight increase of stress beneath the toe. This implies that there is a significant amount of shear stress (τ_{xy}) developed in the soil element as a result of the horizontal arching mechanism, which has redistributed horizontal stress via the shear stress (τ_{xy}) from the centre to the edge of the panel. It therefore has limited

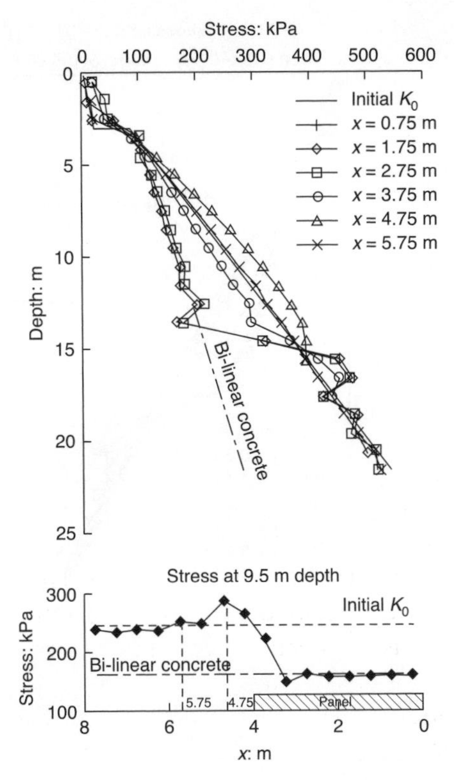

Fig. 2.55 Total horizontal stresses behind the panel after casting of concrete (after Ng and Yan, 1998c)

a substantial reduction of the horizontal stress. The distribution of the shear stress is presented and discussed later.

Figure 2.55 shows a typical distribution of the horizontal stress (σ_{yy}) at 0.1 m behind the trench and at 9.5 m below the ground surface (at the mid-depth of the panel in the clay stratum), normal to the excavated surface. It can be seen that the total horizontal stress distribution at a depth of 9.5 m is fairly uniform behind the panel, except near the edge. Beyond the panel length, a maximum stress is computed at $x = 4.75$ m (0.75 m from the edge of the panel) as a result of the horizontal arching mechanism. The influence of the panel construction on the initial stress conditions becomes insignificant at a distance of $\frac{1}{3}L$ along the length of the panel. This computed distribution of stress behind the panel for a given depth is consistent with the results obtained from a pseudo-3D analysis (Ng *et al.*, 1995b).

Figure 2.56(a–c) shows the variations of the computed horizontal stress (σ_{yy}) with distance perpendicular (i.e. normal) to the diaphragm wall panel at E3, E10 and E14, respectively. At E3, there is a substantial

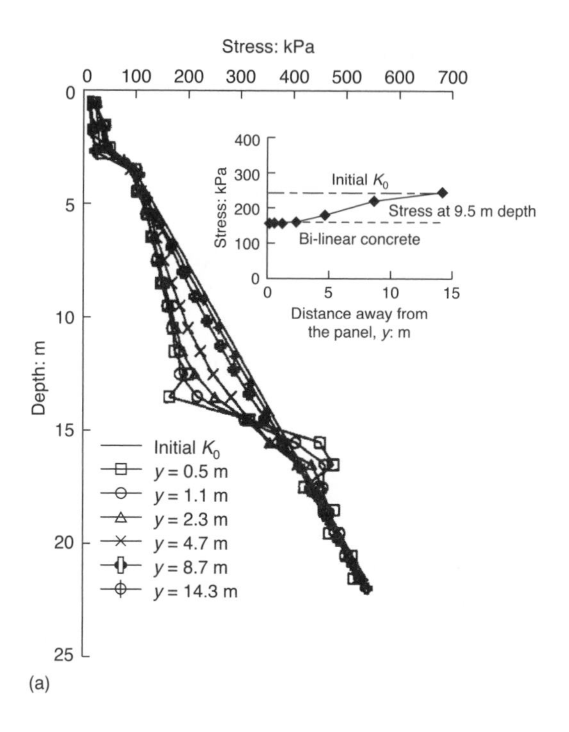

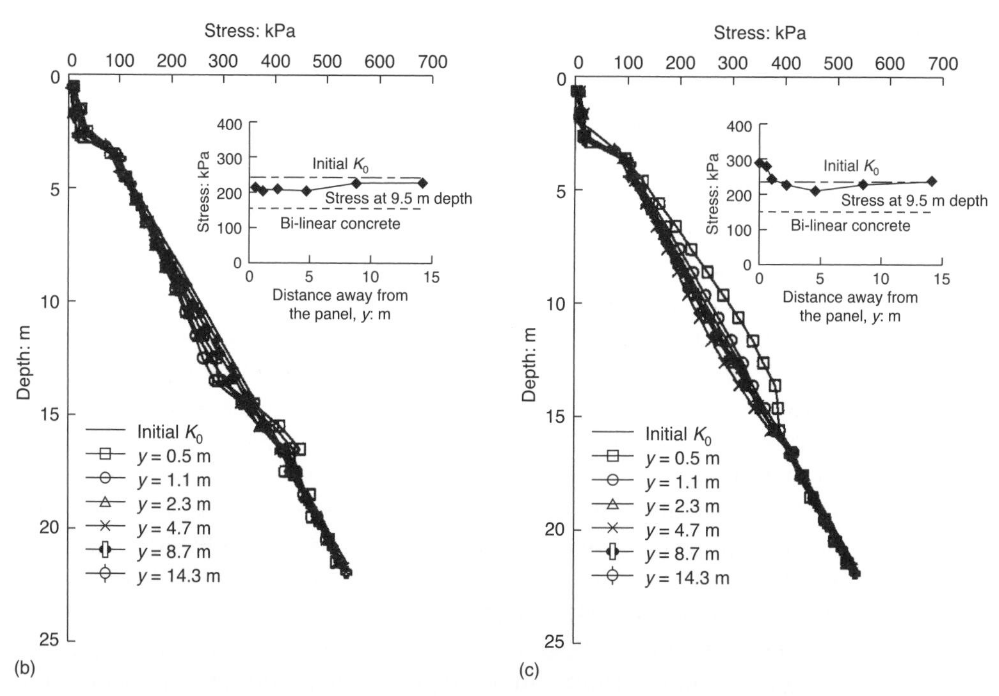

Fig. 2.56 Total horizontal stresses acting with normal distance away from the panel at end of construction for: (a) E3; (b) E10; (c) E14 (after Ng and Yan, 1998c)

reduction of horizontal stress within 3 m (0.38L or 0.2D) behind and above the toe of the panel. Beyond this distance, the horizontal stress increases linearly with distance away from the face of the panel. For the diaphragm wall panel ($L = 8$ m) considered in this study, the influence of the diaphragm wall construction, however, ceases at about 15 m from the face of the trench, that is at a distance approximately equal to one panel depth (1.0D). This finding is consistent with the centrifuge model tests reported by Powrie and Kantartzi (1996).

At E10, only a small stress reduction above the toe and a small increase below the toe are observed. Similar to the stress distribution at E3, the zone of influence due to diaphragm wall construction is also within 15 m from the face of the trench (1.0D). However, instead of a substantial reduction of horizontal stress, there is a stress increase within 2 m behind and above the toe of the panel at E14. In contrast, no stress reduction can be seen below the toe. This clearly indicates that the stress increase above the toe is attributed to the horizontal arching mechanism, which has transferred the horizontal stress behind the panel to the adjacent soils. At the stress distributions at E3 and E10, the zone of influence due to diaphragm wall construction also falls within the distance 15 m from the face of the trench (1.0D). It can be seen that a 3D, non-uniform, stress distribution occurs behind the trench as a result of the diaphragm wall panel construction.

Shear stress distributions behind panel

Figure 2.57(a) and (b) shows the computed distributions of shear stress (τ_{zy}) behind the trench at E3 and E10, respectively, during each stage of excavation. The downward load transfer mechanism is assisted by the shear stress to transfer the initially high horizontal stress (σ_{yy}) progressively downward beneath the toe. This transfer results in a reduction of horizontal stress above the toe, together with an increase of the stress beneath the toe (see Fig. 2.54(b)). Comparing the magnitudes of the shear stress developed at E3 and E10, it can be deduced that the downward load transfer mechanism is more significant at E3 than at E10. For the diaphragm wall panel ($L = 8$ m) considered here, very little shear stress is mobilized beneath the trench during the construction of the diaphragm wall panel at a distance deeper than one-third of the depth of the panel ($\frac{1}{3}D$). This is consistent with the computed normal stress distributions shown in previous figures.

For studying the horizontal arching mechanism during the excavation of the trench, shear stress (τ_{xy}) distributions with depth at E3 and E10 behind the trench at various stages of construction were investigated. As expected, the shear stress distribution with depth at E3 is zero during

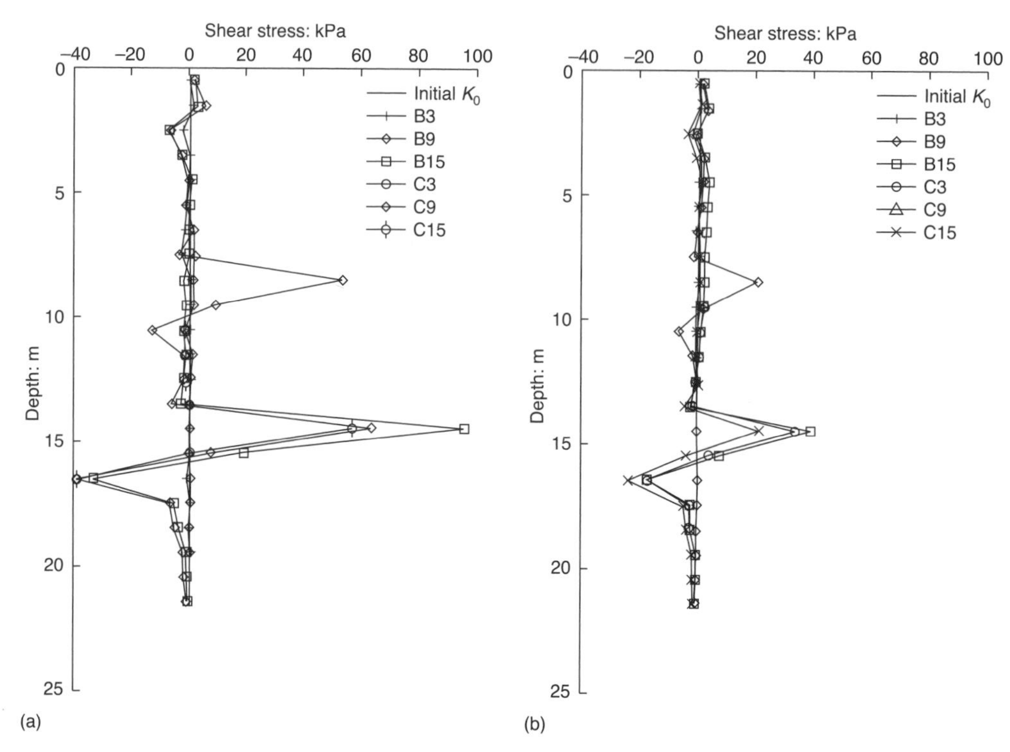

Fig. 2.57 Shear stresses acting at different construction stages for: (a) E3; (b) E10 (after Ng and Yan, 1998c)

each stage of construction. This implies that the directions of the two principal horizontal stresses are parallel and perpendicular to the panel, and hence the horizontal normal stress (σ_{yy}) would be equal to the applied boundary pressures inside the trench (see Figs 2.55 and 2.56(a)). Conversely, there is a substantial shear stress (τ_{xy}) mobilized at E10 at end of the excavation and its magnitude decreases after concreting (Fig. 2.58). The normal stress distribution (σ_{yy}) near the edge of the panel is governed both by the downward load transfer and the horizontal arching mechanisms.

Figure 2.59 shows the distributions of the shear stress (τ_{xy}) with depth and at a distance 9.5 m below the ground surface across the length of the panel. The shape of the distributions is consistent with the normal stress distributions shown in Fig. 2.55. The magnitude of the mobilized shear stress increases with depth and with distance away from the plane of symmetry ($x = 0$). However, for the diaphragm wall panel ($L = 8$ m) considered in the analysis, the zone of influence is fairly limited to a distance less than one-third of D in the z-direction (i.e. depth of panel) and one-third of L in the x-direction (horizontal). The large shear stress

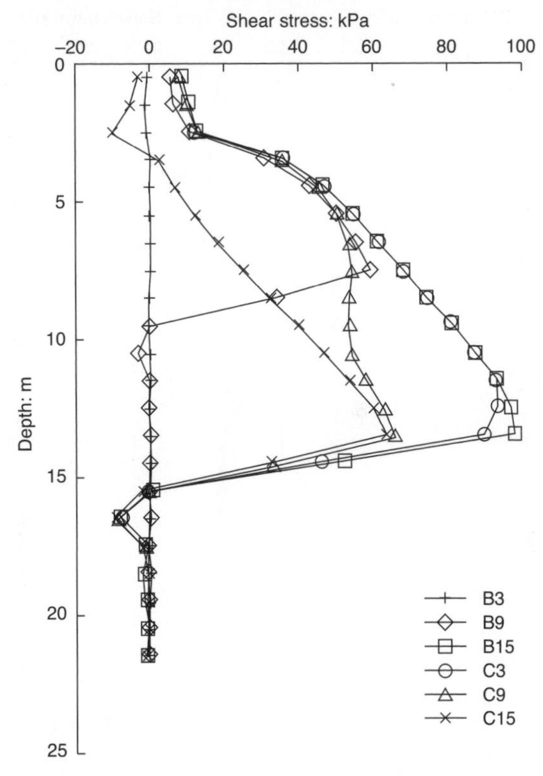

Fig. 2.58 Shear stresses acting on E10 at different construction stages (after Ng and Yan, 1998c)

mobilized at the edge of the panel is to 'bridge' or 'arch' the stress jump from the bi-linear concrete pressure to the initial K_0 pressure line.

Figure 2.60 shows the comparison of the computed total horizontal stresses (σ_{yy}) at E3 and the measured values at the centre of a diaphragm wall panel constructed in Gault Clay at Lion Yard. It should be noted that the theoretical bi-linear concrete pressure envelope has been adopted as an input (initial) pressure imposed on the surfaces inside the excavated trench. From the figure, it can be seen that the computed and the measured values are consistent with each other along the depth of the wall, except near the toe. This is because the modelled depth of the panel is 1 m shallower than the panel in the field. When comparing with the initial K_0 pressure, clearly there is a significant reduction of the total horizontal stress at the centre and above the toe of the panel after concreting, but only a relatively small increase of the horizontal stress below the toe. The small increase of horizontal stress below the toe is contrary to the findings using a pseudo-3D analysis (Ng *et al.*, 1995a), in which the stress reduction was limited only in the vertical plane, that is no simultaneous stress transfers are possible in the truly 3D sense.

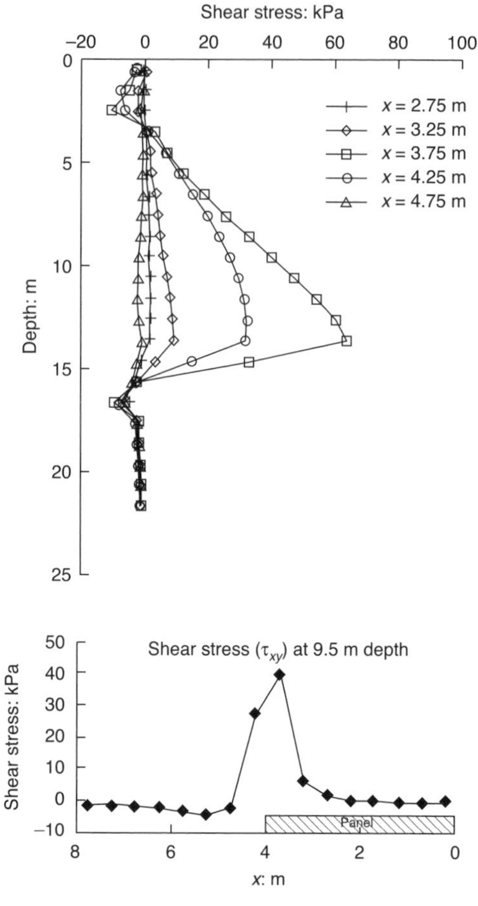

Fig. 2.59 Shear stresses behind the panel after casting concrete (after Ng and Yan, 1998c)

To study how the shear strength is mobilized behind the trench during construction, the computed deviatoric shear stress, $q = 1/2(\sigma_1 - \sigma_3)$, versus depth is plotted in Fig. 2.61 at three locations behind the trench (i.e. E3, E10 and E12). At E3, it can be seen that relatively small shear strength is mobilized along the depth of the trench, except in the region of the toe. This is due to the large shear stress (τ_{zy}) developed near the toe as a result of the downward load transfer mechanism (see Fig. 2.57(a)). Along the length of the panel from E3 to E12, there is a trend of increasing mobilized shear strength. The maximum shear strength mobilized is beyond the edge of the panel at E12 during the excavation stage of the installation of the diaphragm wall (i.e. the bentonite stage), due to the high shear stress (τ_{xy}) developed as a result of the horizontal arching mechanism (see Fig. 2.59). For an isotropic soil model, this suggests that

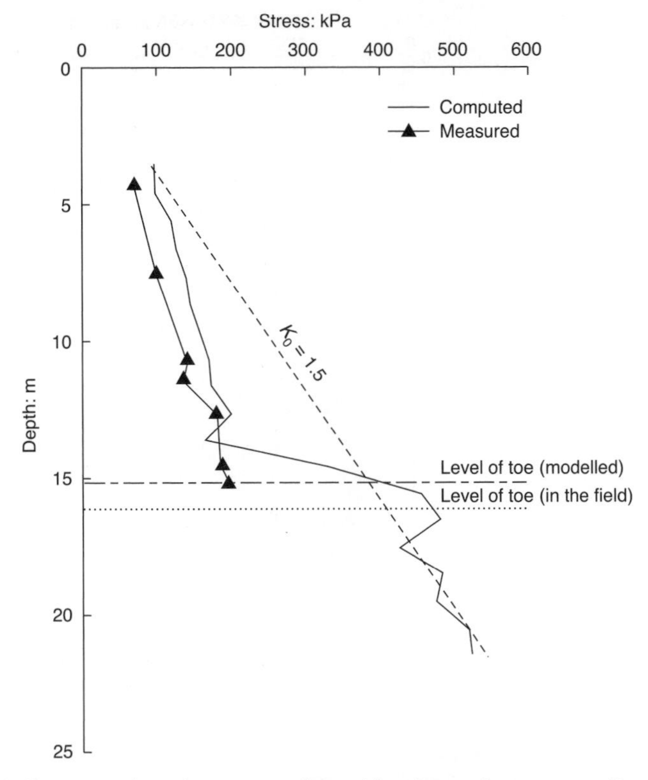

Fig. 2.60 Computed and measured total horizontal stresses after construction of the panel (after Ng and Yan, 1998c)

the soil adjacent to the trench is more likely to yield near the edge than at the centre of the panel.

There is a substantial difference in the mobilized shear strength at both E10 and E12 during the excavation (bentonite) and concreting stages. To the contrary, little difference can be seen between the mobilized shear strength at these two stages at E3, except near the toe of the panel. The computed deviatoric shear stress varies approximately linearly with depth in the clay stratum, except near the toe of the panel. The linear variations of q with depth are due to the linear stress reduction of horizontal stress from the concrete to the bentonite pressures plus the assumed linear increase in soil stiffness with depth for the over-consolidated clay (assumed $K_0 = 1.5$). At the toe, the shear stress increases at the centre (E3), but it decreases near the edge of the panel (E10 and E12) because the different stress transfer mechanism dominates at these two locations.

The conclusions drawn from Fig. 2.61 may only apply for isotropic clays with high initial K_0 values (= 1.5 for this case). It should be noted that the stress paths followed by the soil at E3 and E10 (or E12) are very different.

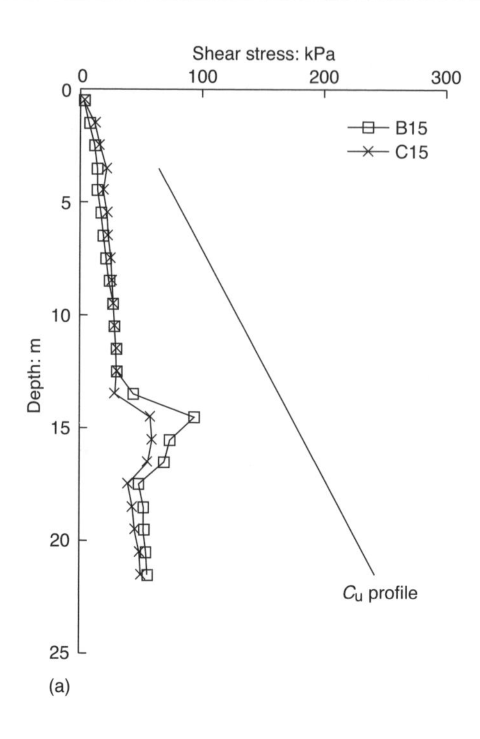

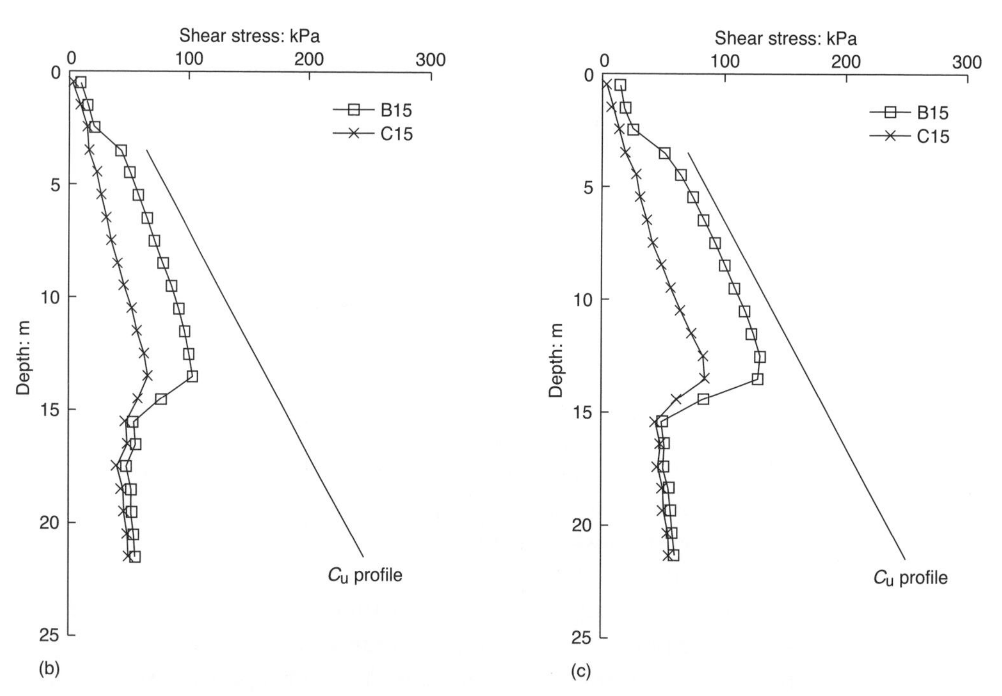

Fig. 2.61 Maximum mobilized shear stress during construction at: (a) E3; (b) E10; (c) E12 (after Ng and Yan, 1998c)

The former has been subjected to an overall unloading stress path during construction (see Figs 2.54, 2.55, 2.58 and 2.59), whereas the latter has been subjected to an overall increasing loading stress path.

Horizontal deformations

Changes of stress during the construction of a diaphragm wall panel will inevitably result in ground deformation. Figure 2.62 shows the computed deformation of the soil behind the trench during each stage of construction. The horizontal ground deformation is largest at the centre ($x = 0$ m) of the panel during the bentonite stage. The deformation at the centre is

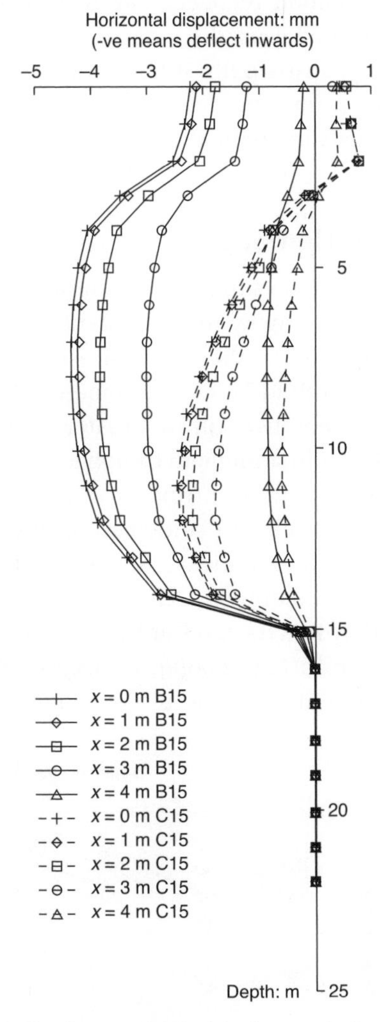

Fig. 2.62 Horizontal displacement behind trench during construction of the panel (after Ng and Yan, 1998c)

about five times larger than that at the edge (x = 4 m). The horizontal deformations along the panel length are clearly not uniform. However, the horizontal deformations are surprisingly uniform along the depth of the trench during the bentonite stage. After concreting, there is some recovery of horizontal deformations, especially near the top of the trench. Recovery of ground deformation has also been observed during the centrifuge model tests reported by Powrie and Kantartzi (1996). The significant deformation recovery computed near the top of the trench is due to the fact that the top 3 m of gravel has a smaller initial K_0 stress than the applied concrete pressure and the stiffness of the gravel is also lower than the underlying stiff Gault Clay. However, the magnitude of the deformation recovery has been over-predicted because the Mohr–Coulomb model used cannot simulate high stiffness response at a 180° reversal of stress path. If the entire panel were constructed solely in the stiff clay, a uniform deformation recovery would probably be expected. This is because both the stiffness and stress change vary linearly with depth.

Surface settlements

Figure 2.63 shows the vertical surface settlement varying with perpendicular distance away from the panel at the edge and at the centre of the panel. Because of the larger stress changes at the bentonite stage than at the concreting stage, settlements that occur at the former stage are approximately two times greater than those computed at the latter stage. A gentle settlement trough appears behind the panel, with the maximum settlement occurring at a distance of approximately 3.1 m (about 0.2*D*) away at all four settlement profiles. At a distance of approximately one trench depth away, however, no significant ground settlements have been computed. This is consistent with the centrifuge results reported by Powrie and Kantartzi (1996) and the computed stress distribution shown in Fig. 2.56(a). As expected, the settlement is larger at the centre than at the edge of the panel because of the horizontal arching

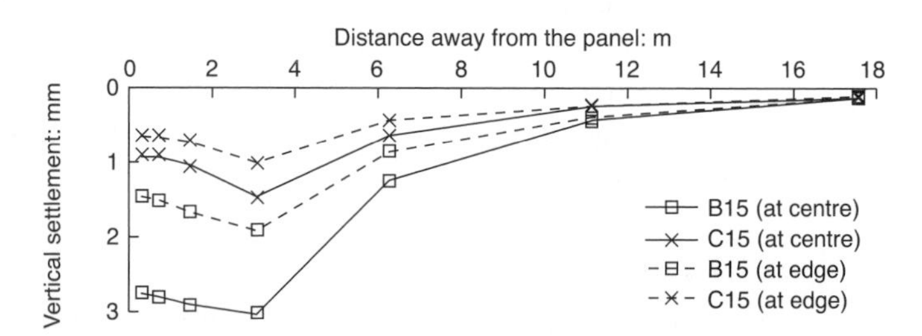

Fig. 2.63 Surface settlement profiles normal to the panel (after Ng and Yan, 1998c)

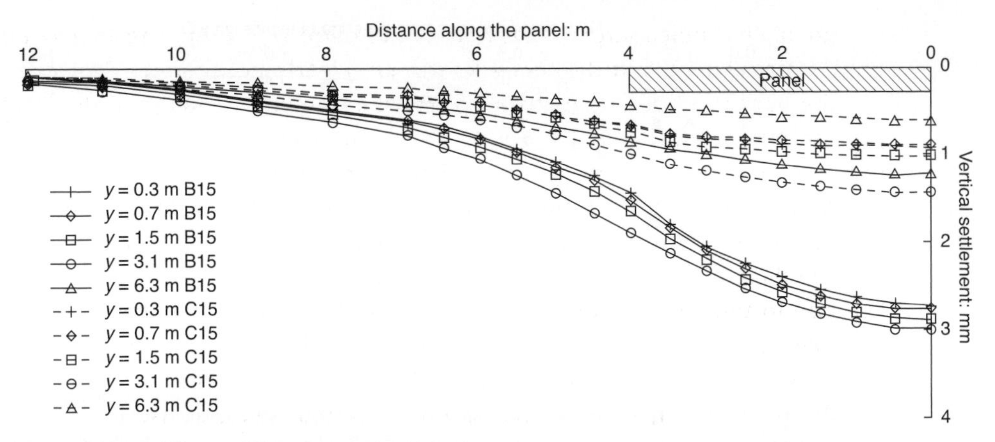

Fig. 2.64 Surface settlement profiles along the length of the panel (after Ng and Yan, 1998c)

effects, which limit the horizontal ground movements (see Fig. 2.59). The settlement ratios between these two locations are computed to be about 2 and 1.5 at the bentonite and concreting stages respectively.

Figure 2.64 shows the surface settlement patterns varying along the length at different distances (in the y-direction) from the face of the panel. When viewed in conjunction with Fig. 2.63, it can be seen that a settlement bowl occurs behind the trench. The shape of the settlement bowl is similar at various distances away from the face of the trench. The radius of curvature of the settlement bowl is smaller at the bentonite stage than at the concreting stage, but the depth of the settlement bowl at the former construction stage is about twice as deep as at the latter construction stage. No significant ground settlements have been computed at a distance (in the x-direction) of a panel length (8 m) from the edge of the panel. Because there is insufficient information available for detailed comparisons between the computed and measured settlements during the construction of the diaphragm wall at Lion Yard, no attempt is made to plot any settlement points on either Fig. 2.63 or Fig. 2.64. However, based on the limited data available, the measured settlement during the construction of the diaphragm wall was in the order of 1–2 mm at about 8 m from the wall. As the Mohr–Coulomb model over-predicted the lateral displacement of the wall during concreting (see Fig. 2.62), ground settlements are likely to be underestimated by the analysis.

Assessment

Although there are differences in soil type and fabric, geological history, groundwater conditions, panel geometry, and construction details and time-scale, a general qualitative comparison of ground settlements

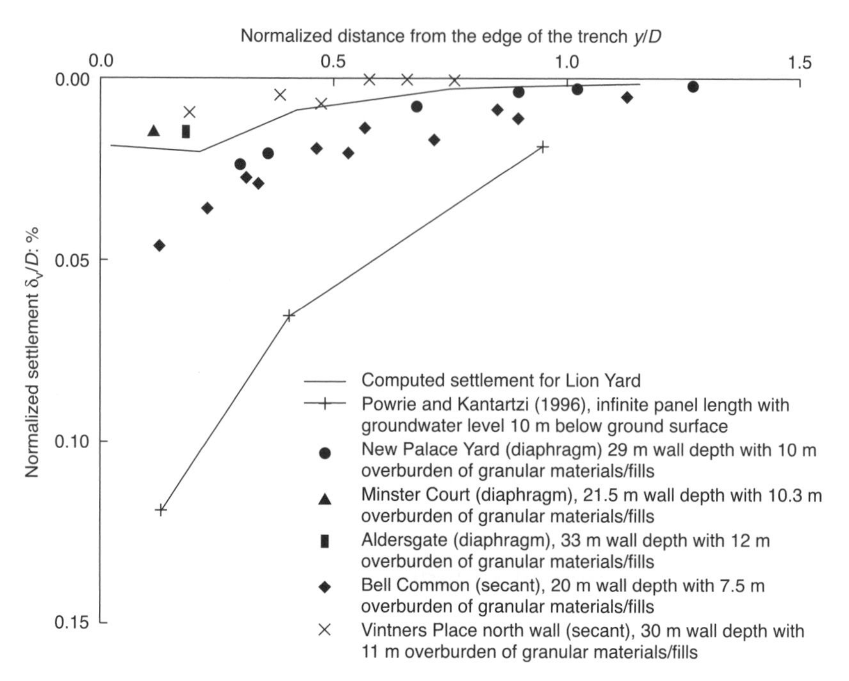

Fig. 2.65 Normalized surface settlement profiles from computation, centrifuge testing and field measurements (after Ng and Yan, 1998c)

between those simulated here and those measured in the field and the centrifuge tests would still be interesting and useful to help engineers and researchers identify common trends and suggest some possible reasons for any major discrepancies seen.

Figure 2.65 compares the computed normalized settlements at the centre of the panel with some relevant field measurements in the stiff London Clay (Thompson, 1991) and with the smallest settlement profiles measured in the five centrifuge model tests reported by Powrie and Kantartzi (1996). The use, in the comparisons, of the computed settlements resulting from a single panel construction, is reasonable because once an excavated panel has been concreted, the construction of adjacent panels has only a very minor influence on the final ground deformation at the centre, behind the first constructed panel (Ng *et al.*, 1995b). In other words, as long as a trench has been concreted before another trench is excavated (not excavating the entire length of a diaphragm wall simultaneously), the settlement profile behind a single panel construction is approximately equal to the final settlements of the construction of an entire diaphragm wall, especially if the length of a panel is long such as the one at Lion Yard (8 m). It can be seen from the comparisons that the computed settlement profile is consistent with field observations. However, the smallest settlement profile recorded in the centrifuge tests is still significantly higher than the measured and computed values, taking into account the adverse effects of infinite panel length and the favourable effects of the low groundwater table used in the tests. In practice, groundwater levels on the sites are within a few metres of the ground surface. Powrie and Kantartzi (1996) suggested that the large ground movements measured in the centrifuge tests are due to the fact that natural clays may exhibit a significantly higher stiffness than the reconstituted kaolin clay used in the experiments because of the ageing and cementing effects in the field. However, another possible explanation for this large discrepancy is that it is due to the creep of the young reconstituted kaolin clay. In the tests, the trench was supported by full concrete pressure ranging from 50 to 70 days before the concrete set. This simulation process in the tests is very different from the one in the field, where a typical concrete panel is completed within one to three days.

Concluding comments

1. Some of the computed results have been compared with some relevant published field records and recent centrifuge model tests. Based on the numerical simulation, significant stress reduction at the centre and above the toe of the panel caused by the installation process is attributed to both the downward load transfer mechanism and the horizontal arching mechanism. These two mechanisms act simultaneously and result only in a gentle increase of horizontal stress below the toe. By comparing the areas of stress reduction and increase above and below the toe of the wall, respectively, it can be deduced that about one-third of the stress reduction above the toe of the panel is attributed to the downward load transfer mechanism, whereas the rest of the stress is redistributed laterally via the horizontal arching mechanism. At the edge of the panel, the stress distributions are complex and the horizontal arching mechanism appears to be dominating, which limits substantial horizontal stress reductions and ground deformations.
2. For the diaphragm wall panel installed in the stiff Gault Clay (assumed $K_0 = 1.5$), the influence of the diaphragm wall construction on the initial stress conditions appears to cease at about 15 m normal from the face of the trench, that is at a distance approximately equal to one panel depth ($1.0D$), at a distance deeper than one-third of the depth of the panel ($\frac{1}{3}D$), and at a distance ($\frac{1}{3}L$) from the edges of the panel.
3. The variation of horizontal ground deformation with depth behind the trench is largest at the centre of the panel during the bentonite stage. The deformation at the centre is about five times larger than that at

the edge. The distributions of horizontal deformations along the panel length are not uniform. However, the horizontal deformations are surprisingly uniform along the depth of the trench during this phase of construction.

4 For the diaphragm wall panel ($L = 8\,\text{m}$) and ground conditions considered here, a gentle settlement trough appears behind the panel, whereas the maximum settlement occurs at a distance of approximately $0.2D$ away from the face of the wall. At a distance of approximately one trench depth ($1D$) away, however, no significant ground settlements have been computed. As expected, the settlement is larger at the centre than at the edge of the panel because of the horizontal arching effects. The settlement ratios between these two locations are found to be about 2 and 1.5 at the bentonite and concreting stages, respectively. Along the length of the panel, a settlement bowl occurs behind the trench. The shape of the settlement bowl is similar at various distances away from the face of the trench. The radius of curvature of the settlement bowl is smaller at the bentonite stage than at the concreting stage, but the depth of the settlement bowl at the former construction stage is about twice as deep as at the latter construction stage. No significant ground settlements have been computed at a distance of a panel length ($1.0L$) from the edge, along the length of the panel.

5 The installation process of a diaphragm wall in stiff clays could result in a significant reduction of horizontal stress above the toe and at the centre of the wall, but with small induced ground movements.

CHAPTER NINETEEN

Effects of modelling soil non-linearity and wall installation on back-analysis of deep excavations in stiff clay

Overview

Practising engineers often work under tight financial restraints. During the design of excavations, engineers would like to know whether a simple linear finite-element analysis is sufficient for a safe design, or whether a complex non-linear numerical analysis provides a more cost-effective solution. This question is further complicated by the recent recognition of wall installation effects, and how and when the effects should be incorporated in the empirical nature of geotechnical design.

This chapter addresses these issues by looking at the case record of a multi-propped excavation in stiff fissured clay at Lion Yard. Numerical simulations using a non-linear model and a linear elastic–perfectly plastic Mohr–Coulomb model, with and without wall installation effects considered, are compared with the field observations. Only those key features in which practising engineers are most interested are discussed.

Numerical analysis

Overview: Mohr–Coulomb and 'brick' models

Three numerical analyses are presented here. These analyses include one non-linear analysis using the 'brick model' (Simpson, 1992), where wall installation has been modelled (WIM), and two linear elastic–perfectly plastic analyses using a Mohr–Coulomb model, one with and one without wall installation being modelled. All the analyses were done using the finite-element program SAFE. Analysis of the excavation using the non-linear brick model with a 'wished-in-place' (WIP) wall was discussed by Ng (1993).

For the simulations presented, the geometry, wall installation and main excavation procedures, and soil parameters except the stiffness were the same for linear and non-linear analyses.

Finite-element mesh

Plane strain conditions were assumed for both Mohr–Coulomb and Brick models. The finite-element mesh adopted in the analyses is shown in

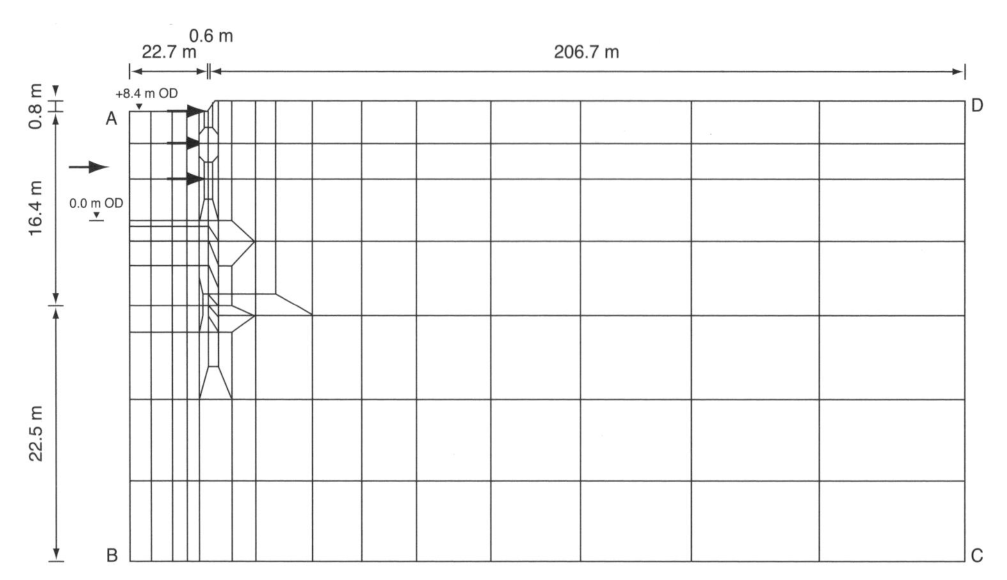

Fig. 2.66 Finite-element mesh for simulating main excavation (after Ng and Lings, 1995)

Fig. 2.66. No slip elements were used at the soil–wall interface, that is, soil elements adjacent to the diaphragm wall were directly connected to the elastic concrete wall elements. The intersection between the Gault Clay and the Lower Greensand was taken as a rigid boundary, BC. The other fixed soil boundary, CD, was assumed at 230 m away from the centre of the site. A vertical sliding boundary, AB, was given to the axis of symmetry in the centre of the site.

Brick model

Simpson (1992) used a physical analogy of a man pulling a number of 'bricks on strings' inside a room (a strain space), to illustrate his idea of modelling the observed recent-history-dependent and path-dependent behaviour of soils at small strains, which are particularly important for stiff clays (Richardson, 1988; Atkinson and Stallebrass, 1991).

The man is taken to represent the point in the strain space (the room) of a soil element, and each brick represents a proportion of the element. The 'strings' are inextensible and their length represents the maximum allowable elastic strain of each 'brick' at a given stress state. Movement of a 'brick' means plastic deformation, and elastic strain developed by a soil element is calculated from the difference between the movement of the man and the sum of the movements of the 'bricks', each weighted by the proportion of the soil it represents. Thus the model simulates the progressive breakdown of elastic behaviour by invoking parallel 'bricks on strings' that slide in the strain space at different shear strain amplitudes when certain proportions of soil reach a fully plastic state.

The brick concept was used in conjunction with a variety of additional equations describing other features of soil behaviour to formulate the model. For instance, elastic volumetric stiffness is assumed to be proportional to current mean normal stress. The incremental elastic mean normal stress $\Delta s'$ is related to elastic incremental volumetric strain Δv_e as follows:

$$\delta s' = \left(\frac{s'}{i}\right)\delta v_e \tag{2.26}$$

where i = initial slope of unload/reload line in $v - \ln s'$ space. By assuming a constant Poisson's ratio μ', the incremental elastic shear stress Δt is related to elastic incremental shear strain $\Delta\gamma_e$ as follows:

$$\delta t = \left(\frac{s'}{i}\right)(1 - 2\mu')\delta\gamma_e \tag{2.27}$$

From Equation (2.26) or (2.27), it can be shown that

$$i = \left(\frac{s'}{G_{max}}\right)(1 - 2\mu') \tag{2.28}$$

where G_{max} is the maximum shear modulus. Full details of the model and its other assumptions are explained by Simpson (1992). In his paper, the use of the model to simulate excavations in Singapore soft clay and stiff London clay is also demonstrated. There is no particular limitation on the use of the model for clays such as Boston Blue clay and Chicago silty clay.

Soil parameters

For the non-linear brick model, an experimental curve that defines the way that shear stiffness varies with shear strain is required. Field measurements of soil stiffness at very small strains on Gault Clay at Madingley (Powell and Butcher, 1991) and results of triaxial stress path tests (Ng, 1992) on Gault Clay from Lion Yard were used in deriving appropriate S-shaped curves relating stiffness and strain level for the numerical analysis. The derived S-shaped curve is shown in Fig. 2.67. A detailed description of the derivation procedures is discussed by Ng (1992). Recent laboratory measurements of small strain stiffness of natural Gault Clay give support to the derived S-shaped curve (Ng *et al.*, 1995a).

The brick model does not require K_0 as an input parameter; instead an estimated preconsolidation pressure for the site is required for simulating geological history, and hence determining the initial lateral stresses in the ground. The actual simulation procedure to derive the initial stress conditions is given later. The angle of shearing resistance ϕ' and the compressibility indices (λ, κ) have been determined from laboratory tests (Ng, 1992) using stress path triaxial and oedometer cells, respectively.

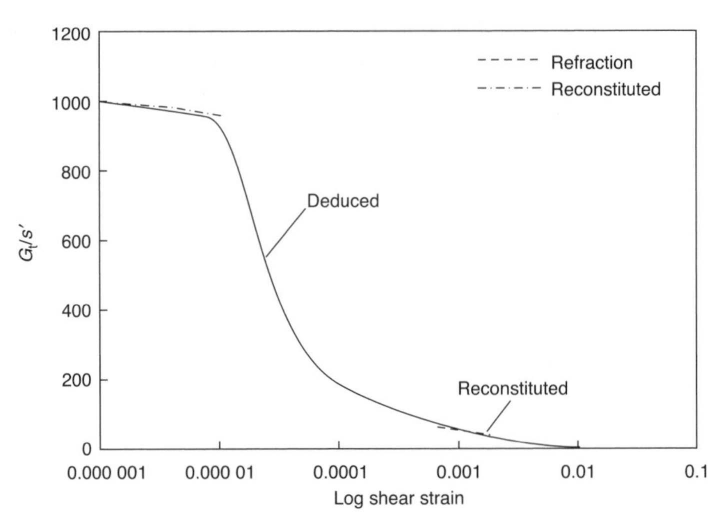

Fig. 2.67 Deduced normalized stiffness–strain relationship for Gault Clay (after Ng and Lings, 1995)

The elastic parameter *i* used in the analysis has been calculated from Equation (2.27) by assuming a Poisson's ratio μ' of 0.2. The maximum shear modulus G_{max} was deduced from the refraction test results (Powell and Butcher, 1991). A summary of the soil parameters used in analysis using the brick model is given in Table 2.4. The initial groundwater conditions were specified according to the results of site investigations.

Table 2.4 Soil parameters for Gault Clay for analysis using the Brick model (after Ng and Lings, 1995)

Description	Parameter
(a) Gault clay	
Slope of ID compression line in v-ln s' space	$\lambda = 0.072$
Slope of unload/reload in v-ln s' space	$\kappa = 0.018$
Slope of initial unload/reload in v-ln s' space	$\iota = 0.0006$
Poisson's ratio for effective stress analysis	$\mu' = 0.2$
Material parameter	$\beta = 3$
Preconsolidation pressure	$s'_{max} = 3200$ kPa
Unit weight of Gault Clay	$\gamma_s = 20$ kN/m^3
Angle of friction	$\phi'_s = 24°$, $\phi'_p = 34°$
Stress–strain characteristics as shown in Fig. 2.67	
(b) Gravel and made ground	
Angle of friction	$\phi' = 40°$
Drained Young's modulus	$E' = 50\,000$ kPa
Unit weight of gravel and made ground	$\gamma_s = 18$ kN/m^3
Coefficient of earth pressure at rest	$K_0 = 0.35$

Table 2.5 Soil parameters for Gault Clay for analysis using the Mohr–Coulomb model (after Ng and Lings, 1995)

Description	Parameter
Unit weight of Gault Clay	$\gamma_s = 20\,\text{kN/m}^3$
Coefficient of earth pressure at rest	$K_0 = 1.5$
Undrained shear strength	$c_u = 60 + 10z\,\text{kPa}$
G/c_u	170
Poisson's ratio for effective stress analysis	$\mu' = 0.2$
Angle of friction for soil elements at the soil–wall interface	$\phi'_p = 34°$
Angle of dilation	0

For analyses using the Mohr–Coulomb model, a normalized stiffness ratio $G/c_u = 170$ was selected for simulating the multi-propped excavation with the aim of matching the measured displacement of the wall at the final stage of main excavation. The selection of the soil stiffness ratio is described by Ng and Lings (1995). The use of the stiffness ratio approach has the effect of modelling soil stiffness increasing with depth, since $c_u = 60 + 10z\,\text{kPa}$ was found at Lion Yard. Other soil parameters used for the Mohr–Coulomb model are listed in Table 2.5.

Drainage assumptions

For both models, soil is assumed to be undrained inside the site and much of the ground outside the site, except near the soil–wall interface (2.9 m wide) where measured pore pressures are explicitly specified for each stage of the excavation simulations. This assumption is considered appropriate for diaphragm wall construction on this site, where bleed channels were observed on the exposed faces of the diaphragm wall and fissures possibly opened in the clay due to active stress relief during excavation.

Simulation procedures

Simulation of geological history at Lion Yard

To model the initial K_0 condition on this site using the brick model, an additional stage was set up to mimic the geological history of the site by consolidating the Gault Clay one-dimensionally to its estimated preconsolidation pressure, followed by erosion, and then reloading by made ground and gravel. This gave an approximately linear variation of K_0 values with depth, except for the first 2 m of clay, where K_0 values less than 1.0 were computed. Below these 2 m, K_0 values increased from 1.0 to 1.75 over the depth of 17 m.

For the analyses with the Mohr–Coulomb model, a constant K_0 value of 1.5 was specified for the Gault Clay before installing the diaphragm wall at Lion Yard. This value was estimated by considering the geological history of the site and measurements from site investigation reports (Ng, 1992), and it was consistent with the average computed K_0 value by the brick model.

Modelling stress reduction as a result of diaphragm walling

Before simulating the main excavation at Lion Yard, the stress reduction due to wall installation was modelled by imposing the measured earth pressures (after wall installation) inside the trench and specifying the measured pore pressures in the surrounding soil. The reliability of the earth pressure measurements has been assessed in the light of measured strut loads and wall rotations (Lings *et al.*, 1993). Six out of seven earth pressure cells appeared to record readings within ±20% of data inferred from other types of instrumentation.

Subsequently the trench was replaced by concrete elements and the imposed pressures were removed for simulating the hardening of concrete in the diaphragm wall.

Main excavation simulation

Diaphragm wall stiffness: the 600 mm concrete diaphragm wall was modelled by a number of linear elastic elements with a Young's modulus of 31 MPa. The Young's modulus used was determined from the test results of concrete cores taken from the actual wall itself (Ng, 1992). No cracking of the wall was modelled, which may not be strictly correct for small parts of the wall at the final stage of excavation (Ng *et al.*, 1992). The use of an uncracked section in this analysis will not alter any conclusions arising from the present investigations.

Prop stiffness: five temporary steel struts at each level were installed in the vehicle opening to support Panel 22 during the main excavation (see Figs 2.14 and 2.15). The strut loads were transmitted to the concrete slab. By considering the force equilibrium between the concrete slab and the steel struts, a simple equivalent stiffness for the whole supporting system can be derived as follows:

$$\frac{1}{K_{\text{equiv}}} = \frac{1}{K_c} + \frac{1}{K_s} \tag{2.29}$$

Concrete creep and shrinkage were allowed for in estimating the slab stiffness in accordance with guidance provided by BS 8110 (BSI, 1985). The estimated equivalent strut stiffness was 36 600 kN/m for the Level 4 support system, and 57 600 kN/m for the rest of the support system at Level 2 and Level 3.

Simulation of construction: the actual construction sequence was simulated by following these steps:

1 Model the effects of geological history and wall installation as described earlier.
2 Excavate to Level 3 by replacing relevant soil elements with air elements which have zero densities and stiffnesses.
3 Install temporary struts at Level 4 and Level 3 by using elastic spring elements (no preload).
4 Continue to excavate to Level 2 by further replacing soil elements with air elements and then install temporary struts at Level 2 ready for the last stage of excavation.
5 Excavate to the final basement level.

Comparison of linear and non-linear analyses with Wall-Installation-Modelled (WIM)

Deflection of wall

When the linear elastic–perfectly plastic model with the Mohr–Coulomb yield criterion was used to analyse the excavation at Lion Yard, it was found that the numerical predictions were mainly governed by the linear elastic component of the model as only a small local zone reached yield. However, for ease of comparison and discussion, the term Mohr–Coulomb model will still be used to distinguish it from the non-linear brick model. Figure 2.68 shows a comparison of the measured and the computed wall displacements using the brick and the Mohr–Coulomb (MC) models. During the three stages of excavation, there is no significant

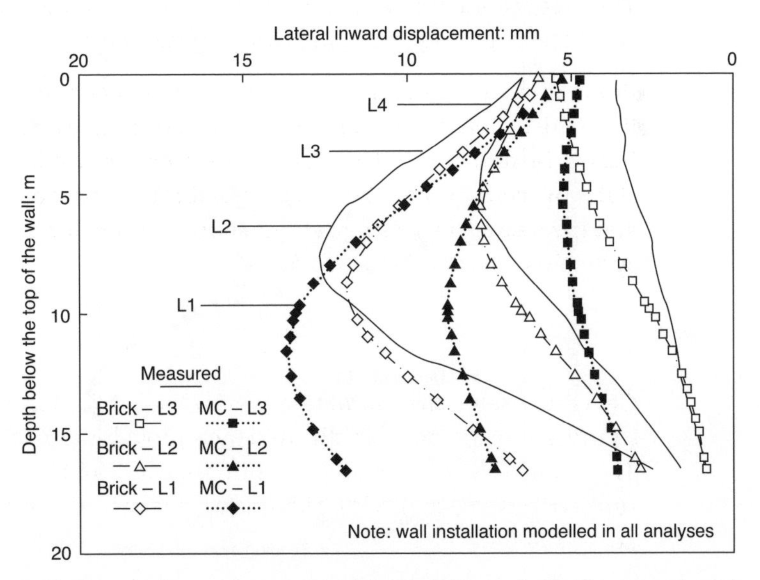

Fig. 2.68 Comparison of computed lateral inward displacements using Brick and Mohr–Coulomb (MC) models (after Ng and Lings, 1995)

difference between the two analyses for wall deformations above each excavated level. However, there is a significant difference in wall deformation computed by the two models beneath each excavated level. Correct computation of the wall deformation beneath the excavated level is vital, since it governs the magnitude of wall curvature and hence the maximum bending moment.

For practising engineers, a major concern is the maximum wall deformation after main excavation. It appears that the simple Mohr–Coulomb model can be a useful tool for obtaining a reasonable calculation of maximum wall deformation, provided an appropriate mean soil stiffness is chosen. This aspect is considered further by Ng and Lings (1995).

Bending moments

Figure 2.69 shows the comparison of the measured and the computed bending moments. Although the maximum wall displacement was reasonably estimated by the simple Mohr–Coulomb model, the curvature of the wall was considerably underestimated and hence so were the bending moments.

In routine design, an uncracked concrete stiffness is normally assumed in order to generate the maximum bending moment for a given curvature of the wall and thus be conservative. Since the curvature of the wall is likely to be significantly underestimated by the Mohr–Coulomb model for a given soil stiffness, the choice of an uncracked concrete stiffness may not by itself lead to a conservative design.

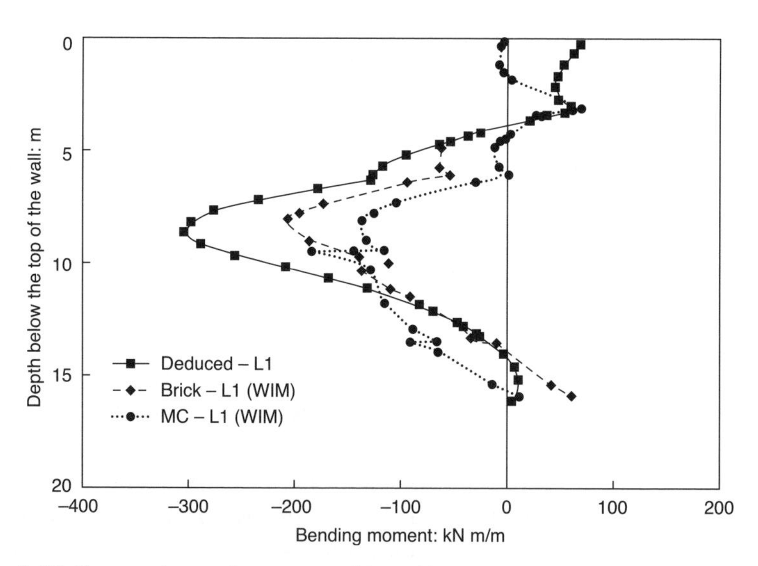

Fig. 2.69 Comparison of computed bending moments using Brick and Mohr–Coulomb models (after Ng and Lings, 1995)

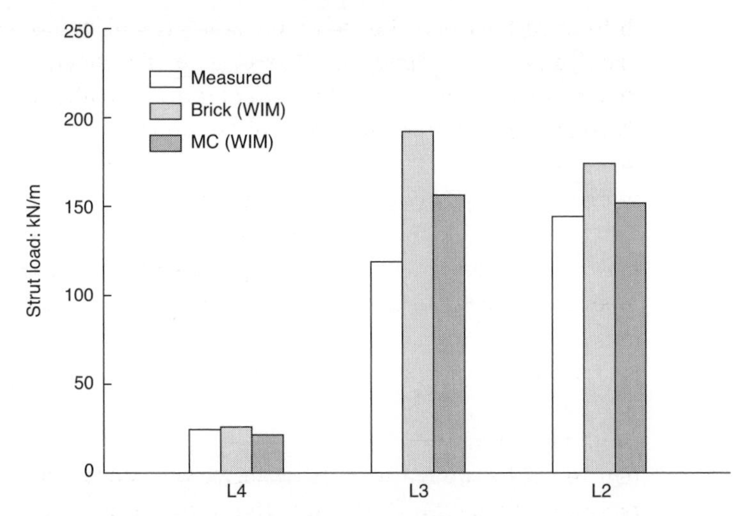

Fig. 2.70 Comparison of computed strut loads at end of excavation using Brick and Mohr–Coulomb (MC) models (after Ng and Lings, 1995)

Strut loads

Figure 2.70 shows the comparison of the measured and computed strut loads at the end of the main excavation. The measured value shown at each level was the maximum during the entire construction period. Lings *et al.* (1991) converted these measurements into apparent pressure diagrams and compared them with the suggested range of design values given by Peck (1969). These showed that the measured maximum loads at Lion Yard approach Peck's suggested lower bound. The strut loads computed by the Mohr–Coulomb model correspond very well with the measured values. As far as strut loads are concerned, there appears to be no advantage in using a non-linear soil model over a simple linear elastic–perfectly plastic one when wall installation effects have been modelled.

Soil movements behind wall

Practising engineers are generally very concerned with the ground movements around deep excavations in urban areas. The consequence of under-predicting the magnitude and extent of ground movements could be very serious, particularly when there are buildings close to the edge of the excavation. In Fig. 2.71 the measured ground movements during main excavation behind the wall are compared with the computed values given by the Brick and Mohr–Coulomb models. It should be noted that ground movements due to wall installation and piling were excluded because the initial reference data set for surveying stations were taken at different times and in many cases after the start of

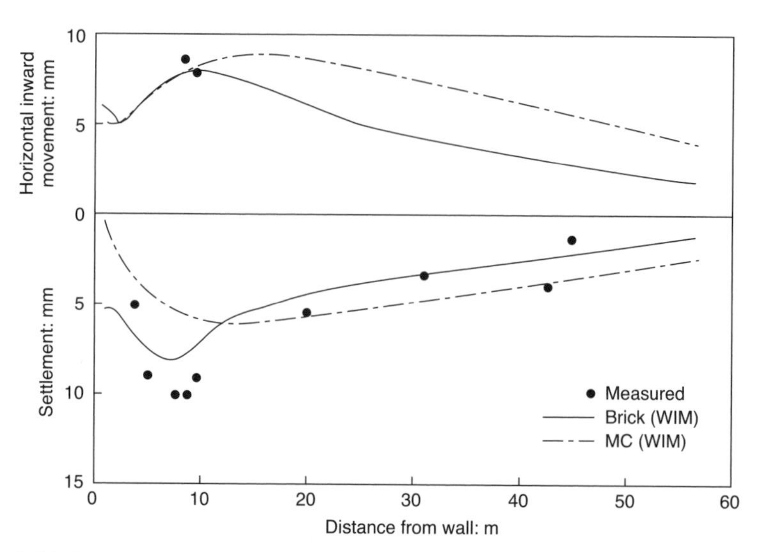

Fig. 2.71 Comparison of computed ground movements at end of excavation using Brick and Mohr–Coulomb (MC) models (after Ng and Lings, 1995)

diaphragm walling. It has thus been rather difficult to make consistent comparisons and the surveying results just before the main excavation have been chosen as a single reference datum for all stations. The general deformation pattern is not significantly altered by this 'shift of origin', since the ground movements during diaphragm walling and piling were small (Ng, 1992). For consistency, the computed ground movements presented in the diagram are only those arising from main excavation.

Although the settlements close to the excavation were underestimated by both models, as shown in Fig. 2.71, it is in this region that Mohr–Coulomb is least satisfactory. The most significant weakness is that the Mohr–Coulomb model failed to predict a 'settlement trough' behind the wall. The result of this is that, between 7 m and 14 m from the wall, the computed slope of the settlement profile is of opposite sign to that given by the non-linear model and the field measurements. In practice this means that if there had been a building 10 m away from the Lion Yard site, the Mohr–Coulomb model suggests that it would have tilted away from the excavation, whereas the brick model and the field data suggest the opposite. This has also been illustrated by Simpson *et al.* (1979), when they back-analysed the tilting of the Big Ben Tower at New Palace Yard.

Another weakness of the Mohr–Coulomb model is that the point of maximum settlement is computed to occur nearly twice as far away from the wall as the brick model simulation. Thus any predictions of settlement close to the edge of the excavation should be treated with caution in design when the Mohr–Coulomb model is used.

In spite of the poor agreement between the two settlement computations within 10 m of the wall, it is surprising to note that the computed lateral inward surface movements in this region are very similar using the Mohr–Coulomb and the brick models. The computed lateral movement curves both pass through the two measured data points. Detailed comparison between the computed and measured lateral inward ground movements is not possible because of insufficient measured data. Further away from the wall there is a general trend for the Mohr–Coulomb model to overestimate both horizontal inward movements and settlements when compared with the brick model.

CHAPTER TWENTY

Comparison of Wall-Installation-Modelled (WIM) and Wished-In-Place (WIP) analyses with Mohr–Coulomb model

Overview

Traditionally, wall installation has been ignored in design because of a lack of understanding of its effects, and also the tendency of designers to prefer conservative predictions. Hence the WIP approach has been used for many years.

In Fig. 2.72 the measured wall deformation at the end of excavation is compared with the computations given by two Mohr–Coulomb (MC) analyses for the Lion Yard case considered in the previous chapter, Chapter 19. The WIP analysis computed larger wall displacements than WIM analysis. More importantly, the WIP analysis computed substantially larger wall curvature than the WIM analysis. This leads to the major

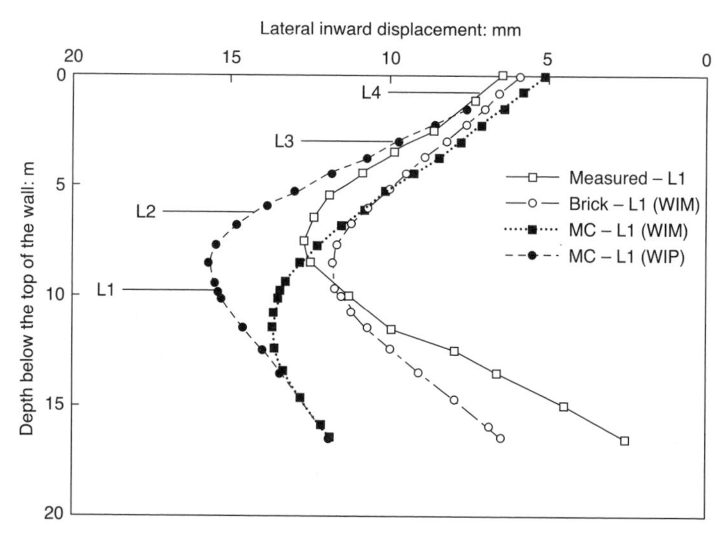

Fig. 2.72 Comparison of computed wall deflections with Wall-Installation-Modelled (WIM) and with wall installation not modelled (Wished-In-Place – WIP) using Brick and Mohr–Coulomb (MC) models (after Ng and Lings, 1995)

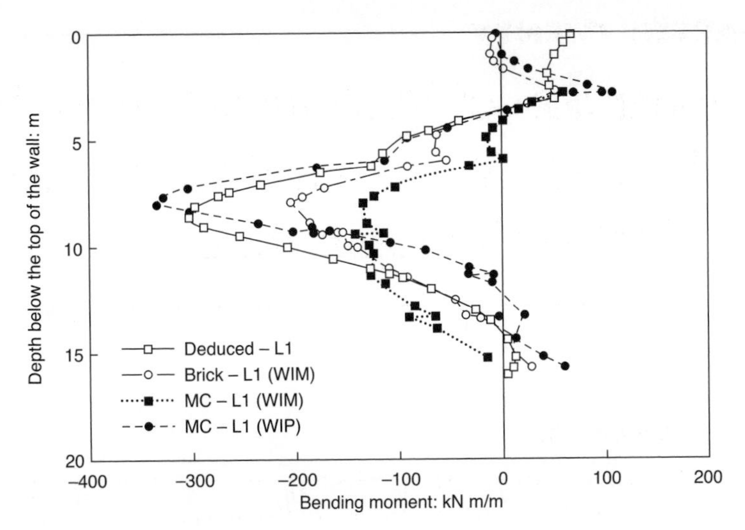

Fig. 2.73 Comparison of computed bending moments with Wall-Installation-Modelled (WIM) and with wall installation not modelled (Wished-In-Place – WIP) using Brick and Mohr–Coulomb (MC) models (after Ng and Lings, 1995)

difference in bending moment computations between the two analyses shown in Fig. 2.73.

A comparison between the measured and computed strut loads is shown in Fig. 2.74. The WIP analysis overestimated the measured strut loads by up to a factor of 2.

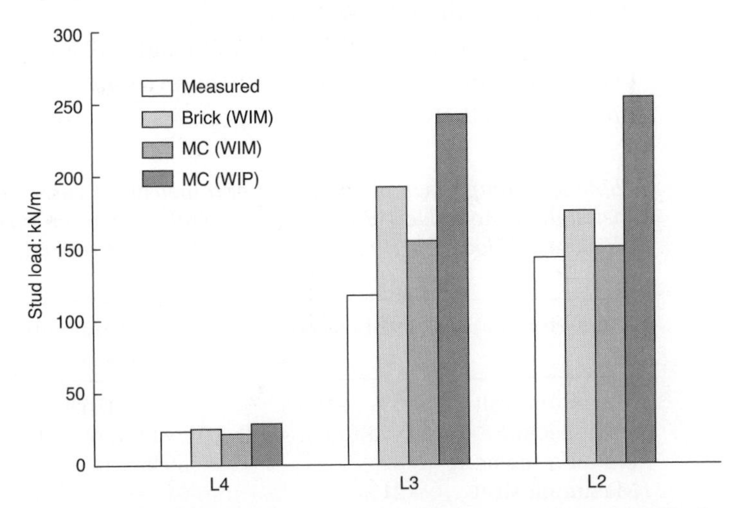

Fig. 2.74 Comparison of computed strut loads with Wall-Installation-Modelled (WIM) and with wall installation not modelled (Wished-In-Place – WIP) using Brick and Mohr–Coulomb (MC) models (after Ng and Lings, 1995)

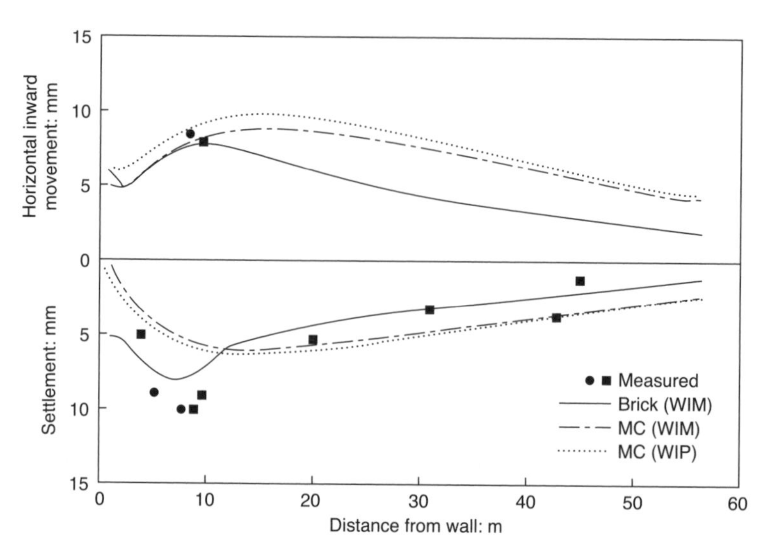

Fig. 2.75 Comparison of computed ground movements with Wall-Installation-Modelled (WIM) and with wall installation not modelled (Wished-In-Place – WIP) using Brick and Mohr–Coulomb (MC) models (after Ng and Lings, 1995)

Figure 2.75 shows the computed and measured ground movements due to main excavation behind the wall. When the Mohr–Coulomb model is used, there is no practical difference between the WIM and WIP analyses, although the WIP analysis computed about 10% larger lateral inward displacements than the WIM analysis.

A summary of the analyses with WIM and WIP is given in Table 2.6. It is evident that the results from the WIP analysis are conservative, in general, provided that there are no buildings close to the excavation. However, the

Table 2.6 Comparison of computed and measured values with Wall-Installation-Modelled (WIM) and with wall installation not modelled (Wished-In-Place – WIP) using Brick and Mohr–Coulomb (MC) models (after Ng and Lings, 1995)

Parameter	Brick–WIM	Mohr–Coulomb v. WIM	Mohr–Coulomb v. WIP
Maximum wall deflection	−6% (+160% at toe of wall)	+8% (+370% at toe of wall)	+24% (+370% at toe of wall)
Bending moment	−32%	−56%	+10%
Maximum strut load	+21%	+5%	+176%
Settlement	Good	Poor ground profile	Poor ground profile

Note: −ve sign means underestimate; +ve means overestimate compared with the measured values.

results may be much too conservative with respect to strut loads as a result of significant overestimation of horizontal earth pressures.

The WIM analysis can substantially improve the computations of strut-loads, wall deflection and earth pressures, but badly under-computes the bending moment of the wall. No improvement is made on the computation of ground movements behind the wall compared with the WIP analysis.

Assessment of mobilized soil stiffness

Selection of single representative soil stiffness

Results from the WIM analysis using the non-linear brick model have been extracted to produce 'stiffness paths' for some typical soil elements around the excavation to give insights into the ground behaviour. Through understanding the ground behaviour, an appropriate average soil stiffness was selected for analysis using the Mohr–Coulomb model as described in this book.

Figure 2.76 shows the variation of normalized stiffness ratio G/c_u with lateral stress reduction ratio $\Delta\sigma_h/\sigma_{h0}$ for some typical soil elements at certain critical locations around the excavation. The lateral stress reduction ratio is defined as the change of total horizontal stress of a particular soil element over its initial total horizontal stress before construction. The diagram effectively illustrates the operational soil stiffness at various locations during construction.

It can be seen that the soil stiffness adjacent to the wall reduced substantially after the wall installation process, whereas the stiffness of

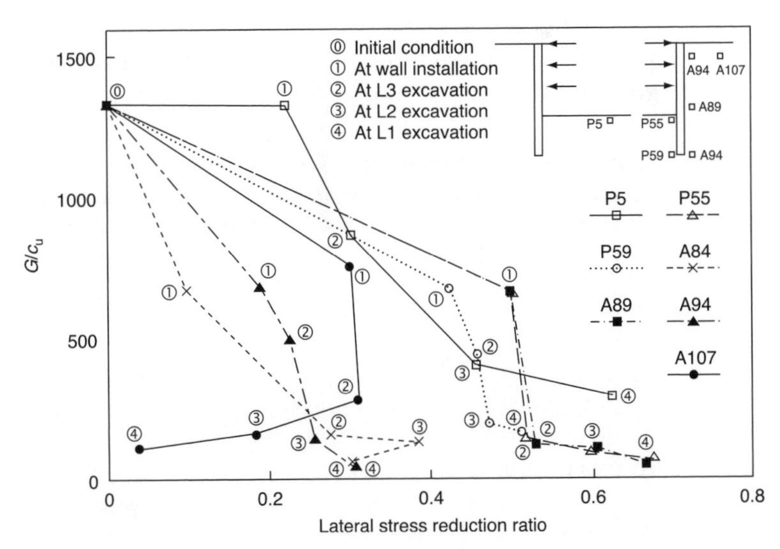

Fig. 2.76 Normalized stiffness paths for soil elements around the site (after Ng and Lings, 1995)

soil elements further away from the wall such as P5 were not affected by this operation. As the main excavation proceeded, the normalized stiffness ratio G/c_u of the soil elements around the excavation continued to decrease. After examination of the soil stiffness operating after the L2 excavation stage of the main excavation, an average normalized stiffness ratio $G/c_u = 170$ (equivalent to $E/c_u = 500$) was selected with the intention of matching wall displacement at the end of excavation. Detailed explanation of the behaviour of the soil elements shown is given by Ng (1992).

Apart from studying the normalized stiffness paths shown in Fig. 2.76, design engineers can also back-analyse relevant case records to obtain an average shear strain developed around an excavation, from which they can then deduce the corresponding soil stiffness for their numerical analysis through laboratory soil element test data. Figure 2.77 shows that the shear strains developed around the Lion Yard excavation were small and generally less than 0.3%, except in front of the wall just below the final excavated level, where the soil was approaching passive failure. A similar pattern and magnitude of shear strains were predicted by Arup Geotechnics for the multi-propped excavation in stiff London Clay at the British Library at Euston (Atkinson and Sällfors, 1991). These indicate that the moduli adopted for any calculations based on the theory of linear elasticity should be appropriate to the average strain level anticipated during excavation. These two case records suggest

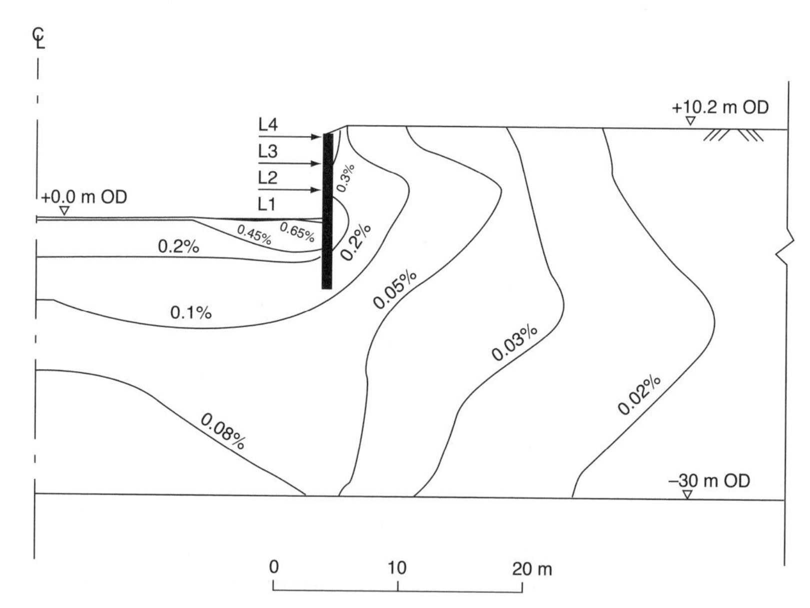

Fig. 2.77 Shear strains around the site at the end of excavation (after Ng and Lings, 1995)

that a laboratory-measured soil stiffness corresponding to an average strain level of 0.1% would give reasonable design predictions for excavations in stiff clays.

Significance of non-linear soil modelling for practising engineers

It is possible to select an appropriate mean soil stiffness in a linear elastic–perfectly plastic analysis, where wall installation is modelled, to predict wall deformation and strut loads that agree reasonably well with the field observations for a particular stage of construction. However, the linear analysis can only produce a good 'snapshot' prediction for a particular construction stage. For instance, good predictions of wall deformation at the end of the excavation inevitably result in over-prediction of the displacements of the wall at the other stages of excavation if the same soil stiffness is used throughout in all stages. More importantly, the bending moment in the wall and ground movements behind the wall are likely to be poorly predicted. This is because the stiffness of soil is not allowed to decrease as shear strain increases. Generally, soil strains become disproportionately greater as shear stresses increase, and hence the deformations become much more concentrated, and form a 'settlement trough' behind a wall.

As shown in Fig. 2.68, the maximum wall deformation was reasonably well computed by the Mohr–Coulomb model at the last stage of excavation using $G/c_u \approx 170$, but the computed lateral movement beneath the final excavated level was excessively large compared with the field observations. A very similar picture in the reported case record at New Palace Yard in London clay has been shown by Burland and Hancock (1977, Fig. 21). The discrepancy between the measured and the computed performance was probably due to the soil at a deeper level operating at a higher mean stiffness ratio than the value assumed. The problem is clearly better accounted for using the non-linear model than the linear elastic–perfectly plastic model.

Of course, non-linear modelling is more involved than the simple linear elastic–perfectly plastic analysis in terms of soil parameters and understanding of numerical modelling and non-linear soil behaviour. Thus the selection of linear or non-linear analysis clearly depends on what resources are available and how good a prediction is required.

Traditional Mohr–Coulomb 'Wished-In-Place' analysis

It is now widely accepted that wall installation in stiff over-consolidated clays does alter the in situ stresses. The results from the WIM–WIP comparison show the effect that modelling installation has on wall deflections, bending moments and strut loads (refer to Table 2.6).

However, a comparison of the computed wall deflections from the brick (WIM) and Mohr–Coulomb (WIP) analyses in Fig. 2.72 suggests the overall deflected shape is surprisingly similar. In fact, it appears that had a higher normalized stiffness ratio G/c_u been adopted in the traditional Mohr–Coulomb (WIP) analysis, quite a reasonable fit for the deflected shape and bending moment might have been obtained. However, compared to the measured field data (and the more sophisticated Brick (WIM) analysis), it would still overestimate the strut loads, and fail to produce the settlement trough behind the wall.

This might explain why the traditional analytical approach has been used with reasonable success in various back-analyses in the past. It appears that, as far as wall deflection and bending moment are concerned, the effects of modelling soil non-linearity and stress reduction due to wall installation largely cancel each other out. For the Lion Yard site it appears that there is no advantage in modelling wall installation, unless at the same time a non-linear soil model is incorporated.

Concluding comments

1. The measured performance of a multi-propped excavation at Lion Yard was back-analysed using the Brick and the Mohr–Coulomb models. The use of a non-linear model with Wall Installation Modelled has generally shown good agreement with the field observations. The Mohr–Coulomb model where wall installation is modelled significantly underestimates bending moment and results in a poor profile of ground deformation behind the wall.
2. The traditional Mohr–Coulomb model with a 'Wished-In-Place' wall can give reasonable wall deflections and bending moments, but will at the same time significantly overestimate earth pressures and maximum strut loads, and also result in a poor profile of ground deformation behind the wall. Unless a non-linear soil model is incorporated at the same time, there appears to be no advantage in modelling wall installation.
3. When using simple linear soil models such as the Mohr–Coulomb, the correct assessment of appropriate mean soil stiffness for the soil is the determining factor in the successful application of the model. This can be done by back-analysing relevant case records to study the development of 'stiffness paths' or shear strain contours.

CHAPTER TWENTY-ONE

Prediction of ground movements due to diaphragm wall installation

Overview

Predictions of ground movement due to the installation of a diaphragm wall are also commonly conducted by adopting a two-dimensional (2D) analysis with plane strain assumptions. The appropriateness of this conventional vertical plane section analysis for modelling diaphragm wall installation effects on ground deformations is very questionable. This is because the complex three-dimensional (3D) installation process cannot be correctly simulated in the conventional 2D analysis.

Elastic solutions

Classic closed form solution

Jaeger and Cook (1969) presented a closed-form solution to examine displacement caused by the formation of a thin elliptical slot in an elastic stressed plate. The displacement (Δh) normal to the major axis of the slot is given by

$$\Delta h \approx \frac{L\Delta\sigma}{2E}\left\{(1-\nu)\left[\left(1+\frac{(2x)^2}{L^2}\right)^{1/2}-\frac{2x}{L}\right]+(1+\nu)\left(1+\frac{(2x)^2}{L^2}\right)^{-1/2}\right\} \tag{2.30}$$

where $\Delta\sigma$ is the stress change associated with forming the slot, L is the length of the slot, x is the distance away from the slot, ν is the Poisson's ratio and E is the Young's modulus of the material.

After simplification by putting $\nu = 0.5$ and $x = 0$, Equation (2.30) can be reduced to

$$\Delta h \approx \frac{L\Delta\sigma}{E} \tag{2.31}$$

During the installation of a diaphragm wall panel, the stress change $\Delta\sigma$ is the difference between the initial K_0 pressure and the hydrostatic pressure of bentonite slurry, which is normally known once the unit weight of the bentonite is measured. The unit weight lies within a narrow range, typically varying from 10.1 to 10.8 kN/m^3. Thus, it can be deduced from the above equation that horizontal displacements induced during the construction of a diaphragm wall panel are governed by the initial K_0 value in the ground and the length of the panel. For a given distribution of Young's modulus with depth, the longer the panel length and the larger the initial stress, the larger the induced displacements are expected to be.

Explicit analytical solution

Lei *et al.* (2001) and Ng and Lei (2003b) have developed a new, simple and explicit analytical solution for calculating horizontal stress changes and displacements caused by the excavation for a diaphragm wall panel. The theoretical solution is obtained by applying the principle of superposition appropriately to model diaphragm wall construction using a basic elastic solution to the problem of an infinite horizontal plate with a rectangular opening subjected to a uniaxial stress at infinity. The basic elastic solution can be obtained by using the method of complex variables with a simplified conformal transformation function. Key parameters governing the magnitude of horizontal stress changes and displacements are identified. Computed results are given in a normalized form in terms of aspect ratio (length to width) of a diaphragm wall panel. Two extreme cases for diaphragm wall panels with dimensions $1\,\text{m} \times 1\,\text{m}$ and $10\,\text{m} \times 1\,\text{m}$ have been analysed to investigate the distributions of stress changes and deformations around the panels during the bentonite stage. By performing a parametric study, calculation charts have been developed for computing horizontal stress changes and displacements for practising engineers and researchers to carry out preliminary designs and for numerical modellers to verify their sophisticated predictions. The ease of use of the charts is illustrated by two examples, and the limitations of the derived solutions are discussed.

Although the above elastic equations can provide some insights to estimate ground movements due to diaphragm wall installation, the formation of a trench for constructing the wall may be expected to cause larger ground movements due to the development of plastic zones. It was therefore more desirable to conduct some numerical analyses to predict ground deformations under various conditions.

Modelling diaphragm wall installation: methodology

Both 2D and 3D finite-difference analyses were conducted by Ng and Yan (1998b) to study the ground movements during the diaphragm wall panel installation. Two similar finite-difference programs FLAC (Itasca, 1995) and FLAC3D (Itasca, 1996) are used respectively for the 2D and 3D numerical analyses. In terms of the numerical modelling, a 2D vertical plane section (with plane strain assumption) and a 3D mesh are adopted to study the construction of a single panel with infinite panel length and the

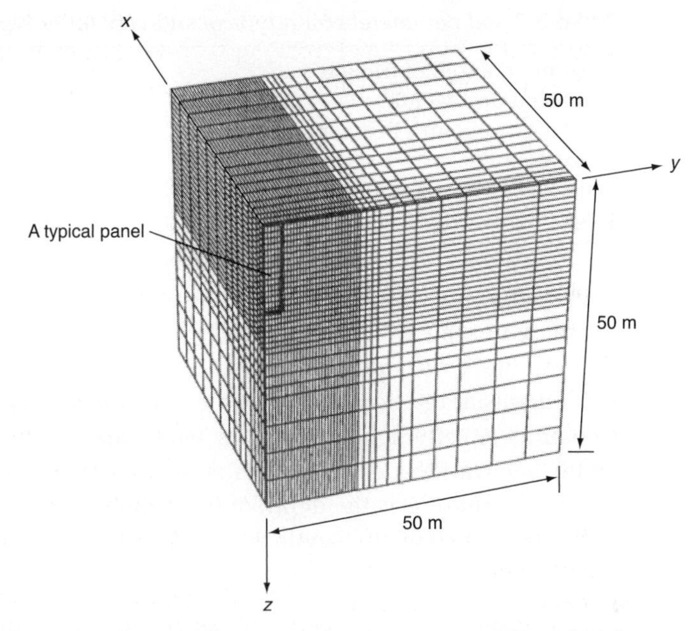

Fig. 2.78 3D finite-difference mesh (after Ng and Yan, 1998b)

construction of panels with different finite lengths separately. Due to the space limitations, only the 3D mesh is illustrated in Fig. 2.78.

The clay is modelled as a linear elastic–perfectly plastic material with a Mohr–Coulomb yield surface. It is well recognized that soil behaves non-linearly even at small strains. The stiffness of soil not only depends on the strain level but also on recent stress history (Atkinson *et al.*, 1990; Atkinson and Sällfors, 1991). Some limitations of adopting the linear elastic–perfectly plastic model for the analyses are discussed later in this book.

By considering mechanical properties for some typical stiff clays (Simpson *et al.*, 1979; Ng, 1992), the model parameters adopted for a stiff clay in the analyses are summarized in Table 2.7. The undrained shear strength and Young's modulus of the clay are both assumed to increase linearly with depth. Since the duration of a panel construction is usually short and the water permeability in the clay is low, the clay is assumed to remain undrained and a total stress approach is therefore adopted for the analysis.

For the 2D analyses, a line of symmetry is used in generating a 2D finite-difference mesh simulating an infinitely long, 0.6 m wide and 15 m deep diaphragm wall panel. The vertical boundary is allowed to move freely in the vertical direction (z-direction). At the bottom boundary, all movements are restrained.

For the 3D analyses, two planes of symmetry ($x = 0$ and $y = 0$) are used in generating a finite-difference mesh (see Fig. 2.78). The planes, $x = 0$

Table 2.7 Soil parameters for a typical stiff clay (after Ng and Yan, 1998b)

Undrained angle of friction	$\phi_u = 0°$
Saturated unit weight	$\gamma_s = 20\,kN/m^3$
Coefficient of earth pressure at rest	Varying ($K_0 = 1.0$, 1.5 or 2.0)
Poisson's ratio	$\nu = 0.495$
Undrained shear strength	$c_u = 30 + 10z$
Undrained Young's modulus	$E_u/c_u = 1000$

and x = 50, are allowed to move freely in the y- and z-directions but restrained in the x-direction. Similarly, the planes, y = 0 and y = 50, are free to move in the x- and z-direction but not in the y-direction. Movements at the boundary plane z = 50 are restrained in all directions. Construction of single diaphragm wall panels with 15 m depth, 0.6 m width and various lengths (L) is simulated ($L = 1/3D$, $2/3D$, $1D$, $4/3D$ and $5/3D$ where D is the depth of the trench).

The construction of a single panel is modelled as follows:

1. Excavate the trench (0.3 m wide, 15 m deep and length varies) and apply hydrostatic bentonite pressure (assuming unit weight = $10.1\,kN/m^3$) on the trench faces simultaneously.
2. Cast the panel with concrete by increasing the lateral pressure inside the trench to a bi-linear wet concrete pressure envelope as proposed by Ng (1992) and Lings *et al.* (1994).
3. Cure the concrete panel by replacing the trench with elastic concrete elements and remove the applied bi-linear concrete pressures on the trench faces simultaneously.

Only the computed results simulating the excavation of a trench under bentonite are presented and discussed in this book. This is because the bentonite stage is believed to be the most critical stage in terms of ground movements. Moreover, ground movements that may occur at the concreting stage are likely to be relatively small as compared with those occurring at the bentonite stage. Soil behaves much stiffer (Atkinson *et al.*, 1990; Ng, 1992) when there is stress reversal during concreting.

Ground deformations

Predicted movements for $K_0 = 1.5$

Figure 2.79(a) and (b) shows the computed ground movements (both vertical settlement and horizontal displacement) varying with perpendicular distance along the centreline away from the trench. The lines marked $L = \infty$ are computed from the 2D plane strain analysis whereas the others are from the 3D analyses. As shown in the figures, the predicted displacement profiles are very similar in shape for 2D and 3D analysis and the 2D analysis always gives the upper bound prediction as expected.

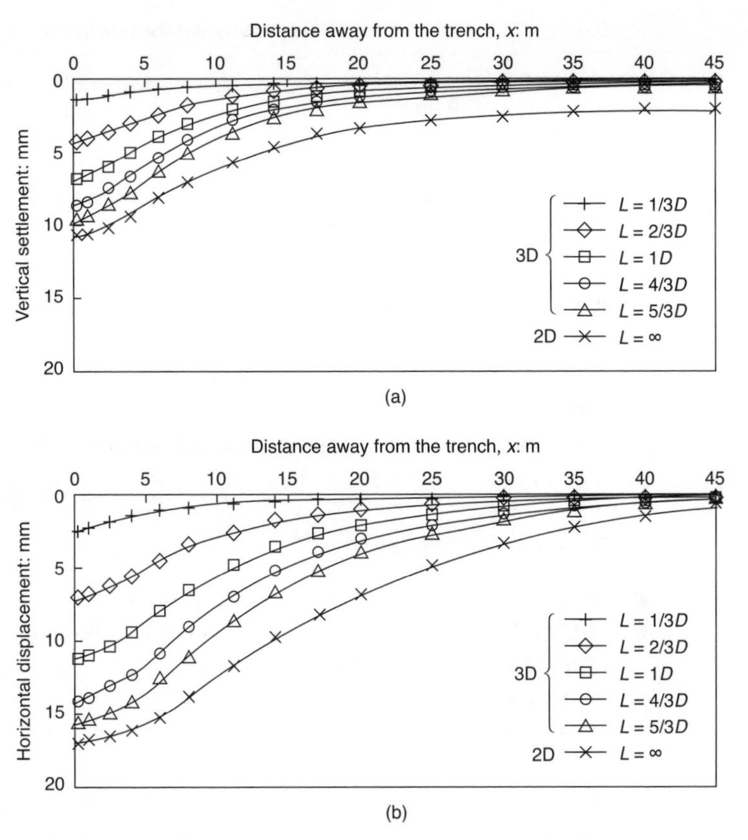

Fig. 2.79 Ground movements at bentonite stage with distance normal to the trench along the centre-line with $K_0 = 1.5$ *(after Ng and Yan, 1998b)*

Both the maximum vertical settlement and horizontal movement occur at the face of the trench. For panels with finite length, the longer the panel, the greater the predicted maximum movement obtained. These agree with the elastic solutions given by Equations (2.30) and (2.31). It is clear that selection of a suitable panel length during construction is crucial for control of ground movements around the panel.

In Fig. 2.79(a), the difference in computed settlements between the 2D and 3D analyses becomes more significant as the distance away from the trench increases. The computed settlements by the 2D analysis are still very significant beyond 30 m (i.e. twice the trench depth) from the trench. This is because the soil arching mechanism, which would reduce soil movements, was completely neglected in the 2D plane strain analysis.

Figure 2.80(a) and (b) shows the computed vertical settlement and horizontal displacement at the corner of the trench varying with perpendicular distance away from the trench. The shapes of the computed

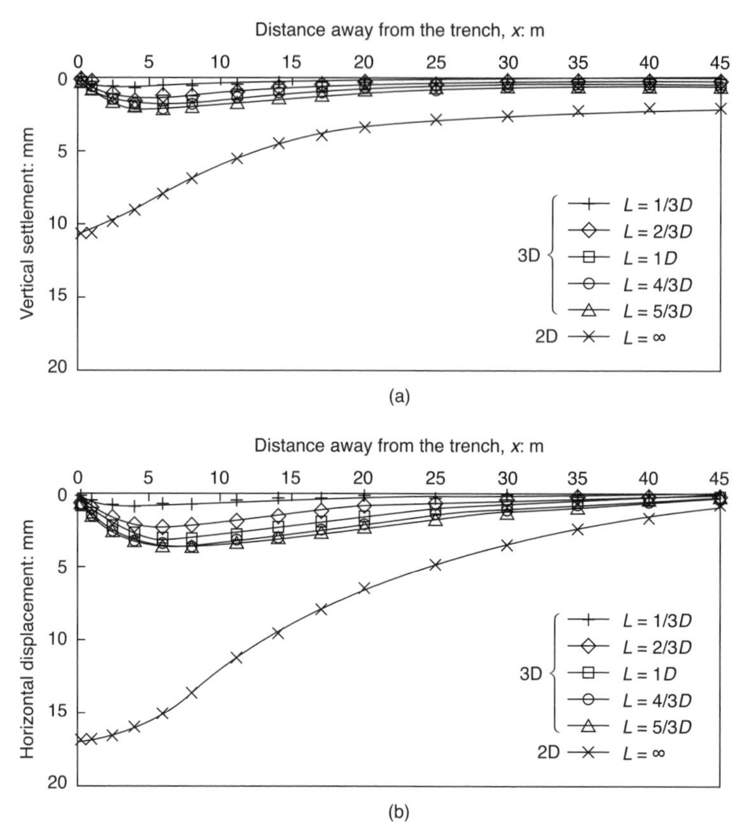

Fig. 2.80 Ground movements at bentonite stage with distance normal to the trench at the corner with $K_0 = 1.5$ *(after Ng and Yan, 1998b)*

displacement profiles between the 2D and 3D analyses are substantially different. For panels with finite length, a settlement trough can be identified, which appears at about 5 m away from the face of the trench. Similarly, the maximum horizontal soil displacement occurs at about 6 m away from the face of the trench. Along the length of a panel, the computed maximum movements at the centre are substantially greater than those at the corner of the trench. Very non-uniform ground deformations behind the trench can be identified.

Mobilized shear strength at $K_0 = 1.5$

The maximum mobilized deviatoric shear stresses $(q = (\sigma_1 - \sigma_3)/2)$ of the soil at the centre and at the corner of the trench during excavation is shown in Fig. 2.81(a) and (b), respectively. It can be seen that the Mohr–Coulomb yielding criterion is not reached at these two sections. At the centre behind the trench, the shear stress induced along the

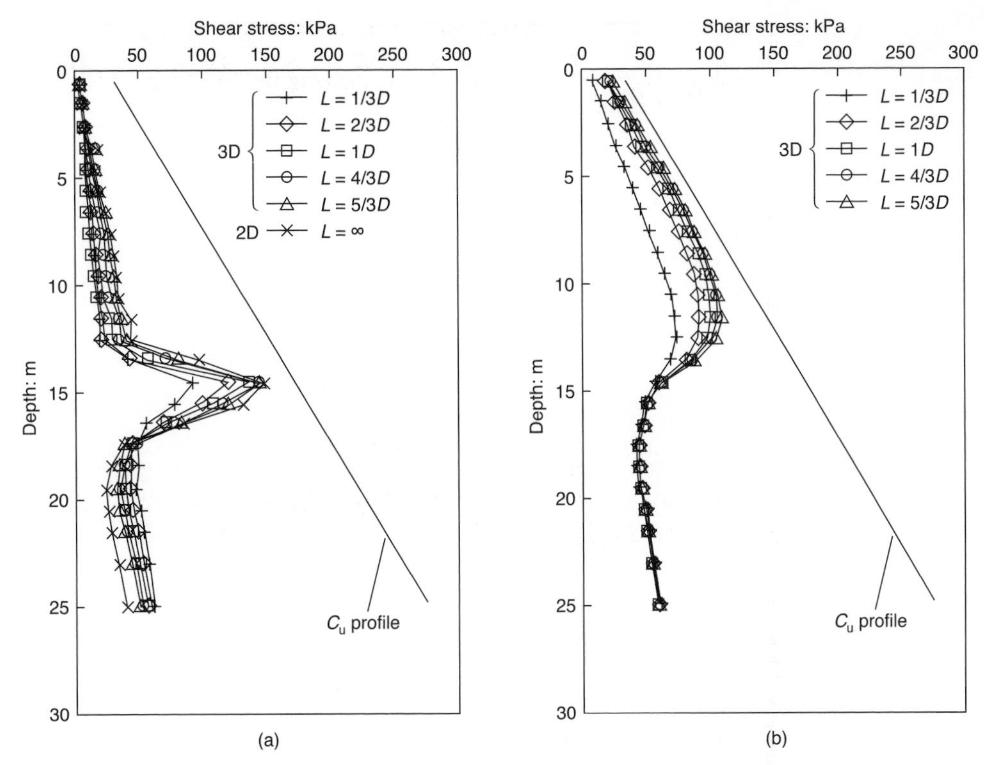

Fig. 2.81 Maximum mobilized deviatoric shear stress at bentonite stage: (a) at the centre of the trench; (b) at the corner of the trench (after Ng and Yan, 1998b)

depth is generally small except near the toe. This is attributed to the downward load transfer mechanism which redistributes the high initial horizontal stress progressively downward via shear stresses during excavation (Ng, 1992; Ng *et al.*, 1995b; Ng and Yan, 1998b). On the contrary, large shear stress is induced along the depth and at the corner of the panel, except near the toe (see Fig. 2.81(b)). The longer the panel length, the higher the shear stress induced. The high shear stress induced at the corner is the result of stress transfer during excavation via the horizontal arching mechanism (Ng, 1992; Ng and Yan, 1998b). Details of the downward load transfer and the horizontal arching mechanisms during the excavation of a single diaphragm wall panel are explained by Ng (1992) and Ng and Yan (1998b).

Influence of initial K_0 values

For stiff clays, the initial horizontal stresses in the ground are likely to be greater than the vertical stresses and the K_0 values may be different from site to site. The influence of K_0 values on soil deformations around a

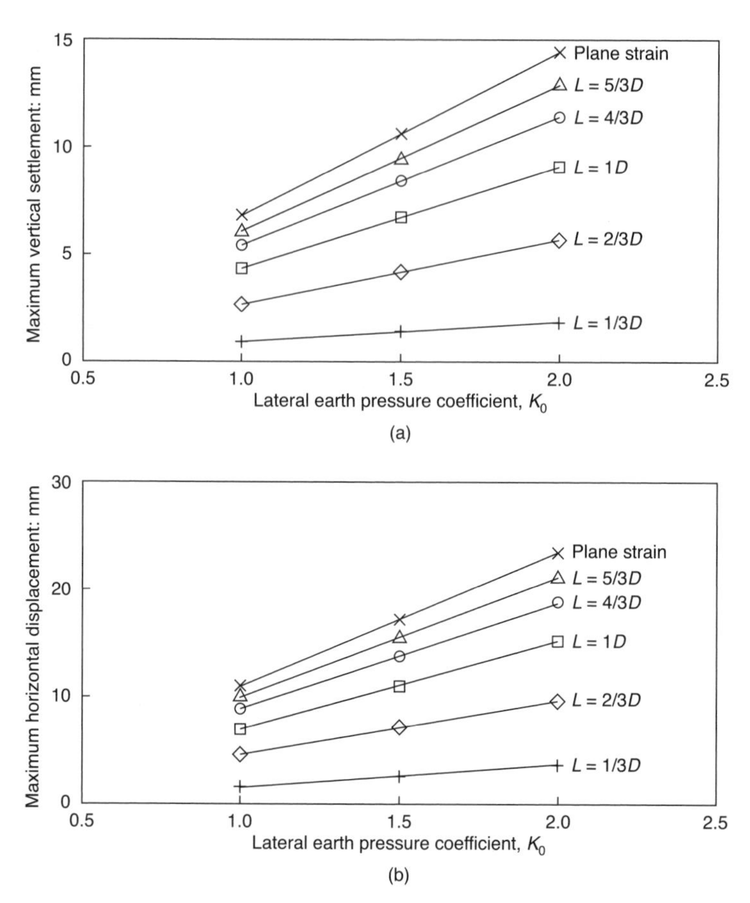

Fig. 2.82 Relationship between initial K_0 *value and ground movement (after Ng and Yan, 1998b)*

diaphragm wall panel has been studied. Figure 2.82(a) and (b) shows the maximum ground movements (vertical settlement and horizontal displacement) obtained from the analyses for different K_0 values. As expected, both the computed settlements and displacements vary linearly with the K_0 value for any panel length. This implies that the soil behaves essentially elastically during the panel excavation and hence the computed results are consistent with the elastic solutions given by Equations (2.30) and (2.31). The higher the initial stress (i.e. higher K_0 value), the larger the stress change and ground movements are expected to be.

Normalization of different panel length (*L*)

Generally, engineers would like to minimize design costs and to avoid performing any complex 3D analysis if possible. In order to assist the engineers, some design charts may be useful. As illustrated by Equation (2.31), the magnitude of ground movements during a diaphragm wall construction is governed by the length of a panel for a given ground condition and bentonite. Normalizing the computed ground movements by the length of a panel seems to be a reasonable approach. However, a problem arises when the panel length is infinite (i.e. 2D plane strain conditions). Since the ground deformations under the plane strain conditions are also proportional to stress change during excavation and soil stiffness, it is possible to normalize the computed 3D soil deformations by the respective computed movements given by a 2D analysis. This enables the engineers to take the 3D effects on ground deformations during diaphragm walling into consideration from a simple 2D analysis. Figure 2.83 shows the normalized ground deformations for the case of $K_0 = 1.5$. The distance away from the trench is normalized by the depth of the

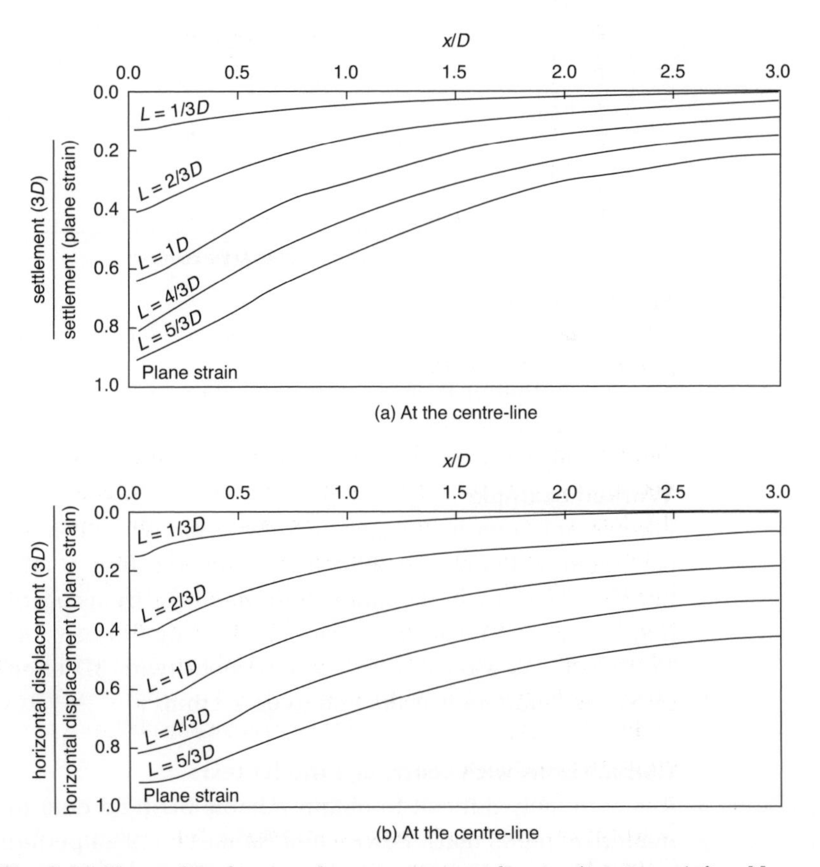

Fig. 2.83 Normalized ground movements at bentonite stage (after Ng and Yan, 1998b)

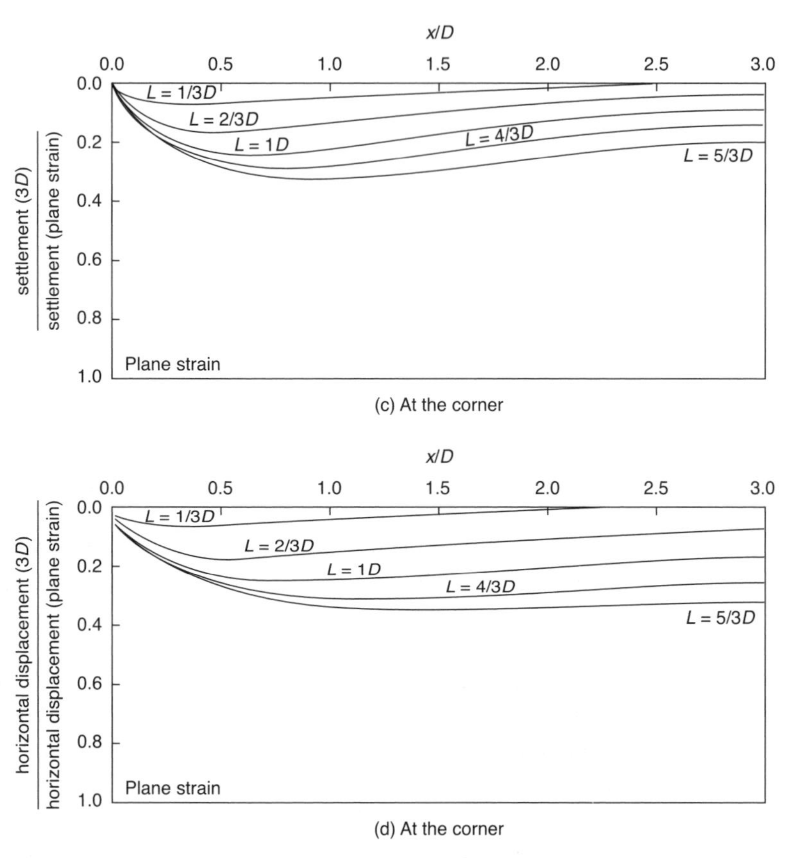

Fig. 2.83 (Continued)

trench (D). Identical charts are also obtained for different K_0 values, as expected (refer to Fig. 2.82).

Worked example

For a diaphragm wall panel with $W = 1$ m, $D = 20$ m and $L = 8$ m, the computed settlement at $0.5D$ away from the trench (along the centre-line) is 10 mm by a 2D plane strain analysis. By using the design chart (Fig. 2.83), a normalization factor can be found to be 0.14 ($x/D = 0.5$ and $L/D = 0.4$). As a result, the 3D settlement (with 3D effects being taken into consideration) is estimated to be 1.4 mm.

Comparisons with centrifuge model tests

It is extremely difficult to obtain reliable 3D field data to compare with numerical predictions. Powrie and Kantartzi (1996) performed a series of centrifuge model tests in an over-consolidated speswhite kaolin to study the effects of a diaphragm wall panel installation on ground deformations

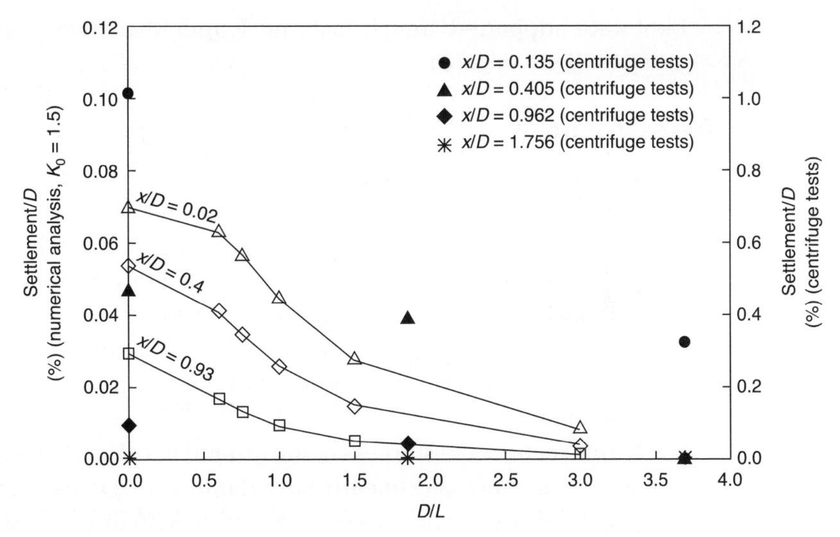

Fig. 2.84 Normalized settlements at bentonite stage along the centre-line of the trench (after Powrie and Kantartzi, 1996; and Ng and Yan, 1998b)

and pore water pressures. The trench modelled was 18.5 m deep and 1 m wide in prototype. A range of panel length was studied, and LVDT displacement transducers were used to measure settlements varying with distance away from the trench both along the centre-line and along the edge. Generally, the longer the panel length modelled, the larger the measured settlements were recorded as expected. Figure 2.84 shows a normalized settlement plot based on the measurements during the centrifuge tests and the numerical predictions presented in this book. It can be seen that the general trend between the two sets of data are consistent. However, the magnitude of settlements reported in the centrifuge tests are about 10 times larger than the corresponding computed values. This seems to indicate that the soil modelled in the numerical simulations is much stiffer than the reconstituted speswhite kaolin used in the centrifuge tests. The undrained shear strength (c_u) of the reconstituted kaolin may be substantially smaller than the assumed values in the numerical simulations. It should be noted that the assumed undrained shear strength in the analyses is based on laboratory and field tests on natural clays which have been aged for millions of years and some bonding may exist in the ground. Moreover, creep of the young reconstituted kaolin clay may be another possible explanation for such a large discrepancy. This is due to the fact that the trench was supported by bentonite slurry for six to seven days (in prototype) before concreting in the centrifuge model tests. The modelled simulation process in the tests is very different from that in the field, where a typical bentonite supported trench lasts no longer than two days. Significant creep is likely to occur.

Assessment

1 It is well recognized that the stress–strain behaviour of soil is highly non-linear at small strains (Atkinson and Sällfors, 1991; Tatsuoka and Kohata, 1994; Ng *et al.*, 1995a; Ng *et al.*, 2000a). The stiffness of soil decreases as strain increases. By using a simple linear elastic–perfectly plastic model, the non-linear behaviour of soils cannot be captured fully in the analyses. It is very important to select an appropriate soil stiffness from laboratory or field tests for numerical simulations.

2 In this section, the stiffness of the ground in both 2D and 3D analyses is assumed to increase linearly with depth with a constant stiffness ratio (E_u/c_u) which is equal to 1000. For soils with different stiffness ratios, the design charts shown in Fig. 2.83 are still applicable, provided that a 2D analysis with appropriate soil stiffness is conducted.

3 The design charts shown in Fig. 2.83 give the design engineer a rapid assessment of how diaphragm wall behaviour assessed by simple 2D analyses can be corrected for 3D effects.

CHAPTER TWENTY-TWO

The prediction and control of displacements around deep excavations in completely decomposed granite

Minimizing displacements

Malone *et al.* (1997) observe that experience in Hong Kong shows that excavations deep below the water table in completely decomposed granite (i.e. material that 'can be broken down between the fingers to its constituent grains and slakes in water') can cause large settlements of the adjacent ground. At sites sensitive to displacements, top-down excavation within a concrete diaphragm wall box is now the favoured construction method for excavation depths greater than around 12 m. However, installation of the diaphragm wall needs to be carried out with care. In the early days of diaphragm walling in completely decomposed granite in Hong Kong, in the late 1970s and early 1980s, large movements were commonly recorded during diaphragm wall construction (Fig. 2.85). Practitioners in Hong Kong have since learnt that, to limit displacements, it is necessary to use short trench panels, take care in excavating and maintain the slurry pressure well above the

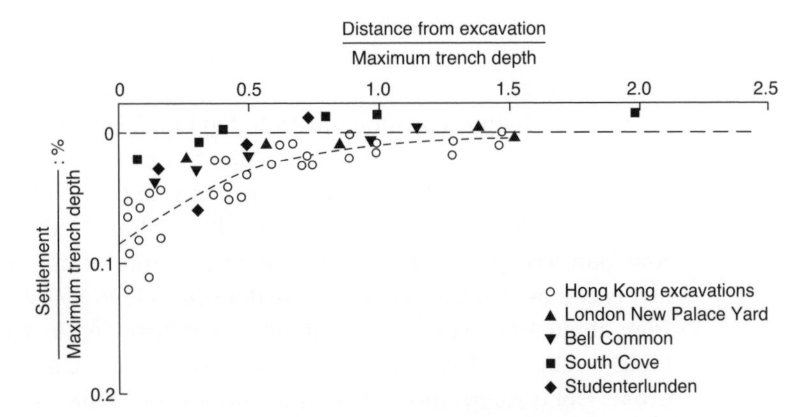

Fig. 2.85 Summary of measured settlements caused by the installation of concrete diaphragm walls (after Clough and O'Rourke, 1990; Malone et al.*, 1997)*

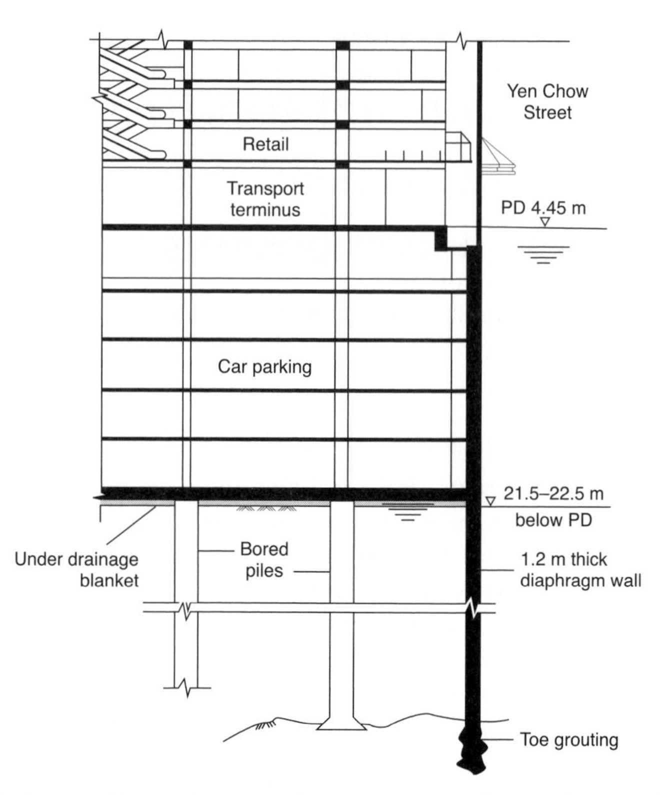

Fig. 2.86 Dragon Centre basement cross-section (after Malone et al.*, 1997)*

groundwater pressure, i.e. rather more than that needed for theoretical static stability. By this means, installation settlements can be limited to say less than about 10 mm. But it remains a challenge to keep displacements due to excavation and de-watering within acceptable limits. A case history will help to illustrate the measures to be taken to minimize displacements.

The 26 m deep excavation illustrated in Fig. 2.86 was carried out by top-down construction within a concrete diaphragm wall box 107 m by 67 m in plan (Lui and Yau, 1995). Ground conditions at Dragon Centre, located in an area of deep weathering in the Sham Shui Po district of northern Kowloon, comprise a deeply weathered granite profile overlain by thin layers of loose sandy fill and marine deposits. Here, SPT N values increase from about 10 at 5 m depth to about 100 at 30 m depth. The groundwater table is about 1.5 m beneath the ground surface. Unlike many sites in urban Hong Kong, this is a 'virgin' site, as it is believed that the ground beneath the site had not experienced any substantial stress changes resulting from construction de-watering activities since it was reclaimed in the 1920s (Lui and Yau, 1995). Adjacent to the site there are sensitive structures, damage to which had to be avoided, and therefore precautions were taken to limit displacements during basement construction. To achieve an effective groundwater cut-off, a deep diaphragm wall was installed, to 'engineering rockhead, Grade 3 rock' at 47–62 m below ground level. Grade 3 rock is defined as 'material that cannot usually be broken by hand, is stained throughout and gives a dull ring when struck by hammer'. Curtain grouting was carried to a depth of some 10 m beneath the wall. A waterstop was installed between trench panels to 30 m depth and a milled hydrofraise formed on panel edges beneath this depth. Care was taken in excavation of the panels, which were 4 m in length, and this clearly paid off as relatively small settlements of the adjacent ground were recorded during diaphragm wall installation (<10 mm), compared with those shown in Fig. 2.85.

On completion of the diaphragm wall, a multiple well pumping test was carried out to test the effectiveness of the groundwater cut-off. A pumping test is now standard practice in Hong Kong for excavations deeper than about 8–10 m at sensitive sites. During the pumping test the groundwater within the diaphragm wall box was drawn down to 24–30 m below

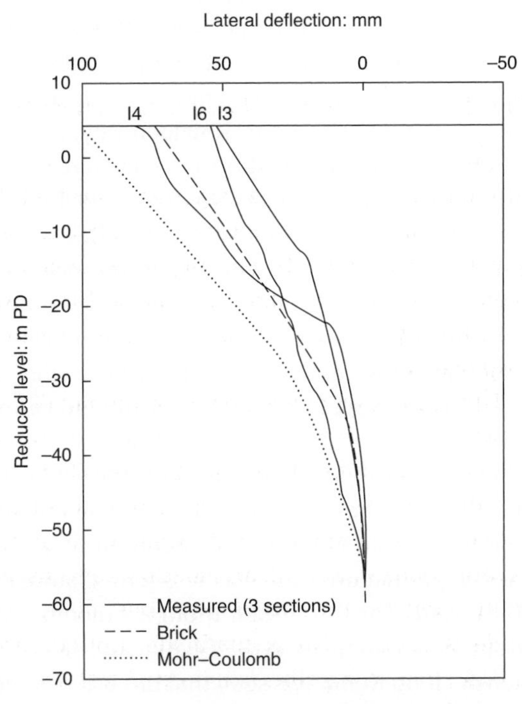

Fig. 2.87 Measured and computed lateral deflection of the diaphragm wall in the multiple well pumping test at the Dragon Centre (after Malone et al.*, 1997)*

ground level. The associated reduction in head in standpipes installed in the adjacent ground outside the box averaged 0.5 m, indicating that a good seal had been achieved over most of the site. As a contingency measure, recharge wells had been installed to maintain water levels, but these were not used during the pumping test; some were operated during excavation. During the works, measurements were taken of surface settlements and lateral wall displacements (Lui and Yau, 1995). Large wall displacements occurred during the pumping test (Fig. 2.87) and these might have been less had the top floor slab been cast before carrying out the pumping test. This procedure is now being introduced for sensitive sites.

Prediction of displacements

The observed displacements were compared with those computed by small strain modelling. The sequence of operations modelled in the finite-element analysis comprised the pumping test, followed by staged excavation to B5 level. The modelling used the SAFE finite-element program with Simpson's 'bricks on strings' soil model (Simpson, 1992). This model has the ability to capture the effect of changes in stress path and strain dependent soil stiffness, from a given modulus decay curve (S-shaped curve – relating soil stiffness with strain level). The modulus decay curves and other material properties were determined from Cambridge self-boring pressuremeter tests carried out in completely decomposed granite at a site in Kowloon Bay, some 5 km to the east of Dragon Centre, and triaxial testing (p' constant with local strain measurement) of soil from the same site (Ng *et al.*, 1998b; Ng *et al.*, 2000a). The completely decomposed granite at Kowloon Bay is a similar soil to that at Dragon Centre, the sites being within the same granite body (a very uniform equigranular medium-grained biotite monzogranite of the Kowloon Pluton) and the soil showing similar fines content (15–30%). The material exhibits pronounced strain-dependent stiffness (Fig. 2.88). The S-shaped curve used in the model is the mean of the measured data, with G_{max}/p' set at 1500.

The measured and computed wall displacements due to the pumping test are illustrated on Fig. 2.87. Inclinometers I3, I4 and I6 are at the midpoints of three sides of the diaphragm wall box. Along with the Brick model predictions, an elastic Mohr–Coulomb model prediction is also illustrated. For the elastic Mohr–Coulomb analysis, Young's modulus is assumed equal to $1 \times N$(SPT) value in MPa, following the normal assumption in Hong Kong. It is seen that the Brick model prediction is good but the elastic Mohr–Coulomb model over-predicts. Wall displacements at the final excavation to the B5 level are shown in Fig. 2.89. The Brick model prediction is again good, in terms of magnitude and curvature, but the

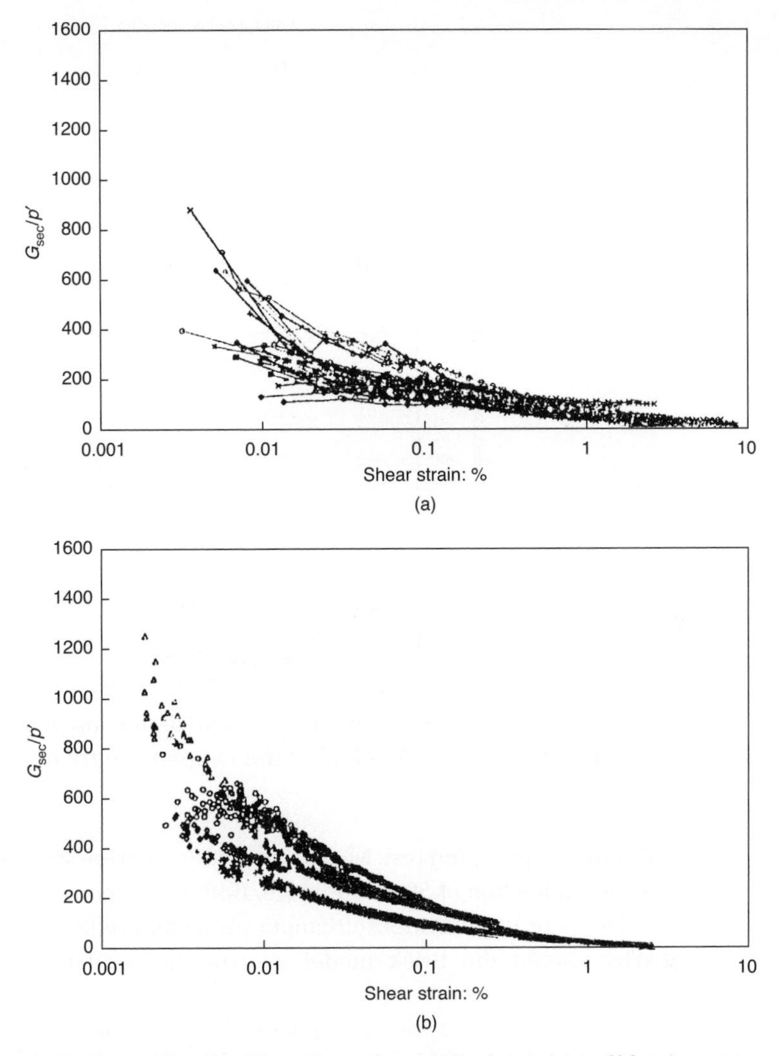

Fig. 2.88 Normalized stiffness–strain results from (a) self boring pressure meter, Kowloon Bay; (b) drained constant p′ *triaxial tests with local small-strain measurement, Kowloon Bay (after Ng* et al., *1998b; Malone* et al., *1997)*

elastic Mohr–Coulomb model is less good. Figure 2.90 shows the associated incremental settlements due to excavation to the B5 level on survey lines normal to the wall at the positions of inclinometers I3, I4 and I6. The measured settlement trough is narrower and closer to the wall than might have been expected for sedimentary sands but the maximum settlement is within the range of expected values, being about 0.15% of excavation depth and about 60% of maximum wall movement. Neither model predicts the settlement very well. Figure 2.91 shows settlements

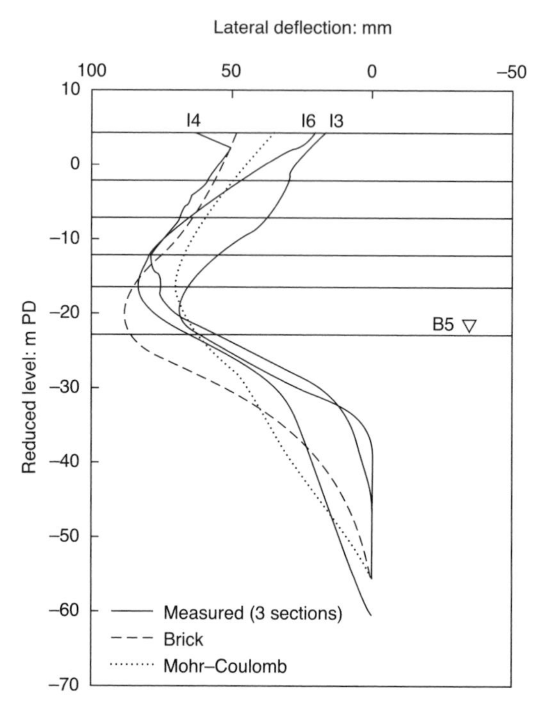

Fig. 2.89 Measured and computed lateral deflection of the diaphragm wall during excavation to level B5 at the Dragon Centre (after Malone et al.*, 1997)*

during the pumping test. Measured settlements are surprisingly low, given a wall deflection of 50 mm or more. Both predictions are poor.

The poor Brick model settlement prediction requires some explanation. Why should the Brick model over-predict settlements but give good

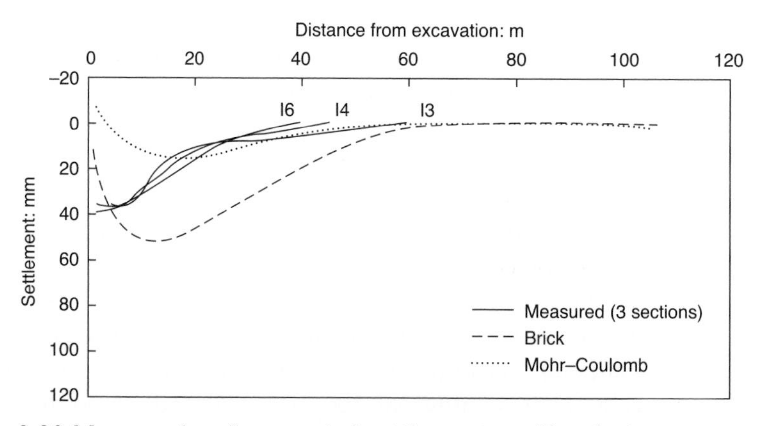

Fig. 2.90 Measured and computed settlement profiles during excavation to level B5 at the Dragon Centre (after Malone et al.*, 1997)*

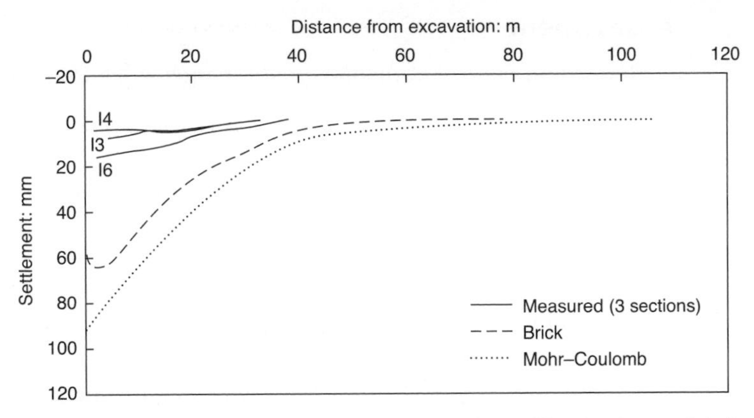

Fig. 2.91 Measured and computed settlement profiles in the multiple well pumping test at the Dragon Centre (after Malone et al.*, 1997)*

predictions for wall movement for both the pumping test and the excavation to the B5 level? The K_0 value at this site is expected to be roughly 0.3 to 0.4, hence one contributory factor may be stress-induced stiffness anisotropy. However, further computation suggests that this factor is unable to fully explain the measurements. Therefore the hypothesis is put forward that the stiff behaviour of the ground in the active zone during the well pumping test is due to remnant structure in the completely decomposed granite – that is, structure present in the soil in the ground but not allowed for in the model. Completely decomposed granite has a remarkably fragile structure. This makes it difficult to take undisturbed samples and to carry out in situ and laboratory soil testing without damaging the soil structure. It is hypothesized that soil structure in the main body of ground in the active zone survived wall installation (<10 mm settlement) and the pumping test (up to 5–15 mm extra settlement) but, adjacent to the excavation, was substantially broken down by staged excavation to the B5 level (40 mm extra settlement). The structure of the soil within the box is unlikely to have remained intact during piling and de-watering during the pumping tests (estimated settlement roughly 75 mm). The resulting low stiffness is represented in the model, which therefore gives a good prediction of wall displacement.

This hypothesis needs to be tested at other sites and by numerical experiments using a modulus decay curve representative of soil with structure intact at small strain and suffering progressive destructuring at larger strain, in a continuum model capable of handling progressive destructuring. These observations, soil tests and computations have yet to be carried out.

Assessment

Regarding control of displacement (conclusions from general practice):

1 'top-down' construction within a diaphragm wall box is favoured for very deep excavations at sensitive sites,
2 precautions need to be taken during diaphragm wall installation, and
3 sophisticated groundwater cut-off systems are required.

Regarding prediction of displacement (conclusions from this case history):

4 completely decomposed granite of the Kowloon Pluton shows small strain stiffness behaviour typical of sedimentary soils,
5 the finite-element predictions with small strain stiffness were good for lateral wall displacement,
6 but less good for vertical settlement, possibly due to the effect of remnant structure not allowed for in the model, and
7 the elastic Mohr–Coulomb displacement predictions were not as good as the small strain predictions.

PART 2

Summary

1 When dealing with excavation problems, it is vital to consider changes in total and effective stress resulting from the excavation process.
2 Idealized plastic geo-structural mechanisms are considered in relation to undrained excavation against a cantilever wall in clay pinned at the toe and in relation to base heave. Possible trends in ground deformations for these two cases are discussed.
3 A set of Short Course Notes is given covering preliminary design for supported deep excavations. These cover the following: pressure envelopes for excavations in sands, in soft and medium stiff clays and in stiff clays; base heave in clays; ground movements around excavations.
4 A most important case study, a deep strutted excavation at Lion Yard, Cambridge, is described in detail. Instruments were installed to measure pore water pressures, ground deformations, strut loads and earth pressures against the diaphragm wall. The observations are presented and discussed and conclusions drawn.
5 The development of lateral pressures exerted by wet concrete against diaphragm walls is reviewed. It is shown that the distribution of pressure with depth is bi-linear; above a critical depth the full fluid concrete pressure applies while below that depth the pressure corresponds to that due to the bentonite under which the diaphragm wall is constructed.
6 Stress transfer and deformation mechanisms around diaphragm walls are investigated. A 3D analysis of the construction sequence of a diaphragm wall panel installed in stiff clay has been conducted using a finite-difference program. Significant stress reduction at the centre and above the toe of the panel caused by the installation process is attributed to both the downward load transfer mechanism and the horizontal arching mechanism. The installation process of a diaphragm wall in stiff clays could result in a significant reduction of horizontal stress above the toe and at the centre of the wall, but with small induced ground movements.

7 A comparison is made between the results obtained using a simple linear finite-element analysis and those given by a complex non-linear numerical approach. As far as maximum wall deformation after main excavation is concerned, it appears that the simple linear analysis leads to a reasonably accurate prediction.
8 Predictions of ground movement due to the installation of a diaphragm wall are generally made by adopting a two-dimensional analysis with plane strain assumptions which cannot accurately reflect the non-linear behaviour of soils. It is suggested however that on a number of occasions the simple approach may yield predictions sufficiently accurate for practical purposes provided that an appropriate value for soil stiffness is selected for the calculations.
9 In order to control displacements around deep excavations in completely decomposed granite it is suggested that the use of 'top-down' construction within a diaphragm wall box is likely to give a satisfactory solution, but precautions need to be taken during the installation of the diaphragm wall and a sophisticated groundwater cut-off system is required. Finite-element predictions with small strain stiffness for lateral wall displacement were reliable, but less so for vertical settlement.

PART 3
Bored and open-face tunnelling below cities

Centre crown of Type 1 section being grouted, Heathrow Express Trial Tunnel, UK (after Deane and Bassett, 1995)

Overview

The major challenges facing geotechnical designers and constructors of bored and open-face tunnels, particularly in crowded urban areas, are:

1. How do we compute the short-term and long-term tunnel stability?
2. What are the best construction techniques for the type of ground being tunnelled?
3. How can the surface and sub-surface settlement 'trough' above a tunnel be estimated?
4. What are the effects and significance of the initial soil effective stress state and anisotropy on ground deformations associated with tunnelling?
5. How can we assess multi-tunnel interaction?
6. What are the benefits of using soil nails to stabilize the open face of a tunnel?
7. Is there removal of support of pile foundations when a tunnel is bored nearby and if so how can we assess it?

This Part 3 – effectively a short course on open-face tunnelling below cities – seeks to answer these questions by first presenting the classic concepts of soil mechanics for item 1 above, showing how easily soil arching supports tunnels – making sprayed concrete linings quite acceptable as used by the New Austrian Tunnelling Method (NATM). Construction techniques are reviewed to cover item 2 above, and the simple estimation of settlement trough via ground loss and Gaussian theory is used to explain item 3 above. The case study of the Heathrow Express Trial Tunnel is presented, showing actual measured ground deformations. We then consider how numerical modelling can give insights into the problems of items 4, 5 , 6 and 7. Finally, we briefly review some of the fascinating aspects of how the Big Ben Tower in London was monitored for the effects of the nearby construction of the Jubilee Line Extension tunnels.

Sources

The main sources for this Part 3 are:

- Mair, R. J. and Taylor, R. N. (1997). Theme lecture: Bored tunnelling in the urban environment. *Proc. 14th Int. Conf. Soil Mech. & Fdn. Engng*, Hamburg, **4**, pp. 2353–2385, Balkema, Amsterdam.
- Deane, A. P. and Bassett, R. H. (1995). The Heathrow Express Trial Tunnel. *Proc. Instn Civ. Engrs Geotech. Engng*, **113**, July, 144–156.
- Bolton, M. D. (1979). *A Guide to Soil Mechanics* (eds M. D. and K. Bolton), Macmillan Press, Cambridge, 439 pp.
- Burland, J. B., Standing, J. R. and Jardine, F. M. (eds) (2001). *Building Response to Tunnelling*. Thomas Telford, London.

We gratefully acknowledge permissions to make verbatim extracts from these sources by Professor Robert Mair, Professor Neil Taylor, Mr Tony Deane, Dr Dickie Bassett, Professor Malcolm Bolton, Professor John Burland, Dr Jamie Standing and Mr Fin Jardine.

Terminology

In the world of tunnelling, terminology is as specialized as in other disciplines. Referring to Fig. 3.1, it can be seen that the tunnel crown, shoulder, springline, knee and invert refer to radial locations on the tunnel cross-section. Other terms that will be useful to define in this chapter are:

Transverse: in the horizontal plane, normal to the axis of the tunnel.

Longitudinal: in the direction of the tunnel axis.

Settlement trough: the three-dimensional surface (and in some cases sub-surface) depression of the ground above the tunnel as the tunnel advances.

Volume loss: the ground volume physically 'lost' by the tunnelling process between the final as-built tunnel section and the original ground outside the final tunnel section.

Tunnel heading: the workface at the front of the tunnel where soil excavation is being made.

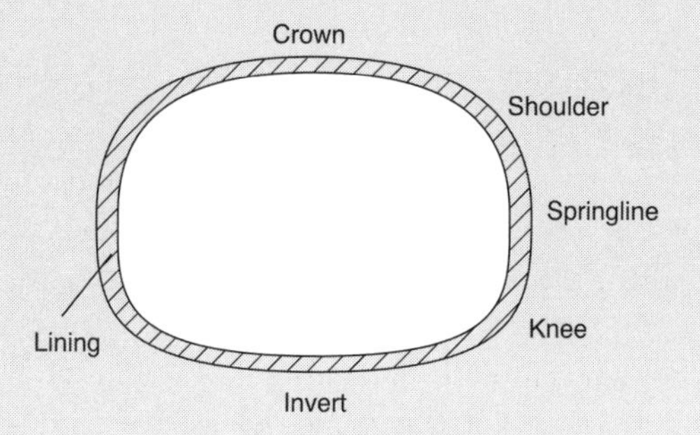

Fig. 3.1 Tunnel cross-section terminology

CHAPTER TWENTY-THREE

Unsupported tunnels: theoretical assessment of the collapse of soil arches

Bolton (1979) points out that it is possible to adopt Rankine's technique of allowing Mohr's circles to touch the envelopes of limiting strength to give a simplified analytical method to illustrate the collapse mechanism of soil arches or unlined tunnels. Bolton considers the collapse of the crown of an underground cylindrical cavity – perhaps part of a tunnel heading, a plastic sewer pipe, a void in some earthwork which must be collapsed by surface compaction, or a plastic bottle deep in a city dump which is to be levelled and developed. Figure 3.2 demonstrates that if the soil conditions and the surcharge pressure σ_0 are uniform there can be no shear stress across the vertical plane AB, by the requirement of symmetry.

Interposing frictionless radial planes CD and EF on either side of the vertical is therefore unlikely to be very pessimistic, if the included angle $d\theta$ is small. It is then possible to analyse pessimistically the collapse of the wedge CDFE into the crown of the cavity.

Since the radii CD and EF are frictionless, and both the surcharge σ_0 and the internal cavity pressure σ_a are perpendicular to the circumferences DF (which is indistinguishable from an infinitesimal arc) and CE respectively, it follows that the principal stresses in the wedge must be radial and circumferential. If the cavity is to collapse, then DC and FE must extend. It follows that the radial stress σ_r must be so small at every radius that the circumferential stress σ_θ is able to cause something like a passive collapse of each of the circular wedge elements represented by VWYX in Fig. 3.2. The stresses must be allowed to change, and they are depicted as σ_r and σ_θ in the figure, which goes on to calculate the forces on a wedge element by summing the stresses over the relevant length of their side. The weight of the element is included, being the unit weight of soil multiplied by the volume of the element. Resolving these forces vertically gives:

$$(\sigma_r + d\sigma_r)(r + dr)\, d\theta - \sigma_r r\, d\theta - 2\sigma_\theta\, dr\, \frac{d\theta}{2} + \gamma r\, d\theta\, dr = 0$$

$$\sigma_r\, dr\, d\theta - \sigma_\theta\, dr\, d\theta + r\, d\sigma_r\, d\theta - \gamma r\, d\theta\, dr = 0$$

$$\frac{d\sigma_r}{dr} = \frac{(\sigma_\theta - \sigma_r)}{r} - \gamma \qquad (3.1)$$

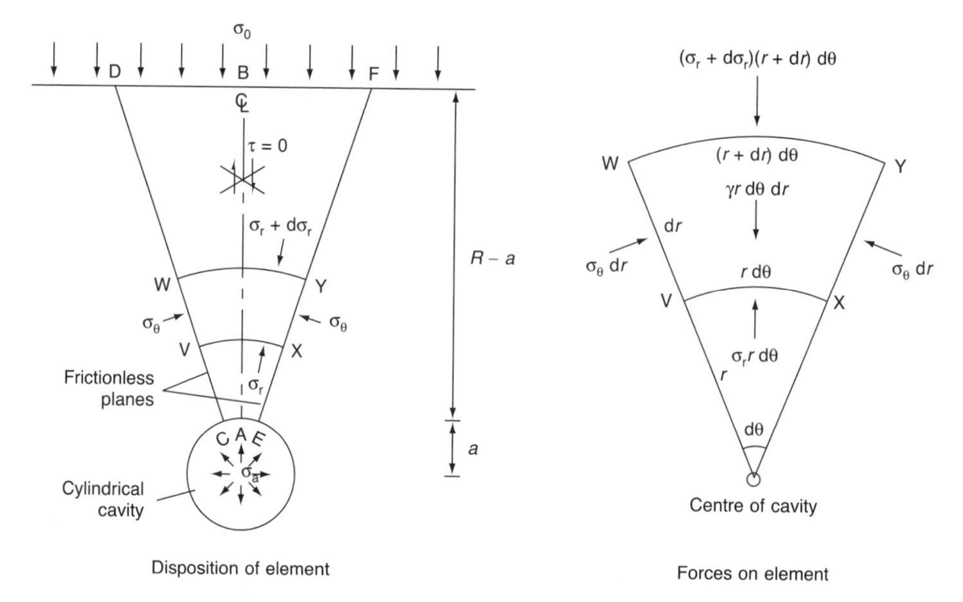

Fig. 3.2 Cross-section through overburden above the crown of a cylindrical cavity (after Bolton, 1979)

Now if the cohesion model is applied with $\sigma_\theta > \sigma_r$ it follows that, in the limit

$$\sigma_\theta = \sigma_r + 2s_u$$

Substituting this into Equation (3.1) we obtain

$$\frac{d\sigma_r}{dr} = \frac{2s_u}{r} - \gamma$$

$$\int_{\sigma_a}^{\sigma_0} d\sigma_r = \int_a^R \left(\frac{2s_u}{r} - \gamma\right) dr$$

where the limits are σ_a at $r = a$ and σ_0 at $r = R$.

On integration

$$\sigma_0 - \sigma_a = 2s_u \ln\left(\frac{R}{a}\right) - \gamma(R - a) \tag{3.2}$$

If the soil were a heavy liquid ($s_u = 0$) then Equation (3.2) reduces to hydrostatics, with

$$\sigma_a = \sigma_0 + \gamma(R - a)$$

But consider what pressure of static compaction σ_0 applied at the surface of a stiff clay of undrained shear strength $s_u = 100$ kPa would be needed to collapse the crown of a 0.01 m radius cavity at 0.2 m depth, assuming atmospheric pressure in the cavity ($\sigma_a = 0$),

$$\sigma_0 = 2 \times 100 \ln 20 - 20 \times 0.99\,\text{kPa}$$
$$\sigma_0 = 600 - 20 = 580\,\text{kPa}$$

Perhaps it is no coincidence that this is the order of size of the tyre pressure of pneumatic-tyred compaction plant, used to compact loose clayey fill in layers up to 0.2 m thick. This result can be generalized by observing that no load could apply a pressure much greater than the $5.5s_u$ of the bearing failure for long strip footings on undrained clay. Ignoring the weight of the soil and equating the cavity pressure σ_a to zero in Equation (3.2), it follows that the greatest possible surcharge pressure is

$$\sigma_0 \approx 5.5s_u \approx 2s_u \ln\left(\frac{R}{a}\right)$$

so that the largest possible ratio of R/a at which collapse occurs is roughly 16.

Now consider what compressed air pressure σ_a would be required in a tunnel heading to prevent the collapse of an unlined 2 m diameter tunnel with a cover of 7 m of firm clay with $s_u = 50$ kPa. Ignoring surcharge

$$\sigma_a = 20 \times 7 - 100 \ln 4$$
$$\sigma_a = 140 - 139 = 1\,\text{kPa}$$

The tunnel is almost safe (temporarily – remember s_u must be the undrained or short-term shear strength) without being lined or inflated. Clearly Equation (3.2) has wide and interesting applications in 'cohesive' ground.

If, on the other hand, the friction model were used with zero pore water pressures, the criterion of collapse for the element would be

$$\sigma'_\theta = K'_p \sigma'_r$$

or, since $\sigma'_\theta = \sigma_\theta$ and $\sigma'_r = \sigma_r$ and because the pore pressure is zero

$$\sigma_\theta = K'_p \sigma_r$$

Substituting this into Equation (3.1), we obtain

$$\frac{d\sigma_r}{dr} = \frac{(K'_p - 1)\sigma_r}{r} - \gamma \qquad (3.3)$$

This is most easily integrated by substituting $\psi = \sigma_r/r$ from which

$$d\sigma_r = r\,d\psi + \psi\,dr$$

so that by Equation (3.3)

$$r\frac{d\psi}{dr} + \psi = (K'_p - 1)\psi - \gamma$$
$$r\frac{d\psi}{dr} = (K'_p - 2)\psi - \gamma$$

$$\int_{\sigma_a/a}^{\sigma_0/R} \frac{d\psi}{(K'_p - 2)\psi - \gamma} = \int_a^R \frac{dr}{r}$$

$$\frac{1}{(K'_p - 2)} \ln \left[\frac{(K'_p - 2)\dfrac{\sigma_0}{R} - \gamma}{(K'_p - 2)\dfrac{\sigma_a}{a} - \gamma} \right] = \ln \frac{R}{a}$$

or

$$\frac{\dfrac{\sigma_0}{\gamma R}(K'_p - 2) - 1}{\dfrac{\sigma_a}{\gamma a}(K'_p - 2) - 1} = \left(\frac{R}{a}\right)^{K'_p - 2} \tag{3.4}$$

Consider what internal supporting stress $\sigma'_a = \sigma_a$ would be required in an otherwise unlined tunnel through dry sand with zero surcharge. From Equation (3.4) with $\sigma_0 = 0$

$$\sigma_a = \frac{\gamma a}{(K'_p - 2)} \left[1 - \left(\frac{a}{R}\right)^{K'_p - 2} \right] \tag{3.5}$$

which is remarkable in that the effect of increasing depth R is swiftly reduced to zero so that

$$\sigma_a \to \frac{\gamma a}{(K'_p - 2)} \quad \text{as} \quad \frac{a}{R} \to 0 \tag{3.6}$$

It is instructive to substitute some typical numbers for a large tunnel, for example $\gamma = 20\,\text{kN/m}^3$, $a = 3\,\text{m}$, $R = 12\,\text{m}$ and $\phi' = 42°$ so that $K'_p = 5.0$. From Equation (3.5)

$$\sigma_a = \frac{20 \times 3}{3} \left[1 - \left(\frac{1}{4}\right)^3 \right] \approx 20\ \text{kPa}$$

so that the (a/R) term is readily seen to approach zero and the required supporting pressure in the tunnel is only 20 kPa, notwithstanding that the average vertical stress at 9 m depth must be 180 kPa. Clearly the stress has 'arched' around the cavity. The requirement for internal support to bolster the friction in the ground is surprisingly little. It should be clear, however, that whereas compressed air is able temporarily to support a tunnel through clay, it would be useless in dry sand since it would increase total stresses and pore pressures equally and leave unchanged the effective stress of the soil in the roof. The effective supporting stress σ'_a in dry sand must therefore be provided by such structural components as, say, sheets of corrugated steel carried over arched ribs. Rabbits on the other hand use both the capillary suction of damp earth and the tensile strength of plant roots to provide the effective stress required in the soil forming the roof of their burrows.

Now use Equation (3.4) to calculate the surcharge pressure needed on the surface of a similar sand fill in order to collapse the crown of a cylindrical cavity 0.10 m in radius and 1 m deep which was so soft that it could only offer a token resistance of 1 kPa (perhaps a plastic bottle)

$$\frac{\dfrac{\sigma_0 \times 3}{20 \times 1} - 1}{\dfrac{1 \times 3}{20 \times 0.1} - 1} = 10^3$$

$$0.15\sigma_0 - 1 = 500$$

$$\sigma_0 = 3300 \text{ kPa}$$

It would require the equivalent of a column of steel 40 m high to crush the 0.2 m diameter cavity only 1 m below its base. The static compaction of sand requires enormous stresses if loose pockets are to be collapsed and densified.

Arching zones cause great difficulties for engineers who want to measure soil stresses. If they introduce any sort of flexible diaphragm into a soil in an attempt to relate its deflection to the pressure acting on it, they may well create a soil arch around the relatively 'soft' diaphragm which could cause a reduction of pressure upon it by a factor of 10 or 100. Consider, for example, the problem of a flexible plate or trapdoor underneath a bed of sand with $\phi' = 42°$ and $K'_p = 5$. Taking Equation (3.5) to refer approximately to a long plate of width $2a$, the vertical pressure on the plate would be roughly $\gamma a/3$ whatever its depth. The equivalent depth $a/3$ of sand 'resting' on the plate could be a tiny fraction of the depth of overburden.

CHAPTER TWENTY-FOUR

Modern tunnel construction techniques

Open-face tunnelling

As pointed out by Mair and Taylor (1997), there is increasing use of sprayed concrete in soft ground tunnelling to form linings, particularly for tunnels of shorter lengths and of non-circular cross-section. These are usually temporary, but may be the sole means of support for significant periods (sometimes more than a year) before the permanent lining is installed. Recent developments have involved composite sprayed concrete linings, in which sprayed concrete is used for both the temporary and permanent linings (e.g. Wittke, 1995; Grose and Eddie, 1996; Negro *et al.*, 1996). Non-circular cross-sections and divided faces are often adopted when using sprayed concrete, as shown in Fig. 3.3, and this allows considerable flexibility in terms of modifying the construction sequence in response to observations. The use of sprayed concrete to form linings is sometimes referred to as the New Austrian Tunnelling Method (NATM).

Ground treatment is more easily undertaken from within tunnels with open faces. Advances have been made in reinforcement of the soil

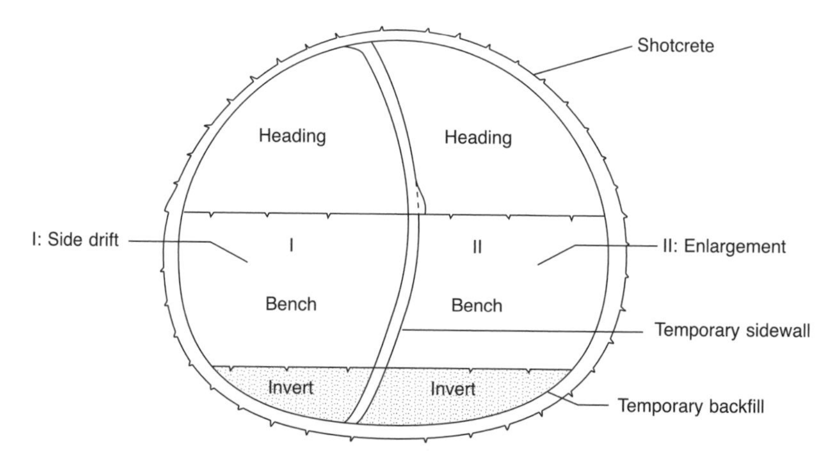

Fig. 3.3 Example of divided tunnel face using sprayed concrete linings (after Institution of Civil Engineers, 1996b; and Mair and Taylor, 1997)

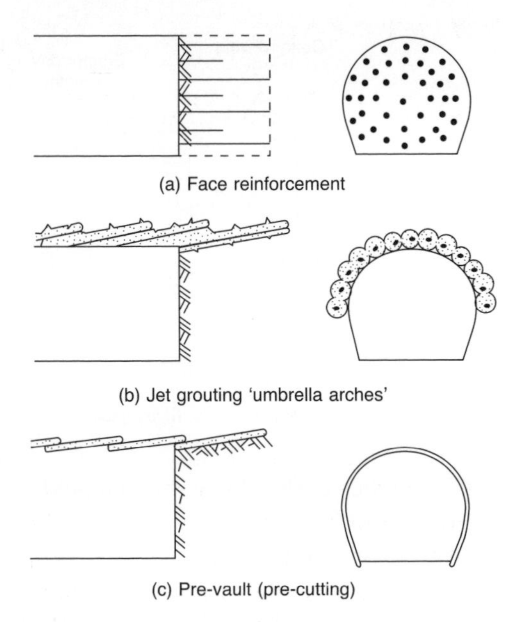

Fig. 3.4 Ground treatment and pre-lining techniques (after Schlosser and Guilloux, 1995; and Mair and Taylor, 1997)

ahead of the face to improve stability and to control ground movements (Fig. 3.4(a)). Improvements in jet grouting techniques are being made to form 'umbrella arches' (Fig. 3.4(b)) as a prelining in difficult ground conditions.

Closed-face tunnelling

Considerable advances have been made over recent years in the use of sophisticated closed-face tunnelling machines which operate on the principle of a pressurized face. These machines are used in unstable ground conditions where the face requires support at all times; this principally applies to permeable ground below the water table (i.e. mainly sands or mixtures of sands, silts and clays) or soft clays. The slurry shield machine, illustrated in Fig. 3.5, is most commonly used in water-bearing granular soils. The face is supported by a pressurized bentonite or polymer-based slurry, which is circulated so that it and the excavated soil are removed to a separation plant.

Earth pressure balance (EPB) machines are being used more universally for all types of unstable ground; the principle is shown in Fig. 3.6. By controlling the entry of soil and water through the cutter face by means of earth pressure balance doors, and by conditioning the spoil so that it can easily be removed through a screw conveyor, it is possible to control

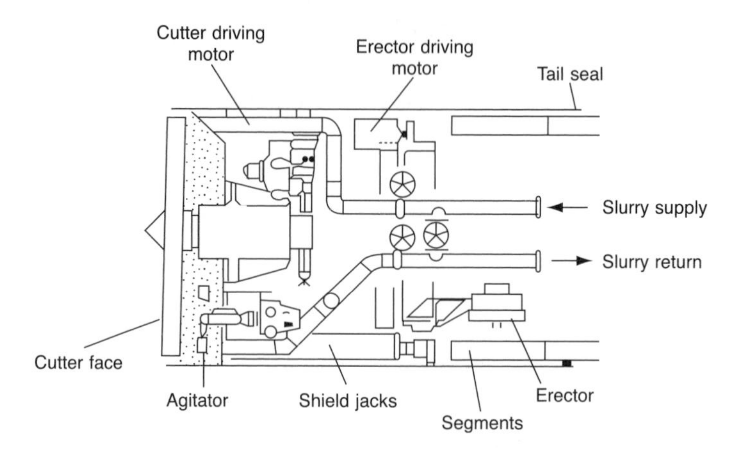

Fig. 3.5 Principle of the slurry shield machine (after Fujita, 1989; and Mair and Taylor, 1997)

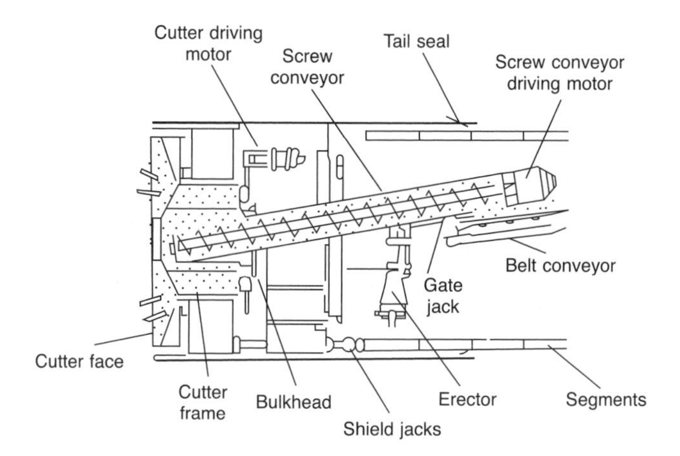

Fig. 3.6 Principle of the earth pressure balance (EPB) machine (after Fujita, 1989; and Mair and Taylor, 1997)

the pressure of the excavated soil in the chamber to balance the earth and water pressures in the ground.

CHAPTER TWENTY-FIVE

Principal design and construction requirements

Overview: stability, ground movements and linings

Mair and Taylor (1997) comment that, as presented by Peck (1969) and developed by Ward and Pender (1981), the three most important requirements for the successful design and construction of a tunnel can be summarized as follows:

- Stability: The choice of excavation and construction technique must be suited to the ground conditions so that it is feasible to build the tunnel safely. Of prime importance is the stability of the opening prior to installation of the lining.
- Ground movements and their effects: Of particular relevance to the urban environment, construction of the tunnel should not cause unacceptable damage to surrounding or overlying structures and services. Prior to construction the ground movements should be predicted and their effects on the structures and services assessed.
- Performance of linings: The tunnel lining, whether it is temporary or permanent, must be capable of withstanding all the influences to which it may be subjected during its design life. This requires predictions of the soil loading acting on the lining and of the deformations of the lining, the latter being of particular significance in the case of external influences such as adjacent tunnel construction.

These three principal requirements, which are closely related, form the basis of this chapter. The first two are of particular importance in the urban environment and therefore more emphasis is placed on these. The subject of this section is restricted to tunnelling in soft ground, but the meaning of 'soft' ground can differ according to the perspectives of the tunnelling engineer and the geotechnical engineer. It is appropriate to define 'soft' ground as that which requires some form of support in the tunnel to prevent instability of the ground, either in the short term or long term. In this context even materials such as stiff to hard clays and weak rocks are classified as 'soft' ground.

Stability

In any tunnelling project adequate stability during construction is clearly of prime importance, and this is particularly the case in urban environments where the consequences of a major tunnel collapse can be catastrophic. In recent years there have been a number of major tunnel collapses associated with the use of sprayed concrete linings (Anderson, 1996; HSE, 1996), although collapses also occur with other forms of tunnelling.

Stability of a tunnel heading (with circular cross-section) can be considered in terms of the idealized geometry shown in Fig. 3.7(a). The heading may be supported by a fluid pressure or compressed air or pressurized slurry (in the case of a slurry shield). It may be excavated in free air in open-face mode, in which case $\sigma_T = 0$. The dimension P represents the distance from the face to the point where stiff support is provided; in the absence of a tunnelling shield, this is the distance from the face to the lining. In most cases when a tunnelling shield is in use in ground of low stability, the ground is in contact with the shield and therefore P can be taken to be zero. An exception is when a shield is being used in ground of higher stability, such as stiff clays; in this case there is often an oversized cutting edge at the front of the shield to ensure a gap between the ground and the length of the shield and

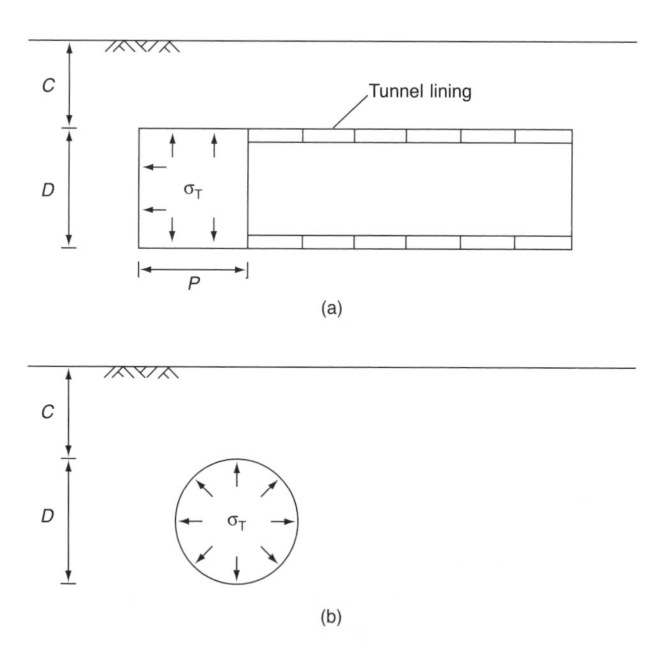

Fig. 3.7 (a) Tunnel heading in soft ground; (b) two-dimensional idealization of tunnel heading (after Mair and Taylor, 1997)

thereby facilitate easy steerage. Determination of P then requires some judgement.

The issue of whether undrained or drained conditions are more applicable to the tunnel stability problem depends principally on the permeability of the soil, the excavation advance rate, and the size of the tunnel. Based on parametric studies of seepage flow into tunnel excavations, Anagnostou and Kovari (1996) concluded that for most tunnels drained conditions are to be expected when the soil permeability is higher than 10^{-7} to 10^{-6} m/s and the excavation advance rate is 0.1–1.0 m/h or less. Hence, in a predominantly sandy soil, drained stability should be considered. In low-permeability clayey soils undrained stability is of more importance during tunnel excavation, but in the case of a standstill, drained conditions could become more relevant. Factors influencing whether undrained or drained conditions are more applicable are also discussed by Negro and Eisenstein (1991).

Short-term undrained stability

Based on the concept introduced by Broms and Bennermark (1967), the stability ratio, N can be defined as:

$$N = \frac{(\sigma_s + \gamma z - \sigma_T)}{s_u} \tag{3.7}$$

where γ = unit weight of the soil, z = depth to the tunnel axis ($= C + D/2$, see Fig. 3.7), σ_s = surface surcharge pressure (if any), σ_T = tunnel support pressure and s_u = undrained shear strength at tunnel axis level.

On the basis of laboratory extrusion tests and field observations, Broms and Bennermark (1967) concluded that the critical stability ratio at collapse, N_c, is about 6. Similar conclusions were reached by Peck (1969).

Davis *et al.* (1980) derived plasticity solutions employing kinematic upper bounds and statically admissible lower bounds for the two-dimensional idealization of a tunnel heading shown in Fig. 3.7(b), assuming constant undrained shear strength profiles. Reasonably close agreement was obtained between the upper and lower bounds. Their results are summarized in Fig. 3.8, which shows the derived values of critical stability ratio, N_c, in terms of the dimensionless ratios C/D and $\gamma D/s_u$. Slight improvements to the upper bound for the case of weightless soil ($\gamma D/s_u = 0$) have been made by Antao *et al.* (1995). A comprehensive set of solutions for the more general two-dimensional case where the undrained shear strength increases with depth was derived by Sloan and Assadi (1993), using finite-element formulations of the upper and lower bound plasticity theorems.

Of most relevance to practical tunnel stability problems is the three-dimensional heading, shown in Fig. 3.7(a). Based on centrifuge model

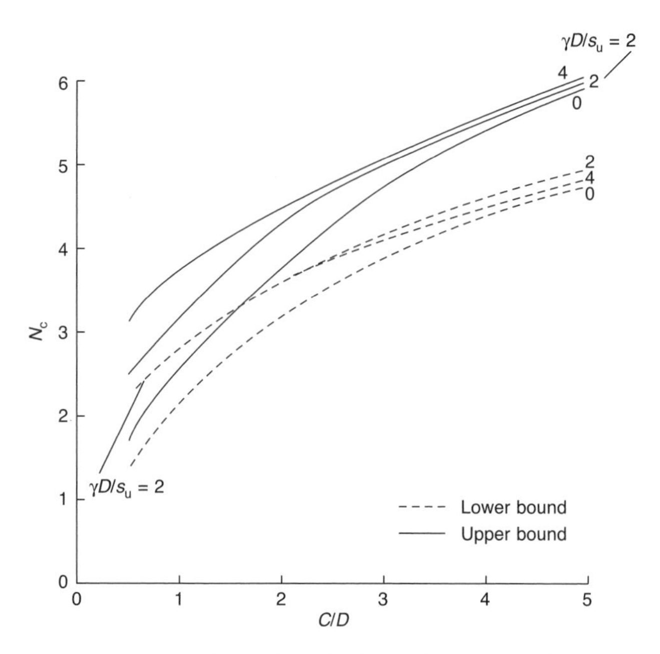

Fig. 3.8 Upper and lower bound critical stability ratios for plane strain circular tunnel (after Davis et al.*, 1980; and Mair and Taylor, 1997)*

tests, the design curves in Fig. 3.9 showing the critical stability ratio, N_c, in terms of the dimensionless ratios P/D and C/D were proposed (Mair, 1979; Kimura and Mair, 1981). The special case of $P/D = 0$ is of particular relevance, and Fig. 3.10 shows part of the design line taken from

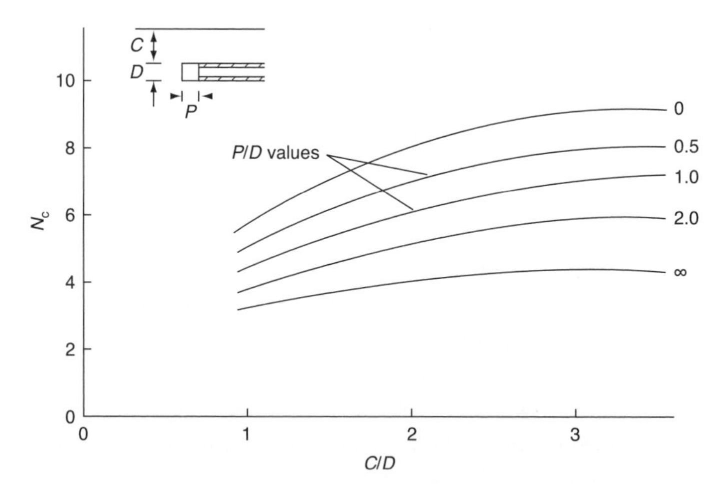

Fig. 3.9 Dependence of critical stability ratio on tunnel heading geometry (after Mair, 1979; Kimura and Mair, 1981; and Mair and Taylor, 1997)

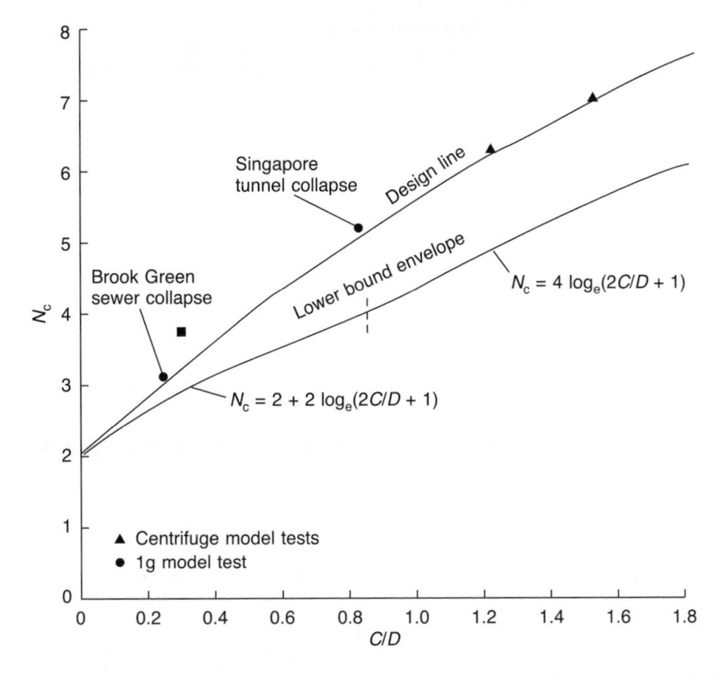

Fig. 3.10 Critical stability ratio values for lined tunnel headings (P/D = 0) with thin clay cover (after Mair, 1993: and Mair and Taylor, 1997)

Fig. 3.9, together with data from laboratory and centrifuge tests, and also from back-analysis of tunnel heading failures in the field (Mair, 1993). Also shown on Fig. 3.10 is an envelope of lower bound plasticity solutions derived by Davis *et al.* (1980); in contrast to the two dimensional case, upper bound solutions (not shown) give N_c values which are much higher than those given by the lower bound solutions. Similar high upper bound values were obtained by Leca and Dormieux (1992).

The stability numbers summarized in Figs 3.9 and 3.10 can also be used to estimate the risk of 'blow-out', which can occur if the tunnel face pressure is too high in soft clays (Mair, 1987; De Moor and Taylor, 1991).

Long-term drained stability

Atkinson and Potts (1977) derived kinematic upper bound and statically admissible lower bound plasticity solutions for the two-dimensional idealization shown in Fig. 3.7(a) for dry cohesionless soils. Their results are presented in Fig. 3.11, for the case of $\phi' = 35°$, where the tunnel support pressure σ_T required to prevent collapse is expressed as the dimensionless ratio $\sigma_T/\gamma D$ plotted against C/D. It should be noted that the upper bound plasticity solution (being inherently unsafe) gives a lower value of $s_T/\gamma D$ than the lower bound solution which is inherently

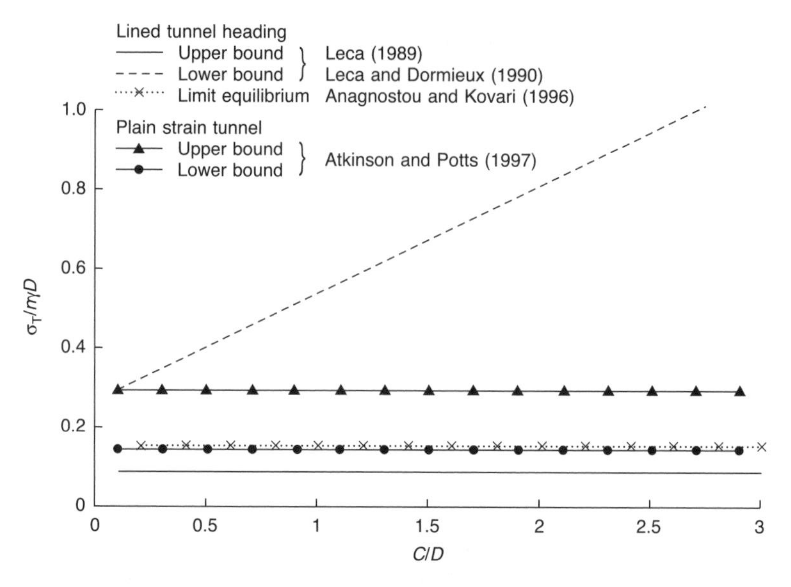

Fig. 3.11 Drained stability solutions for plane strain tunnels (P/D = ∞) and lined tunnel headings (P/D = 0); $\phi' = 35°$ (after Mair and Taylor, 1997)

safe. Centrifuge model tests in dry sands reported by Atkinson and Potts were consistent with these plasticity solutions, and tests on model tunnel headings ($P/D = 0$) showed that the support pressure required to prevent collapse was only slightly lower than for the two-dimensional case ($P/D = \infty$). Their experimental results, together with their plasticity solutions for $\phi' = 50°$ (which was evaluated by the authors for the low stress levels around the tunnel), are shown on Fig. 3.12. The experimental data and the plasticity solutions indicate the support pressure to be independent of the ratio C/D.

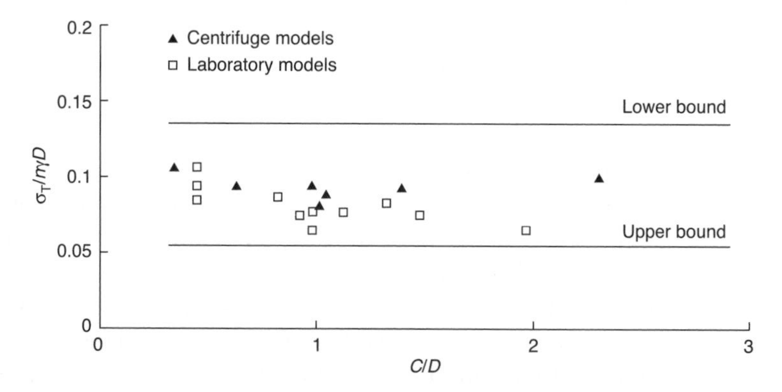

Fig. 3.12 Centrifuge model tests on lined tunnel headings in dry sand (after Atkinson and Potts, 1977; and Mair and Taylor, 1997)

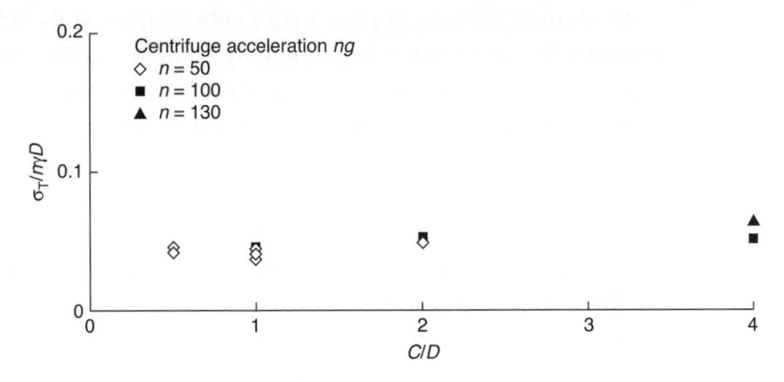

Fig. 3.13 Centrifuge model tests on lined tunnel headings in dry sand (after Chambon and Corte, 1994; and Mair and Taylor, 1997)

Similar findings are reported by Chambon and Corte (1994), who performed centrifuge tests on lined tunnel headings in dry sand; their results are shown in Fig. 3.13. As found by Atkinson and Potts, the required support pressure was almost independent of C/D.

Upper and lower bound plasticity solutions for the stability of the three-dimensional tunnel heading ($P/D = 0$) in dry soils in terms of c' and ϕ' were derived by Leca (1989) and Leca and Dormieux (1990). Their results are also presented in Fig. 3.11 for the case of $c' = 0$, $\phi' = 35°$ for comparison with the two-dimensional results of Atkinson and Potts (1977). Leca and Dormieux found that their upper bound solution was in good agreement with earlier centrifuge model tests reported by Chambon and Corte (1994), the support pressure required to prevent collapse being independent of the ratio C/D. It is of interest to note that, in contrast to the presently available undrained ($\phi = 0$) stability solutions, the lower bound solution for the tunnel heading shown in Fig. 3.11 gives significantly higher support pressures than the upper bound, and these increase with C/D, differing from experimental observations. Anagnostou and Kovari (1996) present a limit equilibrium solution for a lined tunnel heading in terms of c' and ϕ'. Their solution for $c' = 0$, $\phi' = 35°$, as seen in Fig. 3.11, shows the support pressure being independent of C/D, as obtained by Leca and Dormieux (1990) for their upper bound plasticity solution.

The principal conclusion arising from the stability solutions illustrated in Fig. 3.11, and from the centrifuge model test data in Figs 3.12 and 3.13, is that the effective support pressure required to prevent collapse of a tunnel in dry cohesionless soil is very small, irrespective of whether it is a two-dimensional circular tunnel or a three-dimensional heading. It is also independent of tunnel depth. These conclusions are borne out by Bolton's (1979) simple analysis of soil arches presented in chapter 23 at the beginning of Part 3.

All of the foregoing applies to dry cohesionless soils, which are not often encountered in practice in tunnelling. In many cases even soils above the water table contain sufficient moisture to behave as if they exhibited significant values of c'. Then even smaller support pressures are required, as shown by the stability solutions of Leca and Dormieux (1990) for c', ϕ' soils.

In practice, tunnelling is frequently undertaken below the water table, in which case positive measures are necessary to prevent water inflow and ensure adequate drained stability of gravels, sands and silts. Compressed air can be used for this purpose, although it has become less attractive in recent years as more is understood about its potential adverse effects on tunnel workers (especially at high pressures). Its decline in use is also associated with the increasing adoption of full face slurry or Earth Pressure Balance (EPB) machines.

Face stability in slurry and EPB tunnelling is considered by Anagnostou and Kovari (1996). Based on a limit equilibrium approach, they developed computational models which provide a useful framework for quantifying the mechanics of tunnel face failure. For a slurry machine, the stabilizing force capable of being exerted by the slurry depends on the extent of its infiltration into the ground. Under optimum conditions, the slurry seals the face by forming a filter cake and acts like a membrane; the support force then results from the excess slurry pressure over and above the groundwater pressure. When the soils are of high permeability, or when the shear resistance of the slurry is low, the slurry will penetrate the ground and the force it is capable of exerting depends on the depth of infiltration. Stability solutions for face stability of slurry machines, based on limit equilibrium methods, have also been derived by Jancsecz and Steiner (1994).

An example of the approach by Anagnostou and Kovari is illustrated in Fig. 3.14, in which the safety factor against face instability for a 10 m diameter tunnel is shown to be a function of the excess slurry pressure (Δp), the concentration of bentonite (%) and the associated yield strength of the slurry (τ_f), and the characteristic grain size of the soil (d_{10}). The following practical points emerge:

- An increase in safety can be achieved by raising the excess slurry pressure, but only in the finer-grained soils. Above a d_{10} size of approximately 2 mm, increasing the slurry pressure results in deeper infiltration and fluid loss, and is of little benefit.
- Increasing the bentonite content of the slurry for the coarser soils results in more effective support.

In EPB machines, the muck chamber is filled with excavated soil under pressure. Face instability is potentially a problem only when the

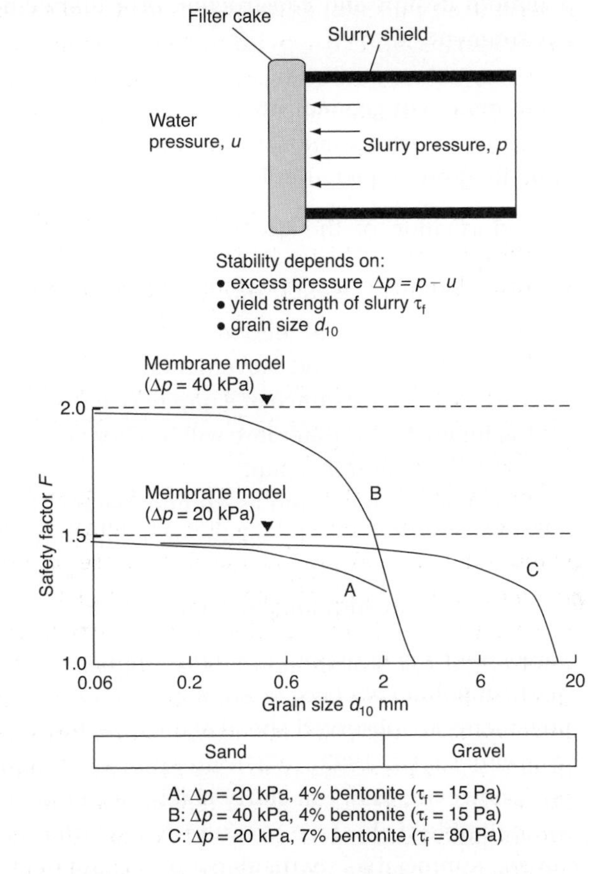

Fig. 3.14 Safety factor against face instability for a slurry shield (after Anagnostou and Kovari, 1996; and Mair and Taylor, 1997)

piezometric head in the muck chamber is lower than the piezometric head in the ground, so that seepage forces act towards the tunnel face. Anagnostou and Kovari provide normalized charts for assessing the tunnel face stability under these conditions.

Ground movements

Introduction

Mair and Taylor (1997) point out that the construction of bored tunnels in soft ground inevitably causes ground movements. In the urban environment these may be of particular significance, because of their influence on buildings, other tunnels and services. The prediction of ground movements and the assessment of the potential effects on the infrastructure are therefore an essential aspect of the planning, design and construction of a tunnelling project in the urban environment.

Components of ground movement

The primary components of ground movement associated with shield tunnelling are depicted in Fig. 3.15. These are as follows:

1. Deformation of the ground towards the face resulting from stress relief.
2. Passage of the shield: the presence of an over-cutting edge (bead) combined with any tendency of the machine to plough or yaw will lead to radial ground movements.
3. Tail void: the existence of a gap between the tailskin of the shield and the lining means that there will be a tendency for further radial ground movements into this gap.
4. Deflection of the lining as ground loading develops.
5. Consolidation: as pore water pressures in the ground change to their long-term equilibrium values the associated changes in effective stress lead to additional ground movements.

Component 1 is of major importance in many cases, particularly in open-faced tunnelling in clays. In London clay, for example, significant ground movements are observed ahead of the face due to stress relief (Ward, 1969; Mair and Taylor, 1993). When pressurized face tunnelling machines are used, however, either in slurry shield or EPB mode, this component can be negligible if the face pressure is carefully controlled. Component 2 can be appreciable, particularly if the over-cutting edge (bead) is of significant thickness, and if there are steering problems in maintaining the alignment of the shield. Component 3 can be minimized by immediate grouting to fill the tail void (or, in the case of expanded linings, by expanding the lining against the soil at the earliest opportunity). Component 4 is generally small in comparison with the other components once the lining ring is completed. Component 5 can be of importance, particularly when tunnelling in soft clays. In cases where there is no tunnelling shield, for

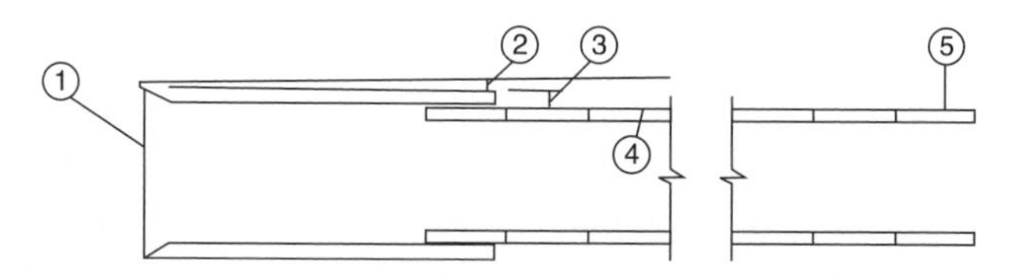

Fig. 3.15 Primary components of ground movement associated with shield tunnelling (after Cording, 1991; and Mair and Taylor, 1997)

example when sprayed concrete linings are used, components 1, 4 and 5 are still applicable.

Transverse surface settlement

For the case of a single tunnel in 'greenfield' conditions, the development of the surface settlement trough above and ahead of the advancing heading is as shown in Fig. 3.16. Following work by Martos (1958) on field observations of settlements above mine openings, Schmidt (1969), Peck (1969) and subsequently many other authors, it has been shown that the transverse settlement trough immediately following tunnel construction is well-described by a Gaussian distribution curve (shown in Fig. 3.17) as:

$$S_v = S_{max} \exp\left(\frac{-y^2}{2i^2}\right) \tag{3.8}$$

where S_v = settlement, S_{max} = maximum settlement on the tunnel centre-line, y = horizontal distance from the tunnel centre-line and i = horizontal distance from the tunnel centre-line to the point of inflexion of the settlement trough.

The volume of the surface settlement trough (per metre length of tunnel), V_s, can be evaluated by integrating Equation (3.8) to give

$$V_s = \sqrt{2\pi} i S_{max} \tag{3.9}$$

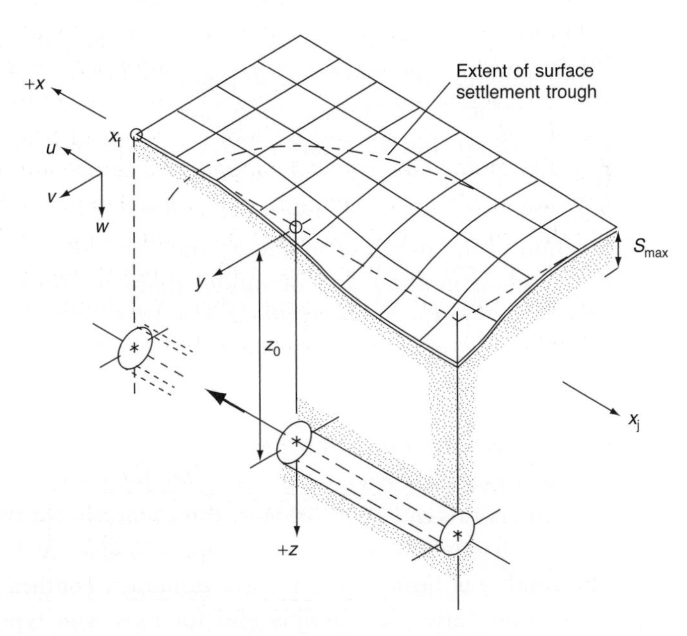

Fig. 3.16 Settlement above advancing tunnel heading (after Attewell et al., *1986; and Mair and Taylor, 1997)*

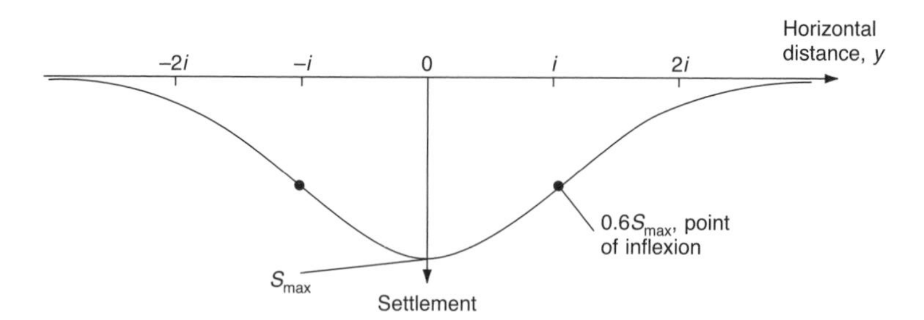

Fig. 3.17 Gaussian curve used to describe the transverse settlement trough (after Mair and Taylor, 1997)

The volume loss, V_t (sometimes referred to as ground loss), is the amount of ground lost in the region close to the tunnel, primarily due to one or more of the components 1–4 in Fig. 3.15. When tunnelling under drained conditions, for example in dense sands, V_s is less than V_t because of dilation (Cording and Hansmire, 1975). When tunnelling in clays, ground movements usually occur under undrained (constant volume) conditions, in which case $V_s = V_t$. Whatever the soil type, it is convenient to express the volume loss in terms of the volume of the surface settlement trough, V_s expressed as a percentage fraction V_l of the excavated area of the tunnel, that is, for a circular tunnel

$$V_s = V_l \frac{\pi D^2}{4} \tag{3.10}$$

Peck (1969) suggested a relationship between the parameter i, tunnel depth and tunnel diameter, depending on the ground conditions, as shown in Fig. 3.18. Many authors have investigated similar relationships (e.g. Cording and Hansmire, 1975; Clough and Schmidt, 1981; Fujita, 1981; O'Reilly and New, 1982; Rankin, 1988). In a survey of UK tunnelling data O'Reilly and New (1982) showed that i is an approximately linear function of the depth of tunnel z_0 and is broadly independent of tunnel construction method and of tunnel diameter (except for very shallow tunnels where the cover to diameter ratio is less than 1). They proposed the simple approximate relationship:

$$i = Kz_0 \tag{3.11}$$

where K is a trough width parameter, and they recommended that for practical purposes K could be taken as 0.5 for tunnels in clays and 0.25 for tunnels in sands and gravels; the database for tunnels in sands was confined to shallow tunnels with depths to axis level in the range 6–10 m. The validity of Equation (3.11) was generally confirmed by Rankin (1988) for a wide variety of tunnels and for most soil types from around the world. Most of the data presented by Rankin (and subsequently by Lake *et al.*, 1992), together with more recent data, are shown in Fig. 3.19 for

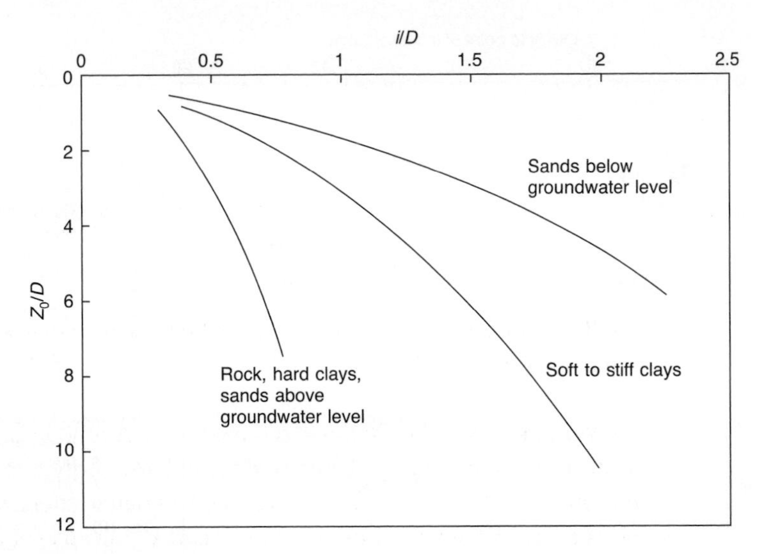

Fig. 3.18 Relationship between settlement trough width parameter and depth of tunnel for different ground conditions (after Peck, 1969; and Mair and Taylor, 1997)

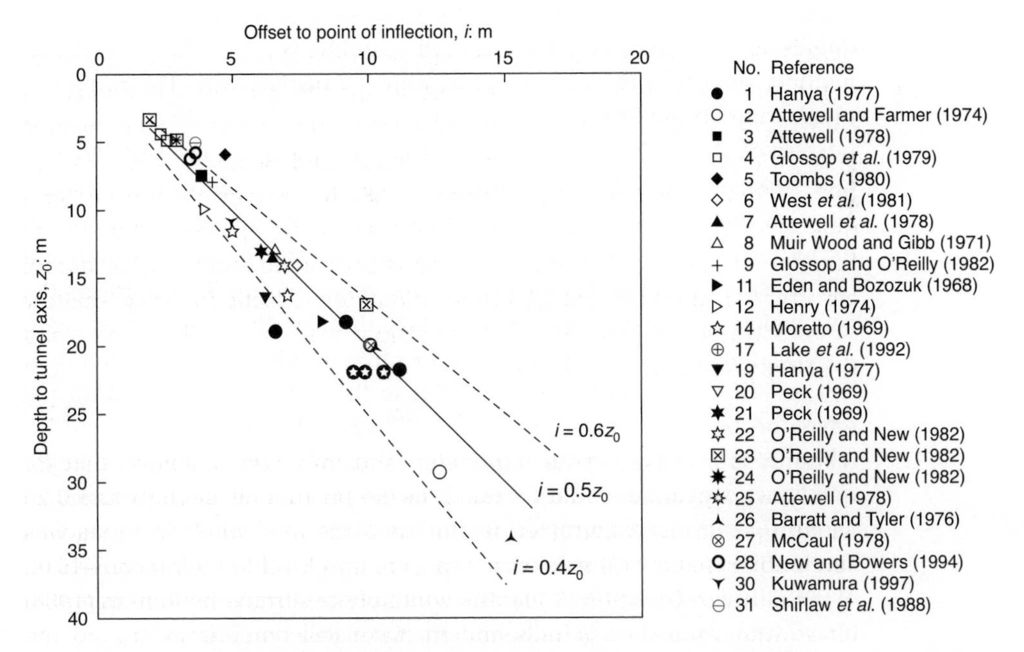

Fig. 3.19 Variation in surface settlement trough width parameter i *with tunnel depth for tunnels in clays (after Mair and Taylor, 1997)*

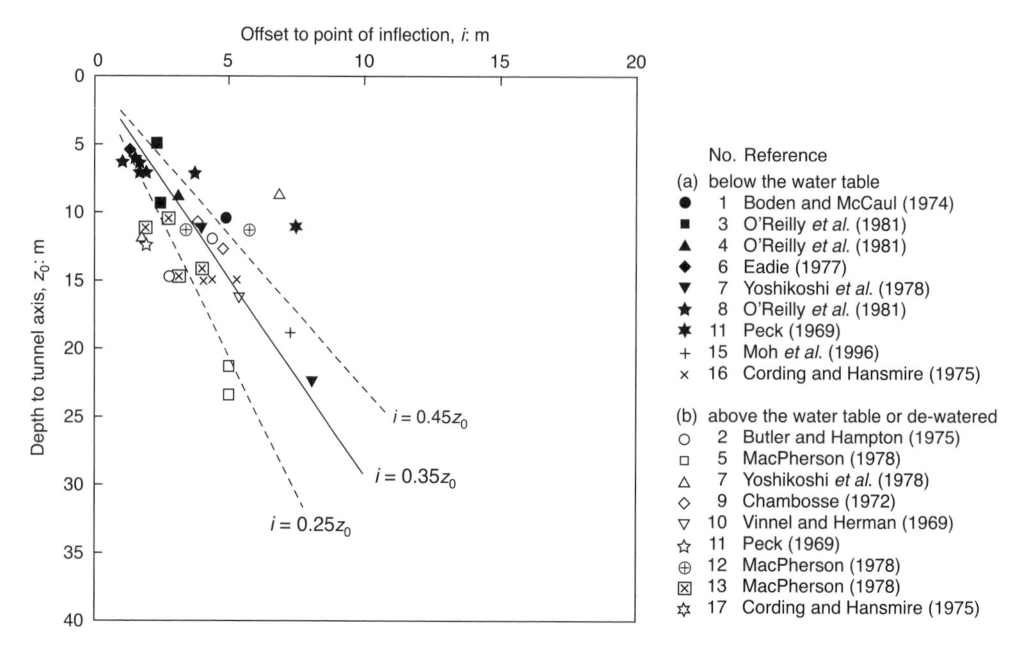

Fig. 3.20 Variation in surface settlement trough width parameter i *with tunnel depth for tunnels in sands and gravels (after Mair and Taylor, 1997)*

tunnels in clays and in Fig. 3.20 for tunnels in sands and gravels. The details of the tunnels, the ground conditions and the excavation methods are given in Table 3.1 and Table 3.2 respectively. The reference numbering on the figures and the tables are consistent with that used by Lake *et al.* (1992), but data have only been included on Figs 3.19 and 3.20 and Tables 3.1 and 3.2 where the ground profile is either predominantly clays or predominantly sands and gravels. Case histories have been excluded where, for example, a tunnel in clay is overlain by significant thicknesses of granular deposits, or vice versa. Figure 3.19 confirms the conclusion of O'Reilly and New (1982) that for the majority of cases $i = 0.5z_0$ for practical purposes, irrespective of whether the tunnel is in soft or stiff clay. There is some scatter in the data, generally within the envelope bounded by $i = 0.4z_0$ and $i = 0.6z_0$. The expression $i = 0.5z_0$ for tunnels in clays is reasonably consistent with the findings of Fujita (1981) who examined data from a large number of case histories in Japan for tunnels constructed using hand-mined shields, blind shields, slurry shields and EPB shields. Fujita generally confirmed the conclusion of O'Reilly and New (1982) that the width of the surface settlement profile above tunnels in clays is independent of construction method.

It should be noted that Equations (3.8) and (3.11) are normally applied to the immediate surface settlements associated with tunnel construction.

Additional post-construction settlement due to consolidation tends to cause wider settlement troughs and this complicates the interpretation of the settlement data. Softer clays are more susceptible to appreciable consolidation settlement, which often develops rapidly and can be difficult to separate from the immediate construction settlement. This may partly explain the observation by Peck (1969) that wider settlement troughs are observed above tunnels in soft clays than in stiff clays.

The data for tunnels in sands and gravels in Fig. 3.20 exhibit rather more scatter than for the tunnels in clays. Two of the data points (Peck, 1969; Yoshikoshi *et al.*, 1978) show considerably wider troughs. Nevertheless, the majority of the data fall within the bounds of $i = 0.25z_0$ and $i = 0.45z_0$ with a mean line of $i = 0.35z_0$. Cording (1991) has noted that the width of the transverse settlement trough above tunnels in granular soils depends to some extent on the magnitude of settlement, with larger settlements tending to cause a narrower overall width of the trough. Figure 3.20 shows that there appears to be no significant difference between tunnels below or above the water table, contrary to the suggestion by Peck (1969).

Tunnels are often constructed in layered strata comprising both clay and granular soils. It has been suggested by Selby (1988) and New and O'Reilly (1991), that the equations for tunnels in clays and tunnels in sands be simply combined taking account of the thicknesses of the different strata, so that for a two-layered system

$$i = K_1 z_1 + K_2 z_2 \tag{3.12}$$

where K_1 is the trough width factor for the soil type in layer 1 of thickness z_1 and K_2 is the trough width factor for the soil type in layer 2 of thickness z_2. Field observations of surface settlement profiles above stratified soils where the tunnel is in sands overlain by clay layers (e.g. Ata, 1996; Atahan *et al.*, 1996) indicate wider profiles than would be obtained if the tunnels were only in sands. There is less evidence, however, of cohesionless layers overlying tunnels in clays causing a narrowing of the surface settlement profile, as implied by Equation (3.12). Indeed, centrifuge model studies by Grant and Taylor (1996) indicate that in the case of a tunnel in soft clay overlain by sand, the surface settlement profile is wider than would be the case if the tunnel were only in the soft clay. This is probably a consequence of the overlying sand layer being significantly stiffer than the soft clay.

Transverse sub-surface settlements

In the urban environment, constraints of existing tunnels and deep foundations often result in new tunnels having to be constructed close beneath such structures. It is becoming increasingly important to predict

Table 3.1 Details of tunnels in clays for which data are plotted in Figure 3.19 (based on Lake et al.*, 1992; after Mair and Taylor, 1997)*

No.	Source	Location	Ground conditions	Excavation methods	Tunnel diameter (m)	Depth to tunnel axis, z_0 (m)	s_{max} (mm)	Inflection offset, i (m)
1	Hanya (1977)	Japan						
1a			Stiff cohesive soil	Slurry shield	7.5	18.3	16	9.2
1b			Stiff cohesive soil	Slurry shield	7.5	21.8	43	11.2
1c			Stiff cohesive soil	Shield, hand excavated	10.7	19.0	22	8.6
2	Attewell and Farmer (1974)	Green Park, UK	Soft OC clay	Shield, hand excavated	4.2	29.3	6	12.6
3	Attewell (1978)	Hebburn, UK	Soft NC clay	Shield, hand excavated	2.0	7.5	8	3.9
4	Glossop *et al.* (1979)	Belfast, Ireland						
4a			Soft silty clay	Shield, hand excavated	2.7	4.9	17	2.7
4b			Soft silty clay	Shield, hand excavated	2.7	4.5	20	2.4
5	Toombs (1980)	Avonmouth, UK	Soft alluvial deposit	Shield, hand excavated	3.4	6.0	13	4.8
6	West *et al.* (1981)	York Way, UK	Stiff OC clay	No shield, hand excavated	4.1	14.1	3	7.5
7	Attewell *et al.* (1976)	Tyneside, UK	Soft silty alluvial clay	Shield, hand excavated	4.3	13.5	23	6.5
8	Muir Wood and Gibb (1971)	Heathrow, UK	Stiff OC clay (London clay)	Shield, hand excavated	10.9	12.9	11	6.6
9	Glossop and O'Reilly (1982)	Haycroft, UK						
9a		section B	Soft silty clay	Shield, hand excavated	3.0	5.5	40	3.6
9b		section A	Soft silty clay	Shield, hand excavated	3.0	6.5	60	3.6
9c		section C	Soft silty clay	Shield, hand excavated	3.0	8.0	68	4.3
11	Eden and Bozoruk (1968)	Ottawa, Canada	Firm OC clay (Leda Clay) extremely sensitive	TBM, rotary	3.0	18.3	6	8.4
12	Henry (1974)	Grangemouth, UK	Soft to very soft laminated silty clay	Shield, hand excavated	2.45 to 2	10.0	25	4.1
14	Moretto (1969)	Buenos Aires, Argentina	Soft to firm clay	Shield, mechanical digger	4.7	16.4	150	7.1
17	Lake *et al.* (1992)	Gateshead, UK	Firm clay, laminated	Hand excavated with timber lagging	1.5 × 1.5	5.3	37	3.7
19	Hanya (1977)	Japan	Firm cohesive soil	Shield, hand excavated	7.3	16.6	63	6.3
20	Peck (1969)	Chicago, USA	Medium clay (Chicago Leda Clay)	Hand excavated	6.1	11.9	23	4.9
21	Peck (1969)	Toronto, Canada	Glacial till	Hand excavated	5.3	13.1	9	6.1

22	O'Reilly and New (1982)	Newcastle, UK	Firm stiff silty clay, glacial till	Partial face machine	5.2	14.2	8	7.0
23	O'Reilly and New (1982)	Sutton, UK						
23a			Stiff fissured clay (London clay)	Hand excavated	1.8	17.1	4	10.0
23b			Firm to stiff weathered clay (London clay)	Hand excavated	1.8	3.4	4	2.0
23c			Firm to stiff weathered clay (London clay)	Full face micro TBM	1.5	4.9	7	3.0
24	O'Reilly and New (1982)	Oxford, UK	Stiff fissured clay	TBM, full face	2.8	11.7	2	5.0
25	Attewell (1978)	Howden, UK	Stiff boulder clay	Hand excavated	3.6	14.2	11	8.9
26	Barrett and Tyler (1976)	Regents Park, UK						
26a		southbound	Stiff OC clay (London clay)	Shield, hand excavated	4.2	34.0	5	15.2
26b		northbound	Stiff OC clay (London clay)	Shield, hand excavated	4.2	20.0	7	10.3
27	McCaul (1978)	Stockton-on-Tees, UK						
27a			Soft to very soft silty clay with sand lenses	Shield, hand excavated	1.3	6.3	44	3.5
27b			Soft to very soft silty clay with sand lenses	Shield, hand excavated	1.3	5.9	56	3.7
28	New and Bowers (1994)	Heathrow, UK						
28a			Stiff OC clay (London clay)	Sprayed concrete lining type I (sidewall only)	–	22.0	21	9.5
28b			Stiff OC clay (London clay)	Sprayed concrete lining type I (complete)	11.3	22.0	28	9.9
28c			Stiff OC clay (London clay)	Sprayed concrete lining type II (sidewall only)	–	22.0	15	10.8
28d			Stiff OC clay (London clay)	Sprayed concrete lining type II (complete)	11.3	22.0	27	9.9
30	Kuwamura (1997)	Chicago, USA	Soft to firm silty clay (Chicago Clay)	Shield	3.7	10.7	18–30	5.0
31	Shirlaw *et al.* (1988)	Singapore, Malaysia southbound tunnel, line 7	Very stiff to hard clay, with boulders	Sprayed concrete lining	6.0	20.0	6	10.1

OC = over-consolidated

Table 3.2 Details of tunnels in sands and gravels for which data are plotted in Figure 3.20 (based on Lake et al.*, 1992; after Mair and Taylor, 1997)*

No.	Source	Location	Ground conditions	Excavation methods	Tunnel diameter (m)	Depth to tunnel axis, z_0 (m)	s_{max} (mm)	Inflection offset, i (m)
1	Boden and McCaul (1974)	London, UK	Sand, medium to coarse, some gravel	Bentonite shield	4.1	10.1	22	5.0
2	Butler and Hampton (1975)	Washington, USA						
2a			Gravel and silty sand	Shield with bucket digger	6.5	14.4	96	2.9
2b		Lafayette Square	Gravel with interbedded silty sand	Shield with bucket digger	6.4	11.6	113	4.5
3	O'Reilly *et al.* (1980)	Warrington, UK						
3a		Lumb Brook sewer	Sand, loose to medium with some gravel	Shield, hand excavated	3.6	4.7	78	2.4
3b			Sand, medium to dense with some clay	Shield, hand excavated	3.6	9.0	19	2.5
4	O'Reilly *et al.* (1980)	Warrington, UK	Sand, loose and silty	Shield, hand excavated, compressed air	2.0	8.4	28	3.2
5	MacPherson (1976)	Washington, USA						
5a		section a	Sand, medium dense silty with interbedded sandy clay	Shield, articulated with digger arm	6.4	20.9	6	5.1
5b		section b	Sand, medium dense silty with interbedded sandy clay	Shield, articulated with digger arm	6.4	23.0	3	5.1
6	Eadie (1977)	Ayrshire, UK						
6a		Ayrshire joint drainage scheme	Sand, silty, fine to medium	Shield, hand excavated	2.9	5.1	14	1.4
6b			loose silty sand with little gravel	Shield, hand excavated	2.9	5.7	16	1.6
7	Yoshikoshi *et al.* (1978)	Tokyo, Japan						
7a		site II	Sand, fine silty	Open shield	3.0	8.5	11	7.0
7b		site III	Sand, fine silty	Blind shield	3.7	22.1	32	8.2
7c		site IV	Sand, fine silty	Blind shield	3.9	10.9	19	4.1
8	O'Reilly *et al.* (1980)	Warrington, UK						
8a		Acton Grange sewer	Sand, fine to medium, fairly uniform, occasional gravel	Bentonite shield	2.9	6.8	20	1.8
8b				Bentonite shield	2.8	6.8	14	2.0
8c				Bentonite shield	2.9	6.0	42	1.1
8d				Bentonite shield	2.9	6.0	19	1.8
8e				Bentonite shield	2.9	5.8	25	1.6
8f			Fully stabilized	Bentonite shield	2.9	6.9	2	3.8

9	Chambosse (1972)	Frankfurt, Germany						
9a		Frankfurt metro T-9	Sand with some limestone and clay marl lenses	Shield	6.5	12.4	70	4.9
9b		Frankfurt metro, Dominikanerga	Sand with some limestone and clay marl lenses	Shield	6.5	10.5	140	3.9
10	Vinnal and Harmon (1969)	Brussels, Belgium	Sand, uniform (top half of face), clayey sand to invert	Shield, hand excavation	10.0	16.0	150	5.5
11	Peck (1969)	Toronto, Canada						
			Sand, dense	Shield, hand excavated	5.4	12.0	85	2.0
			Sand at crown, invert in till	Shield, hand excavated	5.4	10.7	12	7.5
12	MacPherson (1978)	Washington, USA						
12a			Sand, medium dense silty with interbedded sandy clay	Shield, articulated with digger arm	6.0	11.0	46	3.5
12b			Sand, medium dense silty with interbedded sandy clay	Shield, articulated with digger arm	6.0	11.0	56	5.9
13	MacPherson (1978)	Illinois, USA						
13a		Rockford, Illinois	Sand, medium dense with some gravels	Mechanical shield	3.0	10.8	25	2.0
13c				mechanical shield	3.0	14.4	46	3.2
13d				Mechanical shield	3.0	13.8	23	4.1
13e				Mechanical shield	3.0	10.8	38	2.0
13f				Mechanical shield	3.0	10.2	15	2.8
15	Moh *et al.* (1996)	Taipei, Taiwan	Sand, silty (upper 1/3 of tunnel face), lower in silty clay	EPBM	6.1	18.5	26	7.4
16	Cording and Hansmire (1975)	Washington, USA						
16a		A line	Sands and gravels	Mechanical shield	6.4	14.6	76	5.4
16b		B line	Sands and gravels, medium dense, interbedded with silts	Mechanical shield	6.4	14.6	139	4.2
16c		C line	Sands and gravels, medium dense, interbedded with silts	Mechanical shield	5.4	14.6	152	4.5
17	Cording and Hansmire (1975)	Washington, USA	Sands and gravels	Shield	6.4	11.6	280	1.9

how sub-surface settlement profiles develop and how they relate to surface settlement troughs. Mair *et al.* (1993) analysed sub-surface data from various tunnel projects in stiff and soft clays, together with centrifuge model test data in soft clays. They showed that sub-surface settlement profiles can also be reasonably approximated in the form of a Gaussian distribution in the same way as surface settlement profiles. At a depth z below the ground surface, above a tunnel at depth z_0 the trough width parameter i can be expressed as

$$i = K(z_0 - z) \tag{3.13}$$

and the value of K increases with depth as shown in Fig. 3.21. This shows that sub-surface settlement profiles at depth are significantly wider than would be predicted by assuming a constant value of K. Figure 3.21 is

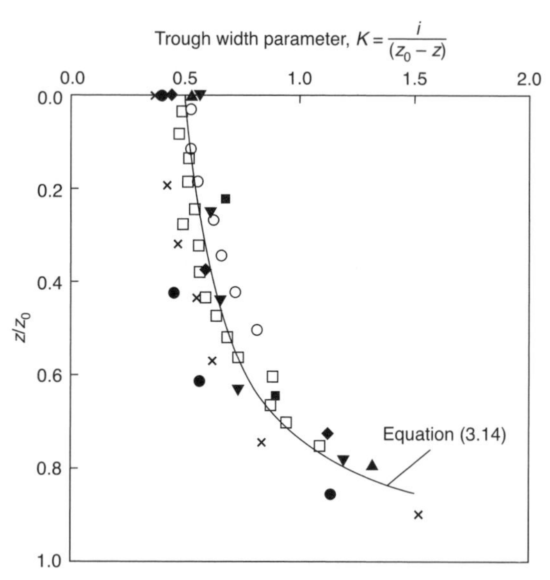

Location	Soil type	D: m	z_0: m	Reference
● Green Park	London Clay	4.2	29	Attewell and Farmer (1974)
▲ Regents Park (northbound)	London Clay	4.2	20	Barrat and Tyler (1976)
▼ Regents Park (southbound)	London Clay	4.2	34	Barrat and Tyler (1976)
■ Willington Quay	Soft clay	4.3	13.5	Glossop (1978)
◆ Heathrow Express	London Clay	11.3	22	New and Bowers (1994)
× St. James's Park (westbound)	London Clay	4.85	31	Nyren (1998)
○ Centrifuge* model 2DP	Soft clay	0.06	0.13	Mair (1979)
□ Centrifuge* model 2DV	Soft clay	0.06	0.22	Mair (1979)

*Models tested at 75 g: equivalent full-scale $D = 4.5$ m, $z_0 = 9.8$ m (2DP), 16.5 m (2DV)

Fig. 3.21 Variation in trough width parameter K *with depth for sub-surface settlement profiles above tunnels in clays (after Mair* et al.*, 1993; and Mair and Taylor, 1997)*

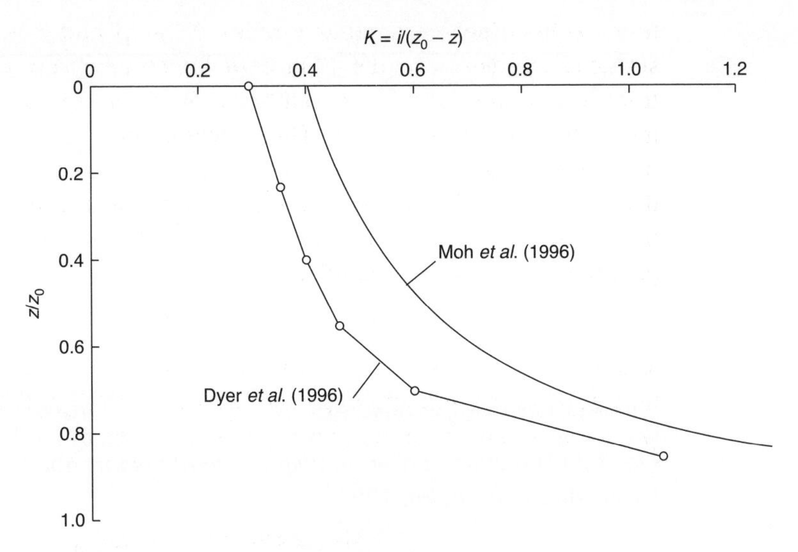

Fig. 3.22 Variation in trough width parameter K *with depth for sub-surface settlement profiles above tunnels in sands (after Mair* et al., *1993; and Mair and Taylor, 1997)*

taken from Mair *et al.* (1993) with some additional data points obtained from the Heathrow Express trial tunnel (New and Bowers, 1994), and the Jubilee Line Extension project in London (Nyren, 1998).

The expression for K proposed by Mair *et al.* (1993), and shown in Fig. 3.21, is

$$K = \frac{0.175 + 0.325(1 - z/z_0)}{1 - z/z_0} \tag{3.14}$$

Similar observations of sub-surface settlement profiles above tunnels in silty sands below the water table in Taipei were made by Moh *et al.* (1996), and above a tunnel in loose sands overlain by a firm to stiff clay layer by Dyer *et al.* (1996). Figure 3.22 shows the variation with depth of K (defined as in Equation (3.13)) presented by Dyer *et al.* together with the variation derived from the data reported by Moh *et al.* A similar pattern of increasing K with depth is evident, as observed for tunnels in clays.

Horizontal movements

Damage to structures and services can arise from horizontal movements (Burland *et al.*, 2001). However, there are relatively few case histories of tunnels where horizontal ground or structure movements have been measured. Attewell (1978) and O'Reilly and New (1982) proposed that, for tunnels in clays, ground displacement vectors are directed

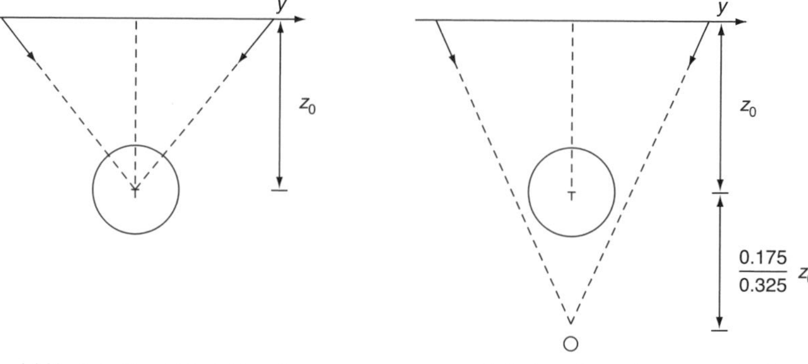

Fig. 3.23 Direction of ground displacement vectors above tunnels in clays (after Mair and Taylor, 1997)

towards the tunnel axis, as shown in Fig. 3.23(a). This leads to the simple relation:

$$S_h = \frac{y}{z_0} S_v \qquad (3.15)$$

This assumption leads to the distribution of surface horizontal ground movement given by

$$\frac{S_h}{S_{h\,max}} = 1.65 \frac{y}{i} \exp\left(\frac{-y^2}{2i^2}\right) \qquad (3.16)$$

The theoretical maximum horizontal movement $S_{h\,max}$ occurs at the point of inflexion of the settlement trough and is equal to $0.61KS_{max}$. This is consistent with field observations by Cording and Hansmire (1975) and Attewell (1978).

Assuming Equation (3.15) to be valid, and if it is also assumed that the trough width parameter K is constant with depth, the vertical and horizontal strains determined by differentiating expressions for vertical and horizontal movement are equal and opposite. This is a necessary condition for undrained (constant volume) movements around tunnels in clays. A variation of K with depth, such as is illustrated in Fig. 3.21 and defined in Equation (3.14), affects the related vertical and horizontal strains. For a constant volume condition, applicable to tunnelling in clays, it turns out that for this variation in K the displacement vectors should be directed towards a point on the tunnel centre-line $0.175z_0/0.325$ below tunnel axis level (Taylor, 1995) as shown in Fig. 3.23(b). This gives horizontal movements of 65% of those that would be obtained by assuming the ground movements to be directed towards the tunnel axis. Deane and Bassett

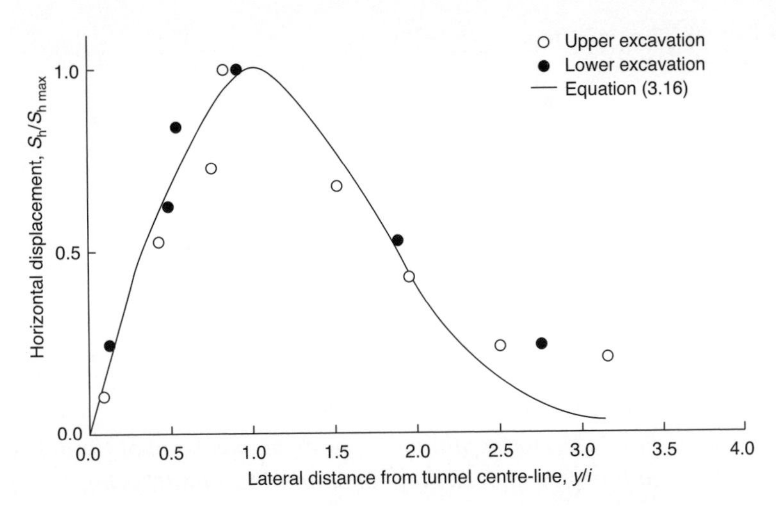

Fig. 3.24 Distribution of horizontal ground surface movements above a tunnel (after Hong and Bae, 1995; and Mair and Taylor, 1997)

(1995) analysed sub-surface movement measurements for two sections of the Heathrow Express trial tunnels in London clay. They concluded that the displacement vectors were directed towards a point midway between tunnel axis level and invert level in one case, and towards a point at (or possibly even below) the invert in the second case.

For tunnels in sands, even the assumption of ground movements being directed towards the tunnel axis may lead to significant underestimates of horizontal ground movement at the ground surface near the edge of the settlement trough (Cording, 1991). This was also observed by Hong and Bae (1995) for a 10 m diameter NATM tunnel in predominantly sandy strata in Korea. Their data are presented in Fig. 3.24, together with Equation (3.16). The distribution is in reasonable agreement with Equation (3.16) except in the region near the edge of the settlement trough ($2i < y < 3i$). However this is usually of little practical significance because the magnitudes of both horizontal and vertical ground surface movements are generally very small near the edge of the settlement trough.

Subsurface horizontal ground movements at tunnel axis level in stiff clays are generally inwards towards the tunnel when open-face tunnelling methods are used. Figure 3.25 shows measurements for five tunnels each approximately 4 m in diameter constructed in London Clay and lined with segments. Four of the tunnels were constructed using a shield, and one of them by hand methods without a shield. For undrained and axi-symmetric conditions, the ground movement δ at a radius r would be proportional to $1/r$.

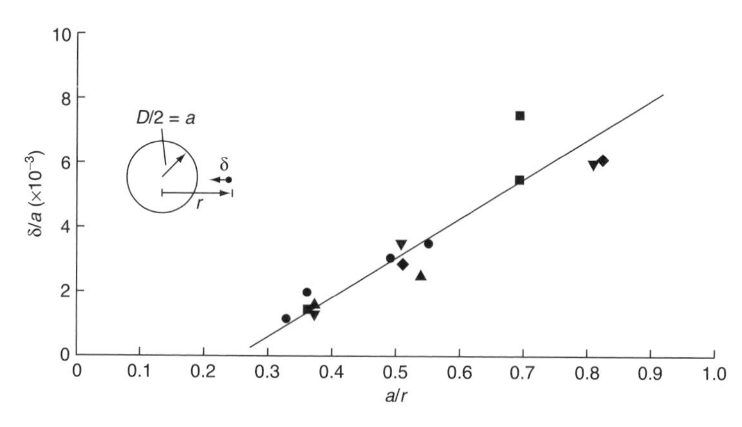

Fig. 3.25 Horizontal movements at axis level of tunnels in London Clay (after Mair and Taylor, 1993; and Mair and Taylor, 1997)

Figure 3.25 shows the movements plotted normalized by tunnel radius (Mair and Taylor, 1993). Zero movement is implied at distances in excess of one-and-a-half tunnel diameters beyond the tunnel boundary at tunnel axis level.

In the case of EPB tunnelling in soft clays, sub-surface horizontal movements at tunnel level may be either inwards or outwards, depending on the bulkhead pressure. Clough *et al.* (1983) measured horizontal movements at four instrument lines for a 3.7 m diameter EPB shield in San Francisco Bay Mud. Where the bulkhead pressures were high, initial outward movements exceeded the subsequent inward movements into the tail void, whereas the reverse was the case elsewhere. Both inward and outward movements around EPB shields in soft clay are also reported by Fujita (1994).

By examining a number of case histories of tunnel construction in clays Attewell and Woodman (1982) showed the cumulative probability curve to be reasonably valid for the longitudinal settlement trough as well as the transverse settlement trough. They found that the surface settlement directly above the tunnel face generally corresponds to about $0.5S_{max}$ for tunnels constructed in stiff clays without face support (see Fig. 3.26). However, for tunnels constructed in soft clays with face support provided by compressed air, the surface settlement directly above the tunnel face was considerably less than $0.5S_{max}$. Pressurized face tunnelling tends to restrict ground settlements developing ahead of the tunnel face (indeed significant heave can be observed in soft clays). Field observations of settlements above EPB or slurry shield tunnelling machines indicate that the majority of the construction settlement is associated with the tail void (component 3 in Fig. 3.15) and that the surface settlement directly above the tunnel face is generally much less than $0.5S_{max}$.

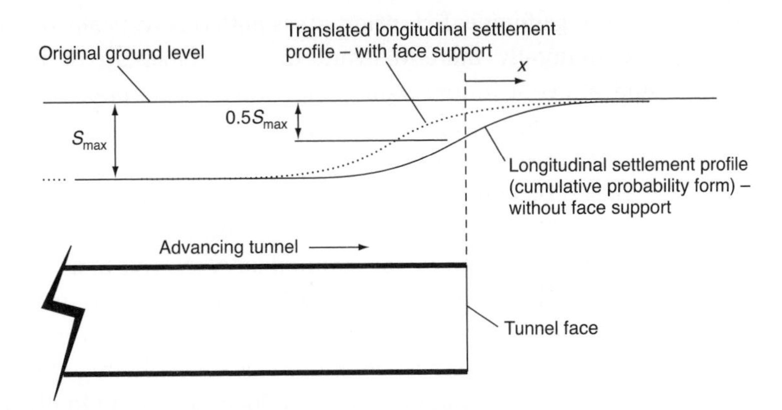

Fig. 3.26 Longitudinal surface settlement trough (after Mair and Taylor, 1997)

It can be concluded that the longitudinal settlement trough having the form of the cumulative probability curve illustrated in Fig. 3.26 is generally reasonable, but it has only been validated for tunnels in clays. The surface settlement equal to $0.5S_{max}$ above the tunnel face is strictly only applicable to open-face tunnelling techniques in stiff clays. Where there is significant face support, as in EPB or slurry shield machines, the major source of ground movement is further back from the face and this leads effectively to a translation of the cumulative curve, as shown in Fig. 3.26.

Volume loss

The magnitude of volume loss V_l (defined in Equations (3.9) and (3.10)) depends principally on the type of ground and on the tunnelling method. Many authors have reviewed volume loss values from tunnelling projects (e.g. Peck, 1969; Cording and Hansmire, 1975; Clough and Schmidt, 1981; O'Reilly and New, 1982; Attewell *et al.*, 1986; Uriel and Sagaseta, 1989; Mair, 1996).

For tunnels in clays, Clough and Schmidt (1981) proposed a relationship between stability ratio (or overload factor) N and V_l based on the closed form solution for the unloading of a circular cavity in a linear elastic–perfectly plastic continuum under axi-symmetric conditions. Attewell *et al.* (1986) and Uriel and Sagaseta (1989) presented field data of volume losses related to stability ratio, based on Clough and Schmidt's proposal; the results show a very wide scatter. The wide scatter is probably associated with many construction details and differing standards of workmanship. Another important factor is the value of undrained shear strength assumed in the calculation of stability ratio N. In a number of cases the influence of sample disturbance and the method of laboratory testing (usually triaxial compression tests) may have led to erroneous values of undrained strength

s_u being adopted. Extension stress paths on vertically orientated specimens are generally more relevant to the unloading of the ground around and above a tunnel than compression stress paths. In soft clays, triaxial extension tests typically give significantly lower undrained shear strengths than triaxial compression tests.

Based on centrifuge model test data and finite-element analyses, Mair *et al.* (1981) and Mair (1989) proposed that V_l should be more properly related to the load factor, defined as N/N_c (N_c being the critical stability ratio), rather than to N alone. Despite the different stress histories and C/D ratios, which meant that the volume losses were markedly different at the same stability ratio, there was a reasonably well-defined relationship between volume loss and load factor. O'Reilly (1988) used the approach to provide good predictions of volume loss at six different tunnelling sites in London Clay.

When closed-face tunnelling methods are employed, using EPB or slurry shields, good control of the face pressure can result in the stability ratio being close to zero, in which case the component of ground movement resulting from stress relief at the face would be very small, leading to smaller volume losses (less than 1%). In such cases in soft clays the principal cause of volume loss is usually the tail void (Broms and Shirlaw, 1989).

Recent experiences with EPB and slurry shield machines in sands and gravels have generally shown small volume losses. Leblais and Bochon (1991) report volume losses in the range 0.2–0.9% for 9.25 m diameter tunnels driven through dense fine Fontainebleau sands at depths ranging from 22 m to 52 m. Values of 0.8–1.3% were observed when the tunnels were very shallow with the tunnel crown being only 4.1–7.2 m below the ground surface. Volume losses reported by Ata (1996) for a 9.48 m diameter slurry shield in Cairo at a depth of about 16 m in medium to dense sands below the water table were in the range 0.2–1.0%, with a mean of about 0.5%.

For settlement associated with bored tunnels, Mair (1996) concluded the following:

- For open-face tunnelling, volume losses in stiff clays such as London clay are generally between 1% and 2%.
- Construction with sprayed concrete linings (NATM) is effective in controlling ground movements. Recent construction in London clay, for example, has resulted in volume losses varying from 0.5–1.5%, which compares favourably with well-controlled shield tunnelling in which there is little or no face support.
- For closed-face tunnelling, using EPB or slurry shields, a high degree of settlement control can be achieved, particularly in sands where

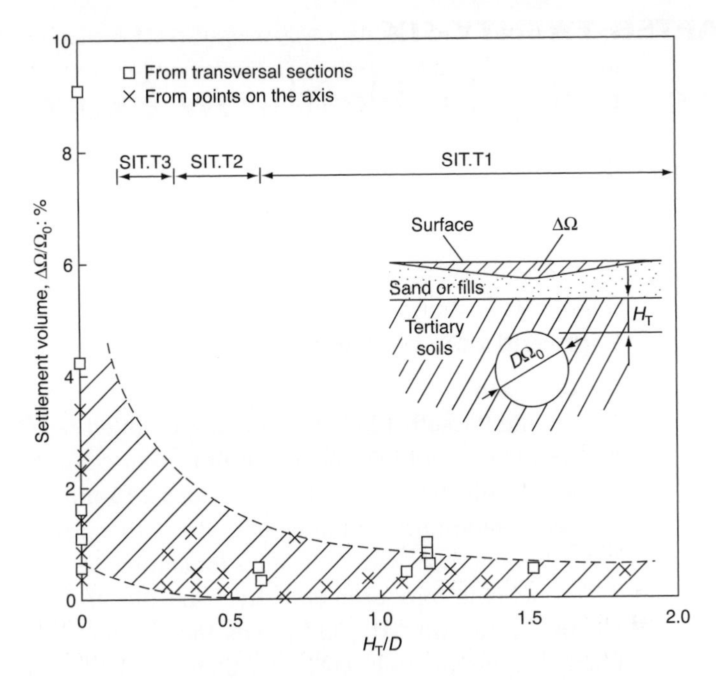

Fig. 3.27 Influence on volume loss of cover of competent soil in mixed ground conditions (after Mair and Taylor, 1997; and Melis et al., *1997)*

volume losses are often as low as 0.5%. Even in soft clays, volume losses (excluding consolidation settlements) of only 1–2% have been reported.

Volume losses may be higher in mixed face conditions for EPB or slurry shields, particularly where sands or gravels overlie stiff clays, or where the cover of competent soil above the tunnel crown is low. This is illustrated by Melis *et al.* (1997), as shown in Fig. 3.27. A 7.4 m diameter EPB shield was used in very stiff to hard sandy clays in Madrid with varying cover (H_T) to overlying sands or fills. The EPB was used in open-face mode when the ratio H_T/D exceeded 0.6, and they state that the observed volume losses were in the range 0.03–1.0%. Larger volume losses were observed for lower H_T/D ratios and significantly larger values (generally 2–4%) were obtained for mixed face conditions.

CHAPTER TWENTY-SIX

Case study: the Heathrow Express trial tunnel

Design and construction

Introduction

Deane and Bassett (1995) describe this detailed case study. The Heathrow Express is now a high-speed rail link from central London to Heathrow Airport. Dedicated rolling stock provides a 15-minute non-stop rail service between Paddington Station and the two current passenger terminal areas at the airport with provisions for connection to the future Terminal 5.

The scheme was promoted by BAA plc, the owner and operator of Heathrow Airport. At 1992 prices the estimated project cost was £260 million, approximately 80% of which was funded by BAA and 20% by British Rail.

The route follows the main Paddington South Wales line out of London before turning south on twin tracks in a retained cut and a cut-and-cover box section. Conventional twin bored tunnels complete the remainder of the running tunnels under the M4 motorway to a station in the centre of the airport. Beyond this point the two tunnels join to provide a single track to a second station on the south side of Heathrow. Both stations consist of twin 200 m long platform tunnels, a central concourse with escalator and lift connections to the terminals, together with the extensive ventilation tunnels and emergency escape shafts required by a modern underground railway.

The station for the London Underground Piccadilly Line in the airport central area was constructed in 1970 using cut-and-cover methods. With the subsequent development in the centre and on the south side of Heathrow it was clear from an early stage in the project's development that many of the facilities for the new stations would have to be constructed in bored tunnels.

Geology

The general geological sequence at the airport comprises

- made ground typically 2 m thick
- terrace gravels between 2 and 4 m thick with standard penetration test valves in the range 20–50 and permeability approximately 10^{-5} m/s

- London clay at least 45 m thick (the strength of the clay varies with depth from 50–250 kPa).

Below the clay are the Woolwich and Reading Beds and then the chalk aquifer.

Scheme design

For the development of the Heathrow Express, NATM (New Austrian Tunnelling Method) offered a number of advantages over segmental tunnelling

- immediate and consistent support of freshly excavated ground
- the flexibility to allow non-circular tunnels
- easier construction of large junctions and complex intersections
- rapid mobilization using standard excavation equipment
- lower capital cost of major equipment
- elimination of shield erection and dismantling chambers.

Settlement

As with any tunnelling scheme in an urban environment, one of the primary concerns at Heathrow is potential ground movement. Many of the tunnels in the two stations are below major structures such as terminal buildings, multi-storey car parks and the London Underground Piccadilly Line running tunnels.

Despite a reasonable track record in soft ground, particularly in southern Germany and Japan, NATM has not been widely used in the UK. In an extensive literature search, Frankfurt and São Paulo were the only locations found with NATM tunnels constructed in ground similar to London clay. Back-analysis of the data from these locations provided initial predictions of likely settlement. These suggested that an acceptable scheme could be developed. However, in view of the many important and sensitive structures above the Heathrow Express tunnels and the uncertainty that still existed with the accuracy of any settlement prediction it was proposed that a trial tunnel should be built prior to the main works going out to bid. The BAA accepted this recommendation and authorized the construction of the trial.

Design of the trial tunnel

The new stations required three similar parallel tunnels each of 9 m excavated width, two platform tunnels and a central concourse. It was therefore decided to use the profile of one of these tunnels as that of the trial. This would reduce the interpretation of the trial tunnel results which would be required to estimate the ground movements in the more sensitive areas.

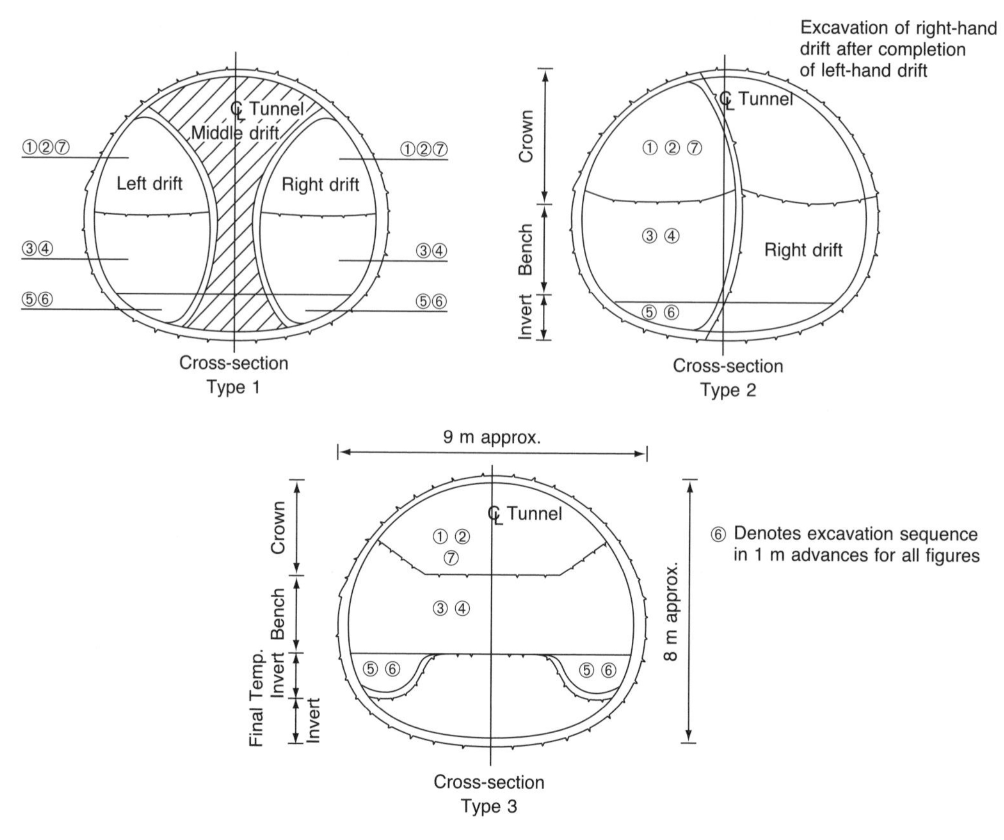

Fig. 3.28 Cross-sections of NATM trial showing construction sequences (after Deane and Bassett, 1995)

The tunnel profile was designed to satisfy the space requirements of the platform tunnels without providing too much unnecessary cross-section; preliminary analysis of a number of outlines confirmed that rapid change in radius around the section would result in unacceptably high local stresses in the shotcrete and this restraint, combined with a decision based principally on engineering judgement that a circular crown would assist in reducing ground movements, resulted in the chosen profile (Fig. 3.28).

Some of the segmental tunnel linings constructed in London Clay were instrumented. The stresses measured showed load from the ground increasing over a long period with the lining finally carrying a high percentage of the full overburden load.

Although the trial tunnel was eventually to be lined with an inner skin of in situ concrete, the overall construction programme required that the shotcrete would have to stand for some considerable time before this support was installed. It was therefore necessary to design the primary support for this long-term load rather than any short-term loading more usually associated with NATM in soft ground.

Using mean and worst-case parameters for the London Clay, finite-element analyses were carried out to estimate the possible effects of soil–structure interaction on the loads in the shotcrete. The analysis consisted of a sequence of steps, starting with the in situ stresses in the ground, followed by the various individual heading excavations which made up the tunnel construction.

The three-dimensional nature of the tunnel excavation, and the stress release that occurs in the ground as the face is advanced, were modelled by reducing the modulus of the ground within the tunnel line in a calculation step prior to the installation of the shotcrete lining. The magnitude of this stiffness reduction was based primarily on experience from previous tunnels.

The resulting design consisted of a 250 mm shotcrete shell with a single layer of steel mesh and lattice girders at 1 m centres along the tunnel. Two 3 m dowels installed in the clay at approximately 40° either side of the crown in each metre of tunnel completed the structural system.

There are a large number of construction methods available with NATM and these could not all be tested in the trial. It was felt that the sequence of excavation was likely to be the most critical in determining settlement. It was therefore decided to drive 100 m of trial tunnel and test three different excavation sequences each over a length of 30 m.

In order to simplify interpretation of the results, the same construction specification for tunnel cross-section, shotcrete thickness, mesh and lattice girder spacing was maintained through all three trial lengths.

The three sequences were chosen to represent the range that might be used during the main construction works. The first and most conservative consisted of two side headings followed by the removal of the central core of the tunnel. The second was to excavate one side of the tunnel and then enlarge it to its full size. The third was a top heading and bench sequence with the bottom of the shotcrete arch of the heading supported on inverted shotcrete arches to try to limit the settlement which might be caused by the feet of the arch pushing into the clay of the bench. The three sequences, noted as Types 1, 2 and 3, are shown in Fig. 3.28.

Location and instrumentation

The trial was located on the line of the running tunnel from the Central Terminal Area to the Terminal 4 station. An escape shaft is required approximately 700 m west of Terminal 4. This shaft was constructed and the trial tunnel driven eastwards towards the Terminal. It has therefore been possible to incorporate these works in the main scheme for the Heathrow Express. The axis of the trial tunnel is approximately 19 m below ground level, 14 m below the top of the clay. The shaft is located in a paved area previously used for staff parking and the tunnel lies under an access road and a second paved parking area. It therefore represents virtually a greenfield site with no significant structures in the area to distort the ground movement measurements taken.

A large amount of instrumentation was installed in and around the trial. Along the line of the tunnel and at three transverse sections levelling pins were positioned just below the sub-base of the road or car park, at 5 m centres. Movements of these points were measured using precise levelling and an EDM theodolite.

To monitor displacements within the ground, rod extensometers and magnetic ring inclinometers were installed in boreholes along two of the cross-sections. Adjacent to these were piezometers and pressure cells in parallel holes.

Within the tunnel, convergence pins were used to measure the movement of the lining. To avoid the problem of interfering with construction progress while a tape was stretched across the tunnel to measure convergence, the pins were fitted with light-emitting diodes and readings were taken using an EDM theodolite fitted with a data logger. The data could then be transferred directly to a PC and convergence plots produced extremely rapidly.

Pressure cells were also installed to measure the stress in the shotcrete and between the ground and the shotcrete lining. The cells were oil or mercury filled and read by way of a vibrating wire transducer. All readings were then taken directly using the PC via purpose-written software. This allowed over 100 cells to be read in the space of 10 min at any chosen time interval. Initially, readings were taken every 6 h, reducing to a daily reading one week after the tunnel was complete at the cell location.

Construction

The trial tunnel contract was let to the Kier–Lilley–Kunz joint venture in January 1992. Work started on site in February and was completed in early June.

The 10.65 m internal diameter shaft was sunk to a depth of 16 m using caisson and underpinning methods. This length of the shaft was lined with bolted precast concrete segments.

The bottom 8 m of the shaft was excavated in 1 m steps and lined with shotcrete. Two tunnel-eyes consisting of stiffened shotcrete rings dowelled into the clay were constructed within this 8 m length. The trial tunnel was driven out of one of these eyes and the second will be used for a running tunnel TBM (tunnel boring machine) to start its drive to the central terminal area during the main construction contract.

From the shaft there is a short transition length to allow the cross-section to change from 6 m circular eye to the 9 m elliptical section of the main trial. Following this transition, the first 30 m of tunnel was driven using the Type 1 excavation sequence. A specific requirement of this sequence was that the removal of the central core lagged at least 20 m behind the two side headings. In practice, the contractor chose to excavate and line the full 30 m of the sides before removing the central core.

Similarly, for both the Types 2 and 3 excavation sequences the first drive, side heading or crown and bench was completed for the full 30 m length before the remainder of the tunnel was excavated and lined. At the end of each trial length a domed shotcrete headwall was constructed to make a distinct break between the two adjacent trial lengths.

The approximate durations for the various phases are given below.

- Type 1 excavation: Side headings – 20 days, central core – 9 days
- Type 2 excavation: First side heading – 12 days, remainder of tunnel – 12 days
- Type 3 excavation: Crown and bench – 15 days, invert – 4 days

The short lengths of each excavation type meant there was a learning curve to be overcome, but it is clear, as expected, that the three types were arranged in order of ascending speed of construction and therefore, it would be expected, in order of descending cost.

All excavation in the tunnel, except for very minor hand trimming, was carried out with a Liebherr 902 excavator. This machine had a second articulation in the boom over a normal back-acter. This allowed great manoeuvrability and a very accurate cut. Although this was an excellent piece of plant for the tunnel, credit must also go to the excavator drivers for the good profile achieved. Particularly in the Type 1 excavation side headings, the space available for the excavator was extremely limited and the large machine was manoeuvred with great skill.

Because of the short length of the trial, the contractor chose to transfer the excavated clay from the face to the shaft using a Volvo front loader. The material was lifted up the shaft in 2 m skips using an overhead gantry crane and then deposited in a stockpile before being loaded onto trucks to be taken to tip. Dry mix shotcrete was batched on the surface adjacent to the shaft and moved pneumatically to the tunnel face through a fixed steel main. Liquid accelerator and water were added at the nozzle (Fig. 3.29).

The dowels used consisted of 25 mm high tensile reinforcing bar threaded at one end. Some dowels, particularly those around the shaft, were placed in drilled holes and grouted but the majority of the dowels in the tunnel were placed in a 1 m deep drilled hole and rammed in the

Fig. 3.29 Centre crown of Type 1 section being grouted (after Deane and Bassett, 1995)

remaining 2 m using the bucket of the excavator. When it was attempted to ram the full 3 m the dowels usually buckled.

The general excavation sequence is given on the cross-sections of each construction type in Fig. 3.28. For the first two 30 m lengths the sequence was designed to provide complete rings of shotcrete as close to the excavated face as possible. It was assumed during design, and proved to be the case in construction, that primary ground movements were arrested by the closing of these stiff rings,

For the third excavation type this procedure was not adopted. A full ring was not completed until the invert was excavated and shotcreted. It was hoped that this or a similar method, which is faster to construct than Types 1 and 2, could be used in areas of the stations where settlement is less critical.

The installation of the tunnel instruments, particularly the pressure cells, caused some disruption to the construction programme but once the instruments were in place the decision to use theodolite reading convergence points and remote reading vibrating wire transducer pressure cells was fully justified. If the tunnel construction had to stop to allow the reading of convergence (at least 18 measurements per trial length) and a similar number of pressure cells, the delay to the programme would have been considerable and comparison of the results with the main works, which would not contain the same quantity of instrumentation, would have been made significantly more difficult.

Review of measured ground movements

Figure 3.30 shows the final settlement data for all three cases reflected about the trough centre-line (not the tunnel centre-line). Based on the measured half-width profiles in Fig. 3.30, the volumes of the surface trough expressed as a percentage of the excavated volume for each method of construction are given in Table 3.3.

For the Type 3 full-face construction it is possible to abstract an acceptable indication of the full 'nose' form of the trough. Work started on 11 May 1992. The construction process had become consistent by 14 May, and thereafter advanced at 2 m/day. Data measured at the instrumented section were therefore plotted ahead of and behind an imaginary face based on this rate of advance.

The potential damage to structures is assessed on the basis of the differential settlement experienced by adjacent foundations and the linear strain between them. Using the definitions in Boscardin and

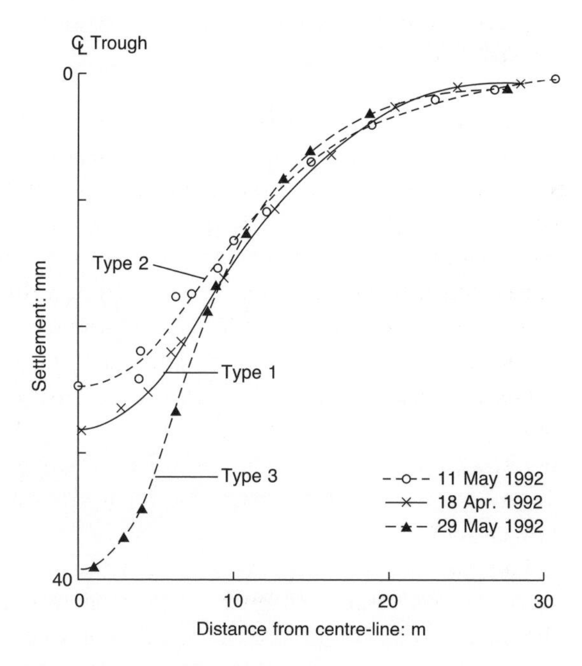

Fig. 3.30 Final settlement data for all Types reflected about trough centre-line (after Deane and Bassett, 1995)

Table 3.3 Volume loss correlated with construction type for the Heathrow trial tunnel (Deane and Bassett, 1995)

Construction	Volume loss: %
Type 1	1.13
Type 2	1.06 (based on mean shape)
Type 3	1.33

Table 3.4 Principal criteria for determining damage for the Heathrow trial tunnel (after Boscardin and Cording, 1989; Deane and Bassett, 1995)

Construction type	Maximum tilt	Maximum relative rotation	Mean tilt	Angle of distortion
Type 1	1 in 500	1 in 550	1 in 1100	1 in 2300
Type 2	1 in 550	1 in 650	1 in 1400	1 in 3500
Type 3	1 in 250	1 in 260	1 in 1000	1 in 620

Cording's work, the principal displacement criteria are listed in Table 3.4.

Figure 3.31 shows the contours for the completed Type 2 construction. The vertical displacement contours show an 'onion' type pattern,

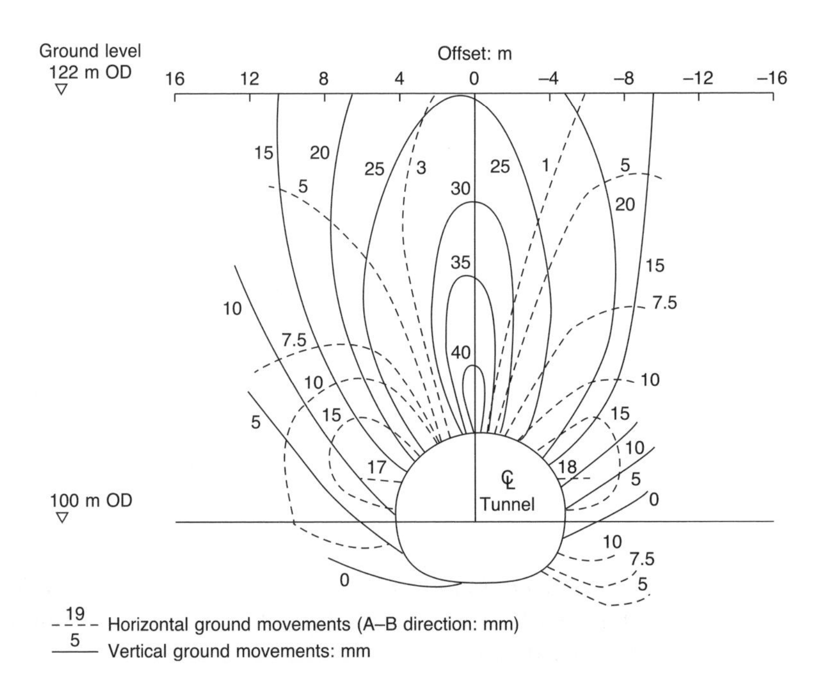

Fig. 3.31 Ground movement contours – Type 2 (completion of left and right drifts 11 May 1992) (after Deane and Bassett, 1995)

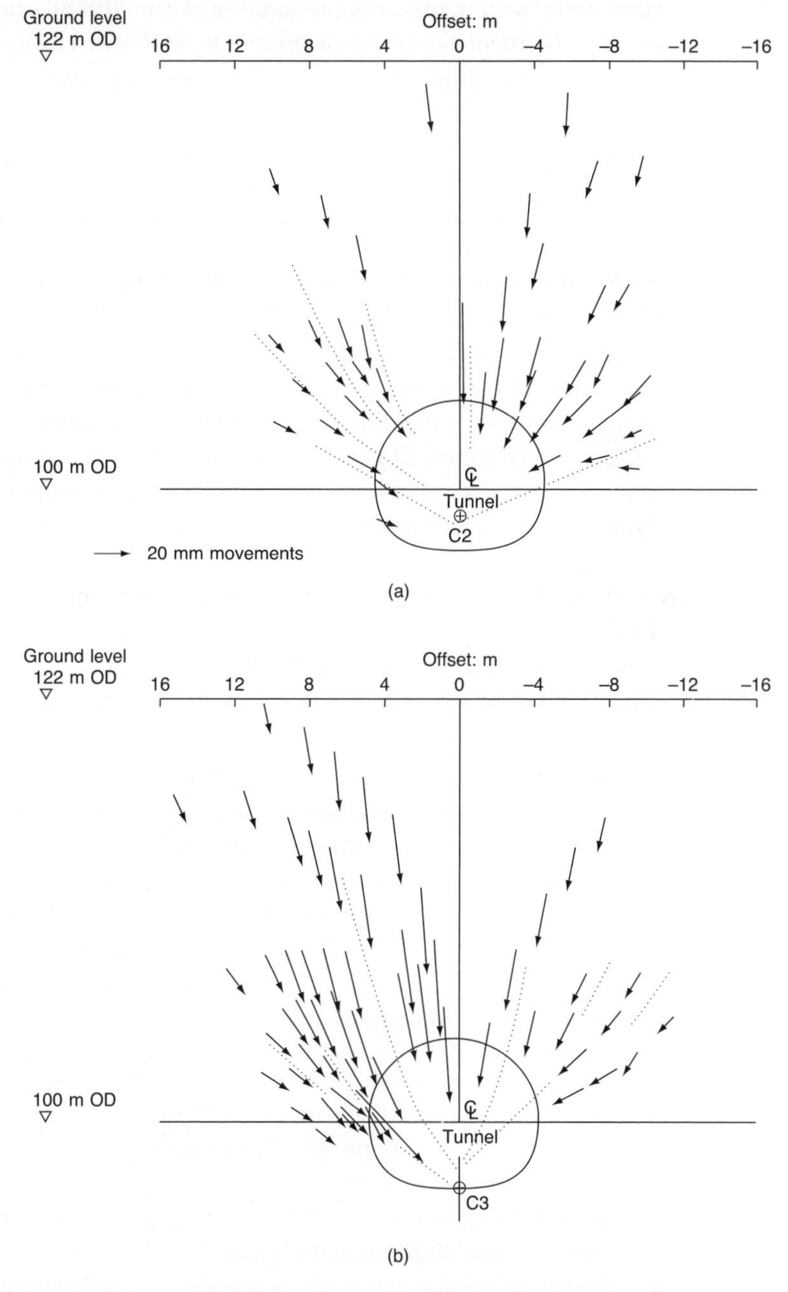

Fig. 3.32 Ground movement vectors: (a) Type 2 (completion of left and right drifts 11 May 1992); (b) Type 3 (completion 29 May 1992) (after Deane and Bassett, 1995)

consistent with experimental data of Mair (1979) and elasto-plastic analysis. Vertical deformations greater than 15 mm for Type 2 construction are concentrated into the crown zone above the centre-line level of the tunnel.

While informative, the horizontal displacement and vertical displacement contours do not give an easily assessed picture of the overall mechanics of deformation. A quantitative representation can be achieved, however, by combining the basic displacement data in vector form. This has been done in Fig. 3.32(a) and 3.32(b) by taking points at which vertical and horizontal displacement contours cross, calculating the angle and magnitude of the full displacement vector and plotting it at an exaggerated scale at the contour intersection point. The full pattern of displacement vectors forms a 'wine glass' flow pattern. The origin of the vector pattern (point C2) appears to be centred above the invert level of the section in Type 2 but quite noticeably at or possibly below the invert (point C3, Fig. 3.32(b)) in Type 3.

Conclusions

All three construction methods were successfully completed without any unforeseen geotechnical problems, such as soil falls or localized shear ruptures. The speed, simplicity and ease of construction increased from Type 1 through to Type 3 and presumably in practice competitive costing would reflect this. The Type 3 construction was dissimilar to Types 1 and 2 as the invert was not closed quickly to form a full ring. It is believed that had this been done settlement on the centre-line would have been reduced to approximately 30 mm, comparable with Types 1 and 2. In fact it is suspected that the settlement magnitude depends less on construction type than on the unsupported time. Without another instrumented Type 3 trial this cannot be definitively proved. In the absence of this proof the conclusions of the trial are as follows:

- Table 3.5 summarizes the significant items of data.
- The Type 2 construction provided the minimum settlement and minimum overall volume loss. On the basis of Boscardin and Cording (1989), the criteria for the most critical location of a surface building would for Types 1 and 2 be in the 'very slight' damage range and for Type 3 in the 'slight damage' range.
- The tunnel centre-line depth is approximately 20 m from mean ground level and the cover therefore approximately 15 m. The visually assessed values of i for the points of inflection suggest $K = 0.41$–0.45. The theoretical values (assuming a Gaussian distribution curve) indicate a K of 0.35–0.45; this assessment is dominated by the central 20 m of the trough which in Type 3 may not be typical or consistent.

Table 3.5 Summary of significant measured parameters from the Heathrow trial tunnel works (after Deane and Bassett, 1995)

Type	Surface settlements					Surface damage	Damage criteria			Sub-surface movements in soil (from inclinometers and settlement units)	
	Max. settlements: mm	Volume loss (average): %	Assessed trough width and point of inflection: m		Theoretical point of inflection: m	Max. tilt	Max. tensile strain: %	Mean tilt in zone of hogging	Angle of distortion	Estimated max. crown displacement: mm	Estimated max. lateral displacement: mm
1	28	1.13	30	8.8	8.4	1 in 500	Not assessed	1 in 1100	1 in 2300	–	–
2	25	1.06	34	9.0	8.8	1 in 550	0.06–0.07	1 in 1400	1 in 3500	42	18
3	40	1.33	30	8.2	7.0	1 in 250	0.06	1 in 1000	1 in 620	55	20

- The overall loss factors were 1.13 and 1.06 for Types 1 and 2 and these can be separated into individual components for the different processes as shown below
  - 1.5% for all initial crown headings
  - 0.8% for all invert closures below a hardened crown
  - 1.2–1.3% for all secondary crown headings, major central excavations and wide inverts.

CHAPTER TWENTY-SEVEN

Finite-element modelling of multi-tunnel interactions and tunnelling effects on adjacent piles

Overview

Finite-element (FE) analysis offers considerable possibilities of modelling many aspects of bored tunnel construction. A comprehensive review of FE analysis of bored tunnel construction was undertaken by Clough and Leca (1989). It is evident that the development of ground movements around an advancing tunnel heading is fully three-dimensional (see Fig. 3.16). Clough and Leca identified a number of reasons hindering the successful development of FE analysis for analysis of this complex problem:[7]

- The cost of a full 3D analysis, which properly simulates construction aspects and the 3D geometry, with a realistic non-linear constitutive soil model, is sometimes difficult to justify for some tunnelling applications [although this is significantly less the case in 2004 than it was in 1989].
- Many of the parameters influencing the results are difficult to define, for example tunnel lining properties, tail void size and soil model parameters (for the particular constitutive model assumed).
- Multiple analyses would often be required for any given project, in view of the usual changes of geology and alignment geometry along the length of the tunnel.
- [Limited] constitutive soil models have been shown to be successful at simulating [all] aspects of soil behaviour important to tunnelling.

A general difficulty in comparing FE analyses with field-data is that discrepancies could be due to one or more of the following:

- deficiencies in the soil model
- deficiencies in the soil parameters adopted

[7] Author's note: to take into account developments in the past 25 or so years, we have paraphrased the words of Clough and Leca (1989) with minor alterations as shown in square brackets thus [. . .].

- idealizations in the modelling, especially with respect to the boundary conditions
- possible uncertainties in the field measurements.

As discussed by Clough and Leca (1989), the tunnel construction process is extremely complex, particularly if shield tunnelling is involved, and it therefore represents almost the ultimate challenge to geotechnical analysts.

There follows some examples of how numerical modelling can assist engineering understanding of some complex problems associated with tunnelling.

Influence of in situ stress state and anisotropy

Overview

Lee and Ng (2002) and Ng and Lee (2004) carried out a series of three-dimensional, numerical, elasto-plastic, coupled-consolidation analyses to investigate the effects of anisotropy (expressed as stiffness ratio $n = E'_h/E'_v$) and the initial in situ effective stress state (expressed as the coefficient of earth pressure at rest K_0) on deformations due to an idealized open-face tunnelling at a given unsupported length. A hypothetical tunnel excavation in a stiff homogeneous over-consolidated London clay was modelled. The diameter of the tunnel (D) was taken as 9 m with a constant cover depth (C) of 18 m. The tunnel was assumed to be an open-face excavation and lined with sprayed shotcrete. The FE program ABAQUS (Hibbitt, Karlsson and Sorensen Inc., 1998) was adopted to model the tunnel excavation and consolidation.

Finite-element mesh

Figure 3.33 shows the FE mesh adopted in the analysis. It should be noted that only half of the tunnel was analysed as a plane of symmetry could be easily identified x = 0 m. The FE mesh was 101.25 m long, 45 m high and 75 m wide. It consisted of approximately 5040 elements and 5642 nodes. A section located at the middle of the mesh (i.e. at y = 0 m, the monitoring section) was monitored during every stage of excavation and construction. Eight-noded brick elements and four-noded shell elements were used to model the soil and the concrete lining respectively. Five soil elements (C – crown, SH – shoulder, SP – springline, H – heel or knee, I – invert) around the tunnel opening at the monitoring section were selected for detailed study.

Transverse surface settlements

Figure 3.34(a) shows the influence of the anisotropy stiffness factor n. As expected, the higher the n value (and so the lower the relative stiffness in

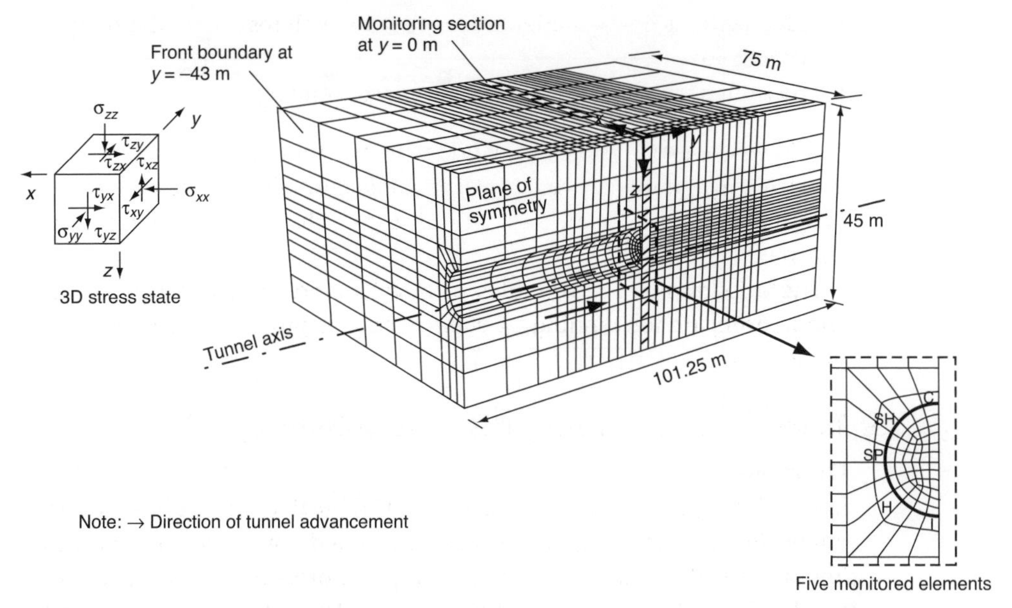

Fig. 3.33 Three-dimensional finite-element mesh for numerical study of influence of anisotropy and K_0 *on the surface settlement trough above a tunnel in clay (Ng and Lee, 2004)*

the vertical direction) the greater the vertical settlement. Figure 3.34(b) shows the influence of the initial K_0 on the immediate transverse surface settlements for $n = 1.6$. For a given n value, the lower the initial K_0 (and so the lower the horizontal effective stress as a proportion of vertical effective stress), the deeper and narrower is the computed settlement trough.

Multi-tunnel interaction

Overview

Ng *et al.* (2004) carried out a series of systematic, three-dimensional coupled FE analyses to investigate the multiple interactions between large parallel hypothetical twin tunnels constructed in stiff clay using the New Austrian Tunnelling Method. Special attention was paid to the influence of the lagging distance between the twin tunnel excavated faces (L_T) and the load transfer mechanism between the two tunnels.

Figure 3.35 shows sections of the FE mesh of the twin parallel tunnels. The FE program ABAQUS (Hibbitt, Karlsson and Sorensen Inc., 1998) was adopted. The modelled tunnel is an oval shape that is 9.2 m and 7.9 m in the horizontal and vertical axes, respectively, that is, similar to the Heathrow Express trial tunnel (Deane and Bassett, 1995). The length of the mesh in the lateral (x-) direction is 178 m (approximately 10.3D from the centre-line to both sides, where D is the equivalent diameter of a

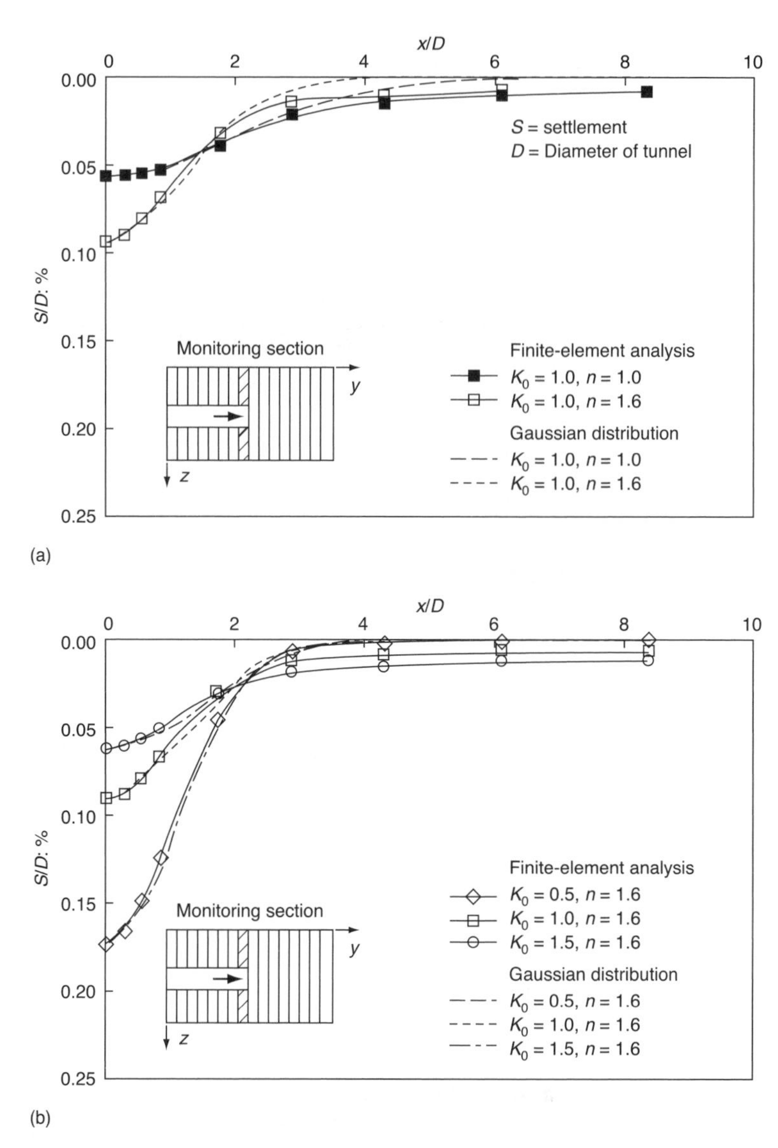

Fig. 3.34 Immediately transverse surface settlements at the monitoring section: (a) $K_0 = 1.0$ *condition; (b)* $n = 1.6$ *condition (Ng and Lee, 2004)*

tunnel equal to 8.64 m). The length in the longitudinal direction and the depth of the mesh was 142.5 m (16.5*D*) and 45 m (5.2*D*), respectively. The centre-line of the tunnels is located at 20 m (2.3*D*) and the pillar width between the two tunnels is 1.0*D*. The entire mesh for the twin tunnels consists of 17 220 three-dimensional, eight-noded solid elements for the soils and 4516 four-noded shell elements for the shotcrete lining and the total number of nodes is 14 868.

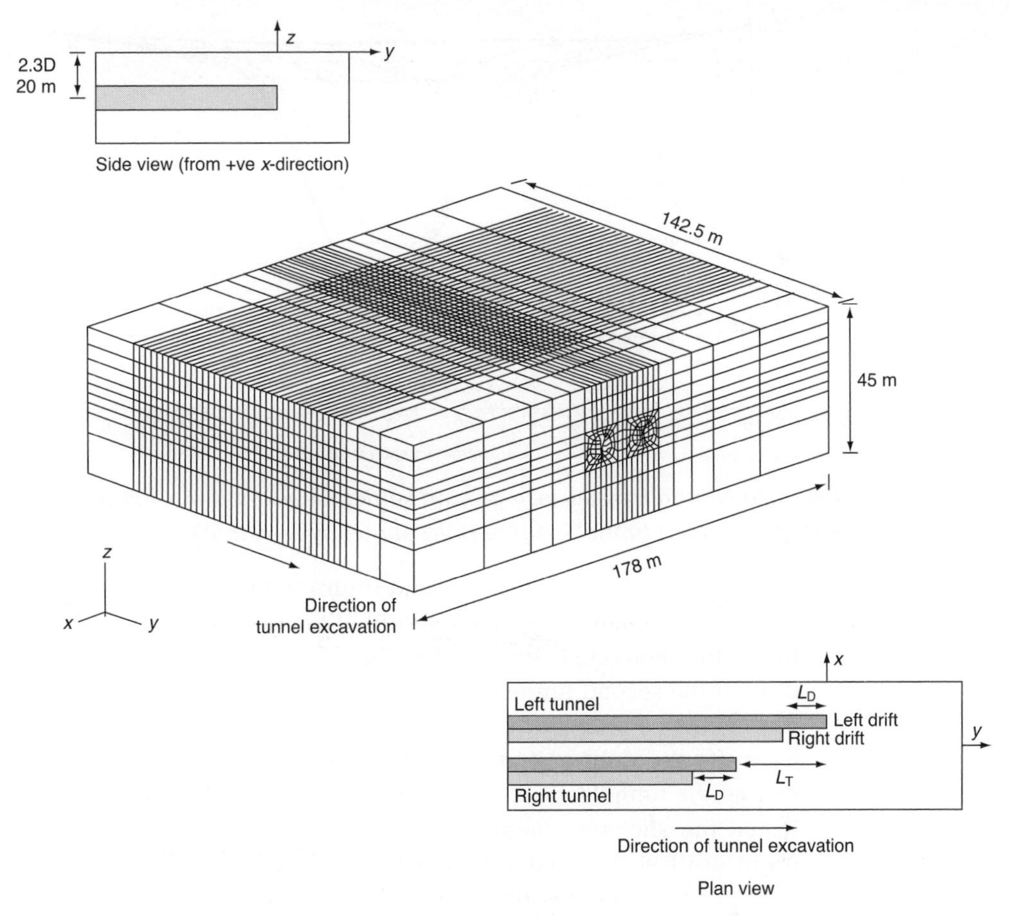

Fig. 3.35 Three-dimensional finite-element mesh for numerical study of twin parallel tunnels in clay (Ng et al., *2004)*

Numerical modelling procedure

The sequence of the tunnel excavation was idealized and modelled as follows:

1. Excavate the left half of the first (left) tunnel.
2. Install the shotcrete lining (250 mm thick) on the exposed surface (left drift) but leave a constant unsupported span of 5 m behind its tunnel face.
3. Excavate the right half of the first tunnel at a lagged distance $L_D = 2.3D$ (or 20 m) behind the tunnel face of the left drift.
4. Install the shotcrete lining (250 mm thick) on the exposed surface (right drift) but leave a constant, unsupported span of 5 m behind its tunnel face.

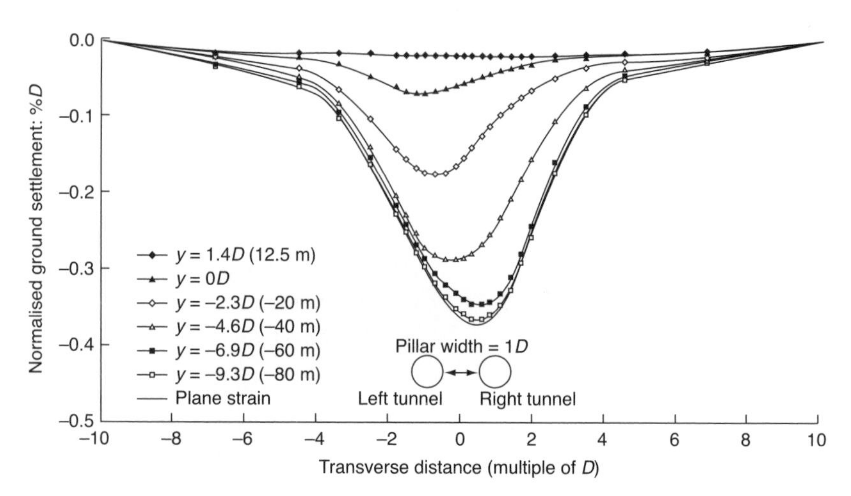

Fig. 3.36 Normalized surface ground settlements at various longitudinal distances for left tunnel lagging right tunnel by $L_T = 3.5D$ *(Ng* et al.*, 2004)*

5 Excavate the left half of the second (right) tunnel at a distance of L_T behind the tunnel face of the left drift of the first tunnel.
6 Install the shotcrete lining (250 mm thick) on the exposed surface (left drift) of the second tunnel but leave a constant, unsupported span of 5 m behind its tunnel face.
7 Excavate the right half of the second tunnel at $L_D = 2.3D$ (or 20 m) behind the tunnel face of the left drift of the second tunnel.
8 Install the shotcrete lining (250 mm thick) on the exposed surface (right drift) of the second tunnel but leave a constant unsupported span of 5 m behind its tunnel face.
9 Repeat (1) to (8) until the two tunnels are completed.

It can be seen from the discrete steps above that it is possible to model numerically numerous and detailed installation and construction procedures.

Surface settlement profiles

The shifting of the settlement trough is clearly illustrated by plotting transverse surface settlements at various longitudinal distances in the y-direction for $L_T = 3.5D$ in Fig. 3.36 . It is clear that the settlement trough shifts gradually from the left to the right and the maximum ground surface settlement offsets from the centre-line of the pillar.

Stabilizing tunnel faces using soil nails

Overview

Open-face excavation is an economical and flexible tunnelling method for constructing tunnels in relatively stable and uniform ground conditions

(Peck, 1969). The tunnel can be excavated in any shape rather than only circular. A concrete lining is sprayed around the excavated surface for temporary support. Modification of the construction sequence and remedial actions can easily be taken if there is a change in ground conditions. Reinforcements such as soil nails can be applied to the tunnel face when excessive deformation is observed or to provide temporary support, and they are removed as the excavation advances.

The effectiveness of applying soil nails for stabilizing the tunnel heading has been studied by Grasso *et al.* (1989) who discussed the use of reinforcement for stabilizing the rock mass during the excavation. They found that the measured tunnel wall convergence was significantly reduced when the reinforcements were used. The excavation rate was also improved. Lunardi *et al.* (1992) reported field monitoring results from the first tunnel excavated in soft saturated sandy-silty and clayey soil with core reinforcements (made of fibre glass nails) in Italy. They concluded that with the use of core reinforcement, the stability of the tunnel face in a poor geotechnical condition is improved and the construction time and cost can be controlled.

In order to find the required soil nail density, Barley and Graham (1997) performed a series of pull-out tests by varying the material (steel or fibre glass) of the soil nail, the borehole diameter, the minimum fixed length and the grouting material. Steel nails were found to have higher pull-out capacity than fibre glass nails.

Pelia (1994) conducted a series of three-dimensional FE analyses to study the stabilizing effects resulting from core reinforcement by soil nails. The tunnel excavation processes were modelled either with or without tunnel lining support and face reinforcement. The tunnel face displacement and stress changes in the core ahead of the tunnel excavation were reduced significantly by the application of soil nails while the radial displacement and the plastic zone around the tunnel were controlled by the lining.

The effectiveness of a soil nail system is influenced by several factors such as nail spacing (or density), nail length, nail diameter, nail stiffness and type of grouting material (Mair and Taylor, 1997). Field monitoring results of tunnel excavations in both rock and soil have indicated that the reinforcing system stabilized the excavation effectively and the advance rate was also improved.

Finite-element idealization

Ng and Lee (2002) conducted a parametric study to investigate the reinforcing effects of soil nails with different axial rigidities ($E_n A_n$) on a tunnel face by adopting the following hypothetical tunnelling case. A circular tunnel 9 m in diameter (*D*) and 18 m of cover depth (*C*) was simulated to be excavated in a stiff uniform over-consolidated London

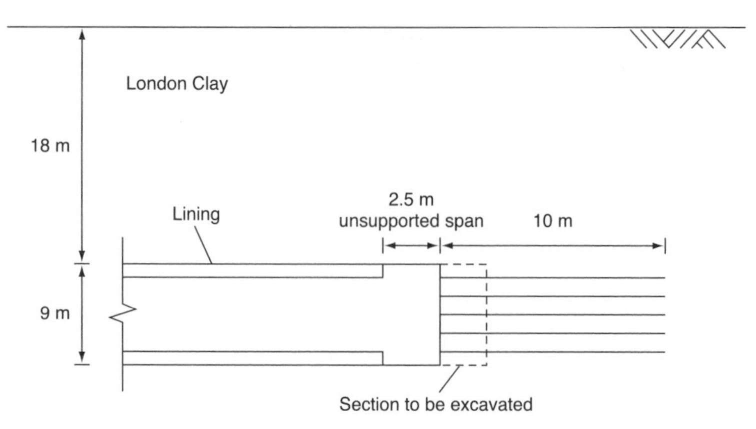

Fig. 3.37 Tunnel geometry and ground conditions modelled (Ng and Lee, 2002)

Clay layer (see Fig. 3.37). The tunnel was assumed to be excavated using the open-face excavation method. A lining was installed to the tunnel wall as the excavation advanced. The tunnelling excavation was idealized and simulated with three-dimensional coupled consolidation analyses using the FE program ABAQUS (Hibbitt, Karlsson and Sorensen, 1998). In coupled consolidation analyses, deformations of the soil medium are related to the dissipation of excess pore water pressure over time.

Tunnel excavation modelling procedures

Here the hypothetical tunnel was excavated to a certain span length and the lining was applied to the excavated surface from a distance behind the tunnel face, leaving an unsupported span behind the tunnel face. For cases where soil nails are required, they were installed prior to the start of the excavation. A portion of the soil nails has been removed (i.e. cut to length) together with the excavated soil during tunnelling. The sequential excavation procedures are simulated by deactivating the soil elements within the proposed tunnel area and activating the lining elements in the FE mesh. To analyse the use of nails, they are activated before the excavation and deactivated together with the soil elements during the excavation.

The effectiveness of the nailing system is evaluated based on short-term ground movement, nail force, pore water pressure and soil behaviour in tunnel heading. These quantities are monitored from the beginning (initial stress is established) to the end (the tunnel face reaches the monitoring section).

The axial rigidities of the soil nails used in each case are summarized in Table 3.6 and the details of the simulation procedures are summarized as follows:

Table 3.6 Summary of parametric cases considered in finite-element analyses of soil nailing the excavation of an open-face tunnel (Ng and Lee, 2002)

Case	E_n (GPa)	σ_y (MPa)	D (mm)	E_nA_n (MN)	E_n/E'_h	Remarks
0	N/A	N/A	N/A	N/A	N/A	No nail
1	15	30	25	7	100	Fibre glass
2	15	30	50	29	100	Fibre glass
3	15	30	75	66	100	Fibre glass
4	60	120	25	29	400	Aluminium
5	60	120	50	118	400	Aluminium
6	60	120	75	265	400	Aluminium
7	105	210	25	52	700	Brass
8	105	210	50	206	700	Brass
9	105	210	75	464	700	Brass
10	150	300	25	74	1000	Cast iron
11	150	300	50	295	1000	Cast iron
12	150	300	75	663	1000	Cast iron
13	200	400	25	98	1333	Steel
14	200	400	50	393	1333	Steel
15	200	400	75	884	1333	Steel

1 Establish the initial stress conditions (i.e. static stress condition with $K_0 = 1.0$).
2 Install soil nails (if required) prior to any excavation.
3 In order to eliminate the boundary effect, excavate the tunnel for 25 m away from the front boundary (i.e. from $y = -25$ m to $y = 0$ m) and install the tunnel lining for 22.5 m simultaneously. Leave an unsupported span 2.5 m.
4 Excavate another 2.5 m span. Remove a length of 2.5 m soil nails along the excavated soil and extend the nails by 2.5 m at the other end. At the same time, apply a 250 mm thick concrete lining to the previously excavated span simultaneously.
5 Advance the tunnel by repeating step 4 until the section at $y = -20$ m is reached.

Tunnel face displacements

The effectiveness of a nailing system can be evaluated by studying the magnitude of the tunnel face's displacement (δ_f). A large tunnel face displacement obviously indicates that the nailing system did not reinforce the tunnel face effectively and vice versa. The magnitude of δ_f is governed by many factors such as soil stiffness, the size of the tunnel, and the amount of stress relief which depends on the initial stress in the ground. A dimensionless parameter, Ω_f (Yoo and Shin, 1999), can be obtained by normalizing the tunnel face displacement

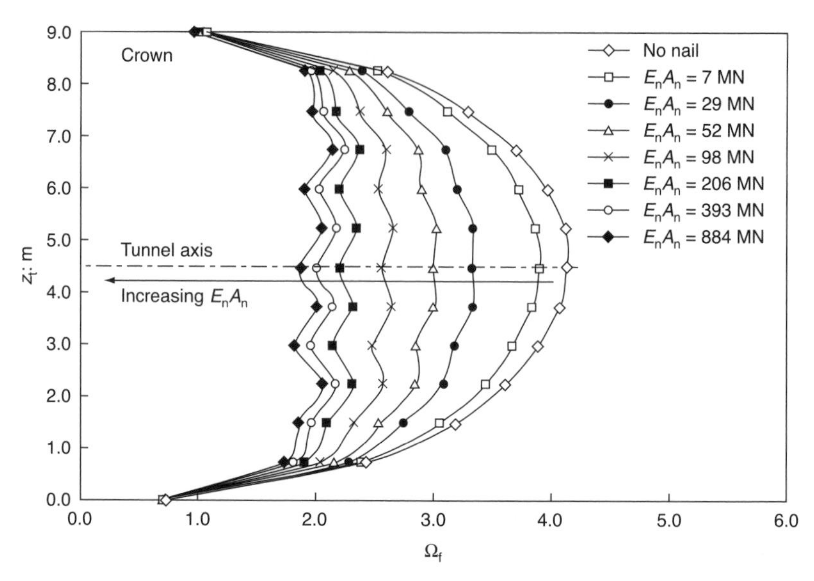

Fig. 3.38 Variation of normalized tunnel face displacement parameter Ω_f with elevation within the tunnel opening (Ng and Lee, 2002)

with the horizontal effective Young's modulus of the soil (E'_h), the tunnel diameter (D) and the initial horizontal effective stress (σ'_0) (i.e. $\Omega_f = \delta_f E'_h / D\sigma'_0$). The variations of the short-term Ω_f against tunnel elevation are shown in Fig. 3.38. Generally, the tunnel face displacement profile is symmetrical about the tunnel axis. This is because the soil stiffnesses (E'_v, E'_h and G_{vh}) used and the vertical and horizontal stress changes that took place due to excavation are all approximately proportional to the depth. As expected, Ω_f decreases with an increasing axial rigidity in the soil nails. For soil nails with low axial rigidity, $E_nA_n = 7\,\text{MN}$ (i.e. a fibre glass nail of 25 mm diameter and $E_n/E'_h = 100$ as shown in Table 3.6), the profile of Ω_f is reduced very slightly as compared with the no nail case.

It can be seen in Fig. 3.38 that the soil nails induce stabilizing effects on the tunnel face and hence minimize the face deformations, especially at and near the longitudinal axis of the tunnel. As E_nA_n increases, the shape of the deformations changes from an elliptical geometry to become almost constant with depth.

Immediate surface settlement

A transverse surface settlement trough obtained immediately after the tunnel excavation can be described by the Gaussian distribution as suggested by Peck (1969) and many other researchers.

Based on the maximum transverse surface settlement S_{max} obtained from each numerical analysis, Gaussian distributions are fitted to each

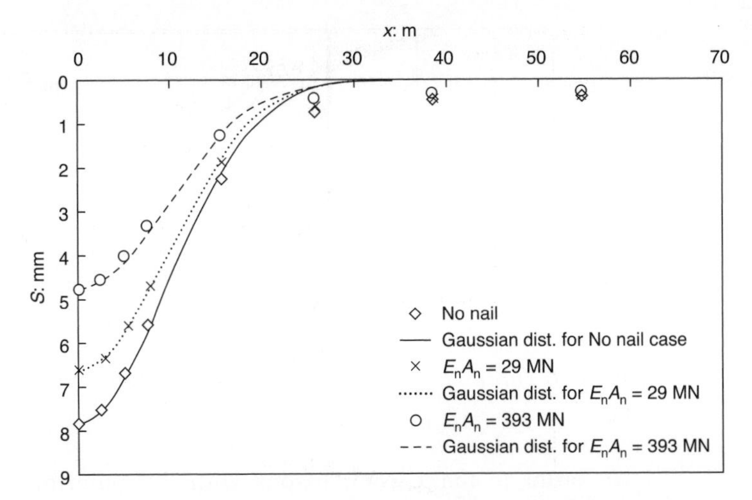

Fig. 3.39 Short term transverse surface settlement (Ng and Lee, 2002)

transverse surface settlement trough from the numerical analysis. As shown in Fig. 3.39 the Gaussian distributions fit well with the transverse surface settlement troughs except that the troughs obtained from the numerical analyses are found to be wider than the Gaussian distributions. The deficiency in the transverse surface settlement trough may result from the simple elastic–perfectly plastic model adopted for modelling the soil. As suggested by Gunn (1993), the prediction of the transverse surface settlement trough due to tunnelling excavation can be improved by using a non-linear elastic soil model.

Volume loss (V_t) is the amount of ground loss in the region close to the tunnel. In an open-face tunnelling in clay, deformation toward the tunnel face is the major component of the ground loss (Mair and Taylor, 1997). Due to the low permeability of clay, it is usually assumed that soil deforms under undrained (constant volume) conditions and V_t is equal to V_s (Mair and Taylor, 1997).

Since the tunnel face displacement is a major component of the ground loss in open-face tunnelling in London Clay, by limiting the tunnel face displacement, the volume loss and the transverse surface settlement can be reduced.

Finite-element modelling of a tunnel advancing on an existing loaded pile

Overview

Lee and Ng (2004) carried out a three-dimensional, elasto-plastic, coupled-consolidation, numerical analysis to investigate the effects of an advancing open-face tunnel excavation near an existing pile at working load.

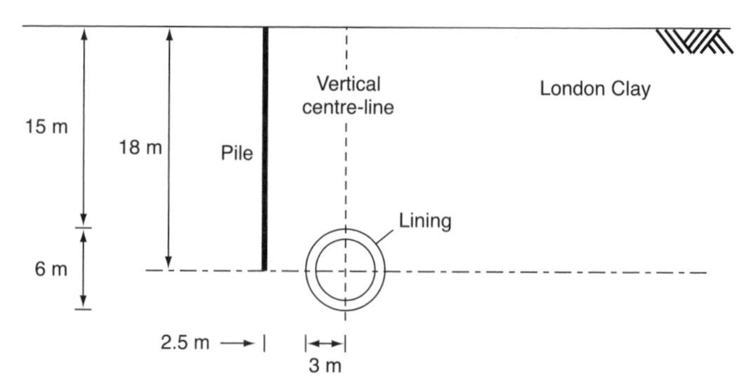

Fig. 3.40 Tunnel geometry and ground conditions modelled (Lee and Ng, 2004)

In order to make comparisons with the published results from the centrifuge model test (test 2) conducted by Loganathan *et al.* (2000), the same tunnel geometry, pile size and location of the pile relative to the tunnel were adopted.

Figure 3.40 shows the geometry of the problem analysed. A 6 m diameter (D) circular open-face tunnel was excavated in an idealized, stiff, homogeneous, over-consolidated London Clay with a cover depth of 15 m ($C = 2.5D$). An 18 m long and 0.8 m diameter pile was assumed to be located 5.5 m from the vertical tunnel centre-line (or 2.5 m from the edge of the tunnel springline). The FE program, ABAQUS (Hibbitt, Karlsson and Sorensen Inc., 1998), was used to model the tunnel excavation. Figure 3.41 shows the three-dimensional FE mesh. As a plane of symmetry

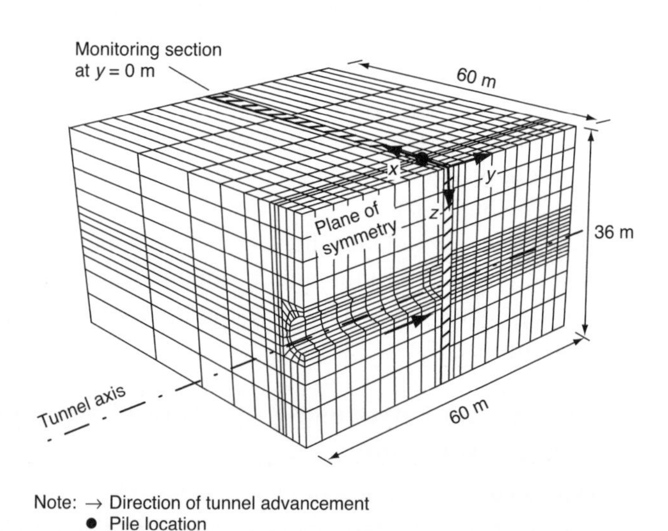

Fig. 3.41 Three-dimensional finite-element mesh for tunnel excavation near an existing pile (Lee and Ng, 2004)

was identified at $x = 0$, only half of the domain was modelled. The mesh was 60 m (10.0D) long, 60 m (10.0D) wide and 36 m (6.0D) high. It consisted of 4209 elements and 4761 nodes. Eight-noded brick elements, four-noded shell elements and two-noded beam elements were used to model the soil, concrete lining and pile, respectively. For ease of comparison and study, a cross-section, called the monitoring section, $y = 0$ m, at which the pile was located, was selected.

A water table, together with a hydrostatic initial pore water pressure profile, was assumed to be located at the ground surface. The tunnel lining was assumed to be impervious.

Numerical modelling procedures

A wished-in-place (construction effects ignored) concrete pile was constructed in the ground and the open-face tunnel excavation with an unsupported span of 3 m (i.e. $D/2$) was simulated. A 250 mm thick shotcrete lining was applied on the circular tunnel 3 m behind the tunnel face as the excavation advanced. The excavation rate of the tunnel was assumed to be 3 m/day and a time increment of 1 day/step was adopted in the analysis. The detailed simulation steps are as follows:

1. Establish the initial stress conditions using $K_0 = 1.0$.
2. Determine the pile capacity (i.e. ultimate load) of the wished-in-place pile and apply and maintain a vertical working load of 1850 kN to the pile head throughout the tunnel construction.
3. Allow full dissipation of excess pore pressure due to the applied load.
4. Excavate the tunnel at 3 m spans (i.e. $D/2$ unsupported length) and apply the shotcrete lining to the exposed surface of the tunnel. At any construction stage, a 3 m unsupported span behind the tunnel face is simulated.
5. Advance the excavation by repeating step 4 until the tunnel is completed.

Pile load capacity

Prior to the simulation of the tunnel excavation, it was important to determine the ultimate capacity of the pile in the modelled ground so that an applied load with an assumed factor of safety (FOS) of 3.0 could be determined. This was done by carrying out a numerical pile load test. An applied load from 0 kN to 6000 kN was imposed gradually on the wished-in-place pile considered in this study. The results of the pile load–displacement curve for the simulated pile are shown in Fig. 3.42.

In order to determine the ultimate load, the failure criterion proposed by Ng *et al.* (2001c) for large-diameter piles was adopted. The displacement

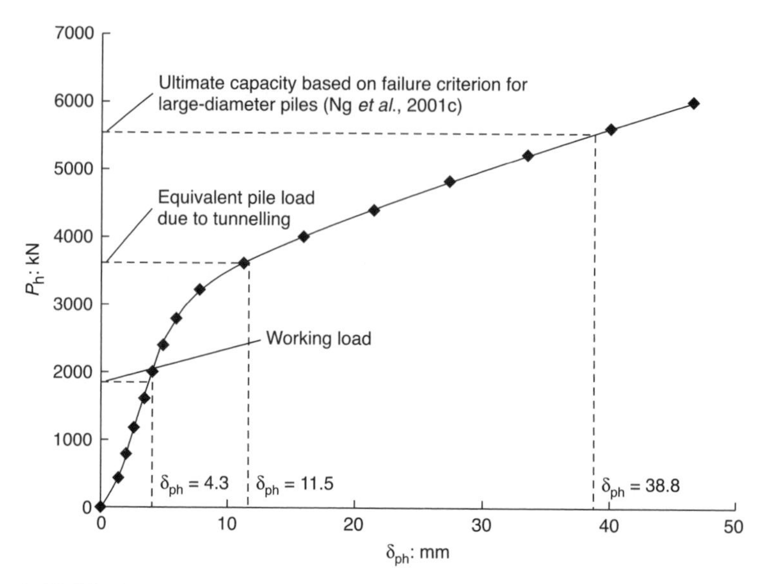

Fig. 3.42 Pile head load versus pile head displacement for a simulated pile loading test (Lee and Ng, 2004)

failure criterion for piles in soil was defined in chapter 4 in Part 1 as follows:

$$\delta_{\text{ph max}} \approx 0.045 d_{\text{p}} + \frac{1}{2}\left(\frac{P_{\text{h}} L_{\text{p}}}{A_{\text{p}} E_{\text{p}}}\right) \tag{3.17}$$

where P_h = pile head load, L_p = pile length, A_p = pile area, E_p = pile shaft elastic modulus and d_p = pile diameter.

As shown in Fig. 3.42, the ultimate pile load for the simulated pile was determined to be 5550 kN.

With the assumed FOS of 3.0, a working load of 5550/3 = 1850 kN was applied to the pile head during tunnelling. Under this working load, an initial pile head settlement (δ_{ph}) of 4.3 mm (0.07%D) was induced (see Fig. 3.42). Any excess pore pressure induced in the ground was allowed to dissipate fully prior to tunnel excavation.

Surface settlements

Figure 3.43 shows the progressive changes in the tunnelling-induced surface settlements in the monitoring section as the tunnel advances. Both the surface settlement (S) and the transverse (x) distance from tunnel centre-line are normalized by the tunnel diameter (D). When the tunnel face is at $y/D = -3$, that is, 3.0D, away from the monitoring section and the pile location ($y/D = 0$), negligible surface settlements are induced in the monitoring section. As the excavation advances further (1.0D away from the monitoring section), the surface settlement

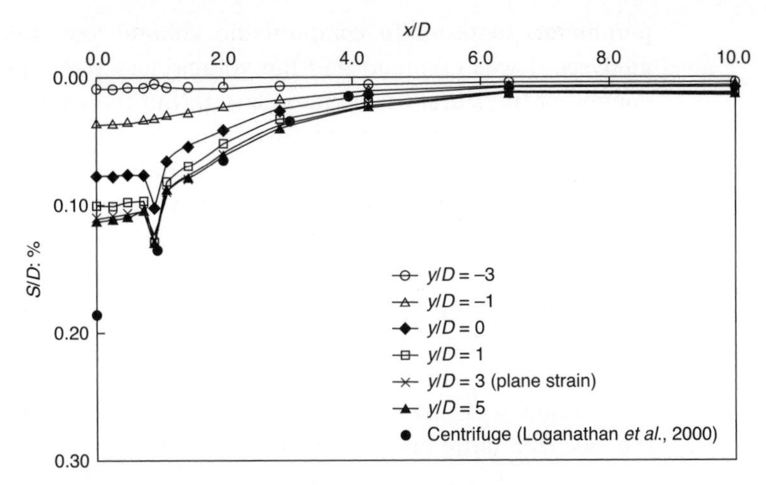

Fig. 3.43 Transverse surface settlements induced at the monitoring section during tunnel advancement (Lee and Ng, 2004)

increases and no significant pile head settlement is observed. However, when the tunnel face reaches the monitoring section (where the loaded pile is located), a large pile head settlement as compared with the surface soil settlement is induced, resulting from the yielding of the soil around the pile toe. This is consistent with results from the centrifuge model tests by Jacobsz *et al.* (2002), who reported larger settlement of the pile head than soil settlement when the pile was located close to the crown of the tunnel.

With further excavation from $y/D = 0$ to $y/D = 1.0$, the surface and pile settlements in the monitoring section continue to increase. As the excavation advances to $3.0D$ beyond the monitoring section (i.e. tunnel face at $y/D = 3.0$), the increase in surface settlement in the monitoring section is negligible. It can therefore be assumed that the monitoring section reaches the plane strain condition when the tunnel face approaches $y/D = 3.0$. Thus, a significant tunnelling influence zone between $y/D = -1.0$ and $y/D = 1.0$ on the loaded pile can be identified.

Surface settlements measured in the centrifuge test (test 2) by Loganathan *et al.* (2000) are also shown in Fig. 3.43 for comparison. As the plane strain condition was modelled in the centrifuge test, the measured surface settlements should only be compared with the computed plane strain surface settlements when the tunnel face reaches $y/D \geq 3.0$. The computed settlements are generally consistent with the centrifuge results, except for the settlements at $x/D \leq 1.0$. In the centrifuge test, a predefined 1% volume loss was imposed to cause ground movements. However, in the three-dimensional finite-element analysis, an actual excavation sequence was modelled. Volume loss was not used as a controlling parameter. In order to compute the volume loss from the numerical analysis, it was assumed that the volume loss was equal to the surface volume of the surface settlement trough but the excessive localized surface settlement around the pile was ignored. The computed plane strain volume loss was 0.9%, which is nearly the volume loss imposed in the centrifuge test. The numerical simulation underestimates the maximum measured surface ground settlements near the tunnel axis, that is, at $x/D = 1.0$ or smaller. The underestimation of the maximum surface settlement may be due to neglecting the reduction of the soil stiffness with strain in the numerical analysis. Obviously, the computed plane strain surface settlements cannot be fitted with a normal Gaussian distribution with the maximum settlement at the tunnel centre-line, if the localized large pile settlement is included.

Plastic strains around the tunnel and pile

Figure 3.44 shows plastic strains (in percentages) induced around the tunnel and the pile during the tunnel advancement. In order to identify zones of compression and extension strains, the compression plastic strain (+ve) is defined when the average of the lateral strain (ε_{xx}) and

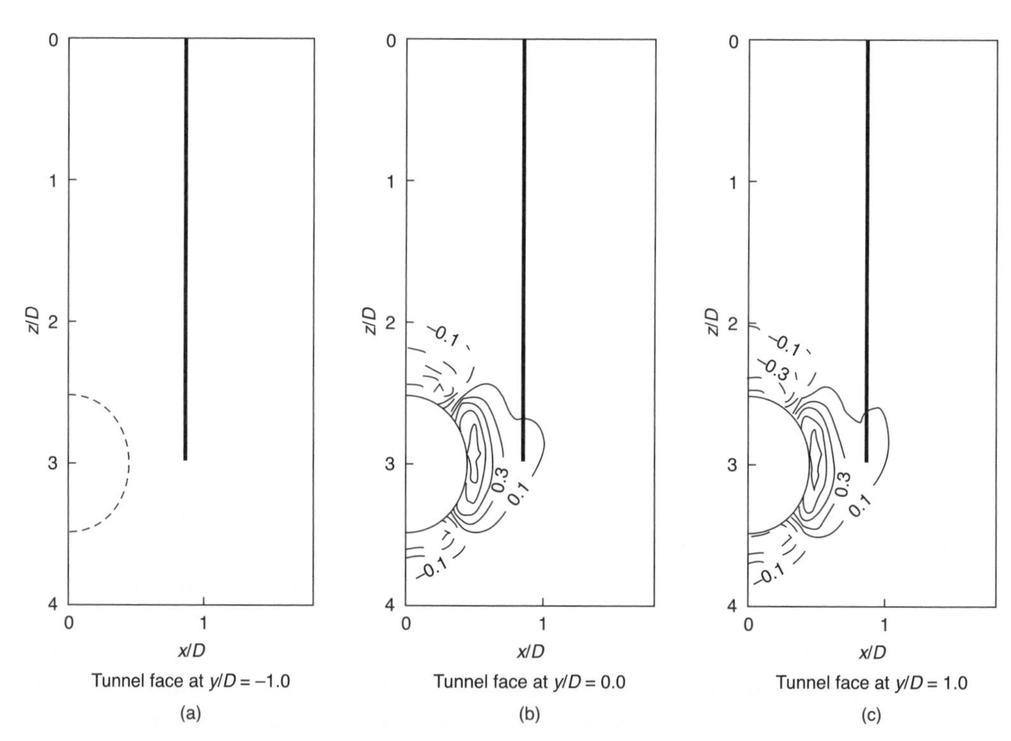

Fig. 3.44 Plastic strains (%) induced around the tunnel and pile as the excavation advances (Lee and Ng, 2004)

the longitudinal strain (ε_{yy}) is less than the vertical strain (ε_{zz}), as indicated with solid lines in the figure. On the other hand, if the average of ε_{xx} and ε_{yy} is greater than ε_{zz} the plastic strain is defined as an extension strain (−ve), as indicated with dotted lines in the figure.

As shown in Fig. 3.44(a), when the tunnel face is 1.0*D* or more away from the monitoring section and the pile, that is, located at $y/D = -1.0$, no significant plastic strain is induced in the monitoring section. As the tunnel face reaches the monitoring section (see Fig. 3.44(b)), significant plastic compression and extension strains are induced during the face excavation at the tunnel springline and crown, respectively. Since the pile is located close to the tunnel, development of a plastic strain can be clearly found around the pile tip, suggesting yielding of the soil and resulting in large pile settlement as illustrated in Fig. 3.45. When the tunnel face advances to 1.0*D* beyond the monitoring section (i.e. the pile location), the plastic strain zone near the pile tip further expands slightly (see Fig. 3.44(c)). The development of plastic strains reveals a significant tunnelling influence zone between $y/D = -1.0$ and $y/D = 1.0$ on the loaded pile, which is consistent with the evaluation of the surface ground settlements as the tunnel advances (see Fig. 3.43).

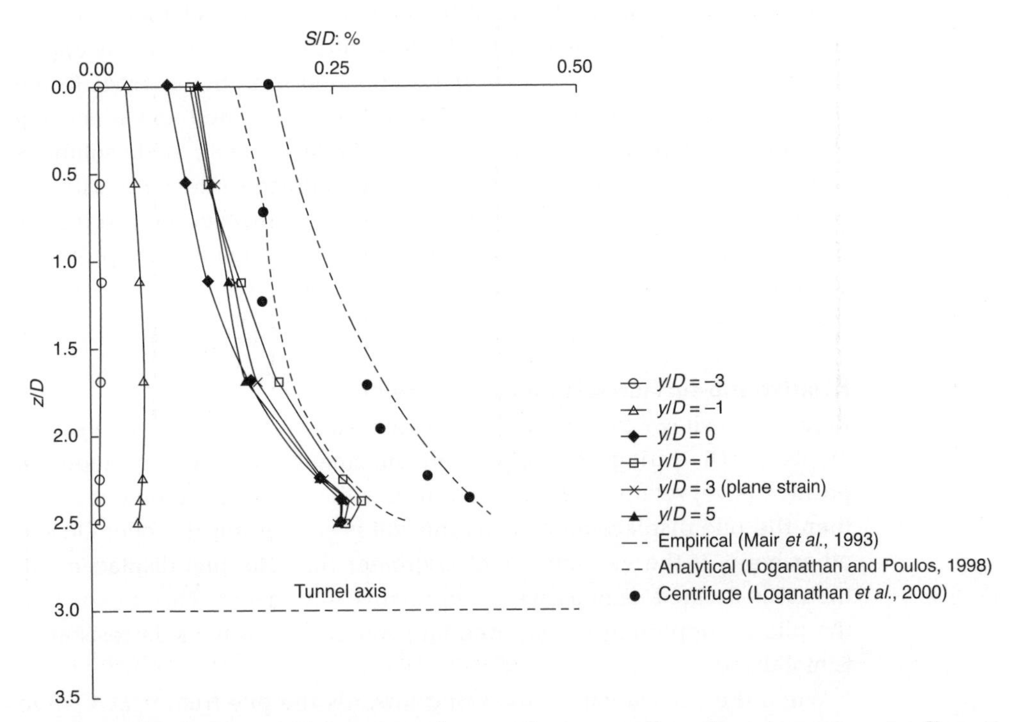

Fig. 3.45 Subsurface settlements at the monitoring section along tunnel centre-line as excavation advances (Lee and Ng, 2004)

Subsurface settlements

Figure 3.45 shows the development of normalized sub-surface settlements (S/D) in the monitoring section (along the tunnel centre-line at various depths) during tunnel advancement. The magnitude of the sub-surface settlements is small and the distribution is almost constant with depth as the tunnel face advances from $y/D = -3$ to $y/D = -1$. However, when the tunnel face reaches the monitoring section, the sub-surface settlements increase significantly and non-linearly with depth due to the yielding of the soil around the tunnel (see Fig. 3.44(b)) and the sub-surface settlements reach their maximum values as the excavation face reaches $1.0D$ beyond the monitoring section. Although there is a slight decrease in the sub-surface settlement above the tunnel crown when the plane strain condition ($y/D = 3$) is reached, no obvious sub-surface settlement can be found when the tunnel face continues to advance away from the monitoring section. Consistent with the development of the surface settlements and plastic strains, a significant tunnelling influence zone between $y/D = -1.0$ and $y/D = 1.0$ on the loaded pile can be identified.

By comparing the computed plane strain sub-surface settlements in the monitoring section with results of the centrifuge model test No. 2 reported by Loganathan *et al.* (2000), it can be seen that the finite-element analysis (FEA) results show a similar trend to the centrifuge test results. By applying the deduced volume loss of 0.9% obtained from the analysis to the empirical formula proposed by Mair *et al.* (1993) and the analytical solution developed by Loganathan and Poulos (1998), both solutions show a similar trend and distribution of sub-surface settlements with the computed plane strain sub-surface settlements. However, both solutions predict higher sub-surface settlements than those computed, probably because the variation in the soil stiffness with strain is not considered in the analysis.

Relative sub-surface settlements of the pile

Figure 3.46 shows the normalized relative sub-surface soil to pile settlements (S/D) during the tunnel advancement. At a given depth, a positive (+ve) relative settlement indicates that the soil settlement is less than the pile displacement. Thus the soil is supporting the pile. On the other hand, if the soil settlement is greater than the pile displacement, then it is defined as negative (−ve) relative settlement. This means that the pile is supporting the surrounding soil and negative side resistance is mobilized.

When the excavation is advancing towards the pile from $y/D = -3$ to $y/D = -1$, no significant relative sub-surface settlements are induced. As the tunnel face reaches the pile at $y/D = 0.0$, significant and highly

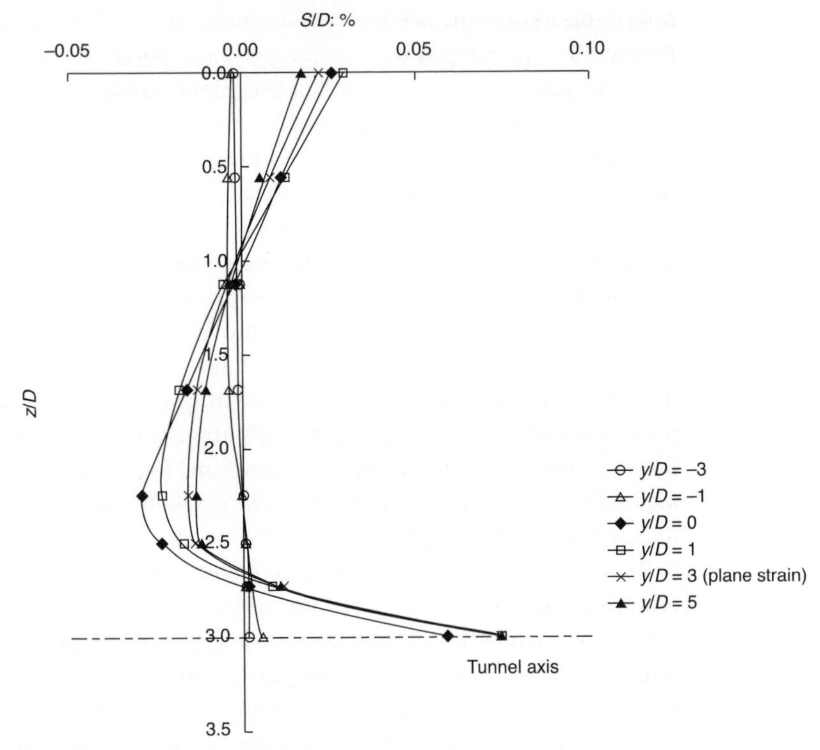

Fig. 3.46 Relative subsurface soil to pile settlements during tunnel advancement (Lee and Ng, 2004)

non-linear distributions of relative sub-surface settlements with depth are induced. For the top 1D of the pile (i.e. $z/D < 1.1$), a positive sub-surface settlement is mobilized, suggesting that the soil is supporting the pile within this range of the pile depth. On the other hand, between the depths $1.1 < z/D < 2.75$ of the pile, the relative sub-surface settlements are negative, implying that negative side resistance is induced along the shaft of the pile as a result of stress relief from the tunnel excavation. However, for the pile section at $z/D \geq 2.75$ depth, large positive relative sub-surface settlements are induced, indicating a substantial mobilization of toe resistance to support the vertical load acting on the pile. Obviously, the tunnel excavation induces complex distributions of relative sub-surface settlements along the depth of the pile.

As the tunnel excavation advances further to $y/D = 1.0$, the relative movements between the pile and the surrounding soil in the monitoring section increase slightly near the ground surface, but the mobilized negative skin friction decreases marginally whereas the relative movements near the toe increase to mobilize the toe resistance. The distributions of the relative movements along the pile length remain essentially unchanged.

When the tunnel face reaches the plane strain condition at $y/D = 5$, there is a continuous reduction of both the positive and negative relative movements along the pile from $0 < z/D < 2.75$, with essentially constant relative movements near the pile toe.

Change of factor of safety of pile due to tunnelling

To investigate the reduction of an equivalent factor of safety (FOS) of the loaded pile due to tunnelling, an additional pile head settlement of 7.2 mm (or $0.12\%D$) at the plane strain condition ($y/D = 3.0$) is considered as the cause of the reduction in pile capacity when a displacement-controlled failure load acceptance criterion is used (Ng *et al.* 2001c). Due to the initial applied working load (displaced vertically by $0.07\%D$ or 4.3 mm) and the tunnelling effects, the pile is displaced by 11.5 mm in total.

Referring to the pile–load displacement curve in Fig. 3.42, when the working load of 1850 kN (i.e. the ultimate load of 5550 kN divided by the FOS of 3) is applied a displacement of 4.3 mm results. From the analysis an additional 7.2 mm of pile displacement is caused by tunnelling. The total displacement is therefore $4.3 + 7.2 = 11.5$ mm. Referring to Fig. 3.42 this reads off as an equivalent total pile load of 3600 kN. This means the FOS of the pile drops from 3 (5550/1850) to nearly 1.5 (5550/3600) – an alarming halving of the Factor of Safety!

CHAPTER TWENTY-EIGHT

Effect of ground movements on buildings

Classification of damage and assessment criteria

Mair and Taylor (1997) observe that Burland (1995) and Mair *et al.* (1996) categorized building damage in terms of deflection ratios Δ/L for sections of the building in hogging or sagging mode, as shown in Fig. 3.47. A damage category chart for buildings in a hogging mode with $L/H = 1$ is shown in Fig. 3.48; this takes into account strains due to both shear and bending modes and is broadly equivalent to the chart produced by Boscardin and Cording, since generally β approximately equals 2–3 times (Δ/L). The evaluation of β from settlement measurements is not always straightforward because the tilt of the building needs to be identified; in practice the evaluation of Δ/L is easier. Figure 3.49 shows the more general results

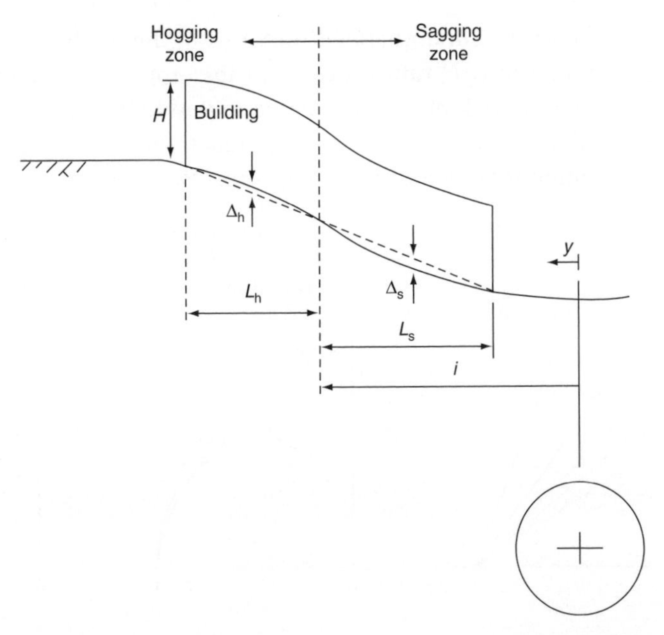

Fig. 3.47 Deformation of building above a tunnel (after Mair et al.*, 1996; and Mair and Taylor, 1997)*

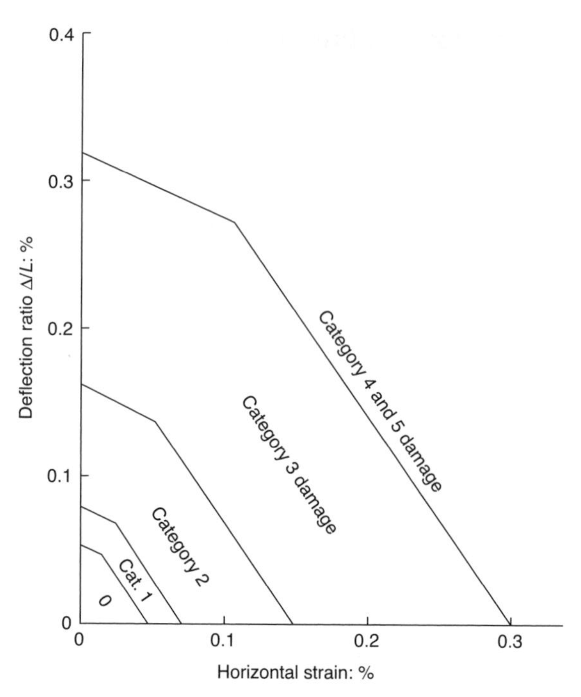

Fig. 3.48 Building damage categories relating to horizontal strain and deflection ratio for L/H = 1*, hogging mode (after Burland, 1995; Mair* et al.*, 1996; and Mair and Taylor, 1997)*

given by Burland (1995) (in terms of the limiting tensile strain, ε_{lim}) for a range of L/H ratios. Based on this approach, a methodology for assessment of risk of building damage due to bored tunnelling is described by Mair *et al.* (1996). This approach has been used successfully for a number of major tunnelling projects.

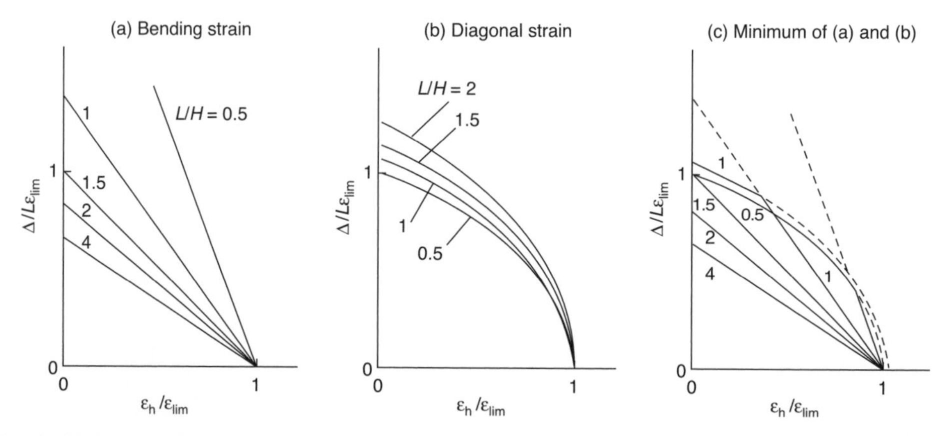

Fig. 3.49 Generalized building damage categories relating to horizontal strain and deflection ratio for different L/H *values (after Burland, 1995; and Mair and Taylor, 1997)*

Influence of building stiffness

Potts and Addenbrooke (1996, 1997) used finite-element analyses incorporating a non-linear elastic plastic soil model to study parametrically the influence of building stiffness on ground movements induced by tunnelling, as shown in Fig. 3.50.

About 100 analyses were undertaken of different configurations of tunnel and building dimensions. Two important dimensionless parameters were introduced: the relative bending stiffness ρ^* which expresses the relative stiffness between the building and the underlying ground, and the corresponding relative axial stiffness α^*. These are defined as:

$$\rho^* = \frac{EI}{E_s H^4} \tag{3.18}$$

$$\alpha^* = \frac{EA}{E_s H} \tag{3.19}$$

where H is the half-width of building ($= B/2$), EI and EA are the equivalent bending and axial stiffness of the building, and E_s is a representative soil stiffness taken by Potts and Addenbrooke to be the undrained secant stiffness at 0.01% axial strain in a triaxial compression test on a soil sample at a depth of z/2. The expression for ρ^* is similar to that used by Fraser and Wardle (1976) and by Potts and Bond (1994) in soil–structure interaction analyses of rafts and retaining walls respectively. The expression for α^* is similar to that used by Boscardin and Cording (1989).

The results of the study by Potts and Addenbrooke are summarized in Fig. 3.51. Modification factors to the deflection ratios (Δ/L) that would be obtained from the 'greenfield site' settlement profiles are shown as different curves for sagging and hogging deformation modes for different e/B ratios (e being the eccentricity of the tunnel from the centre-line of the

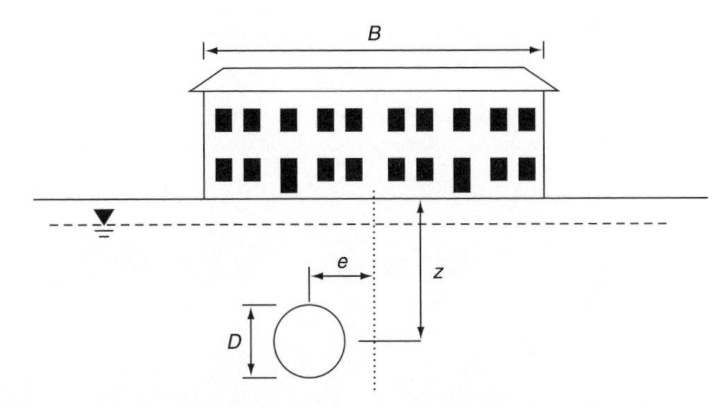

Fig. 3.50 Geometry of problem analysed by Potts and Addenbrooke (1996) (after Mair and Taylor, 1997)

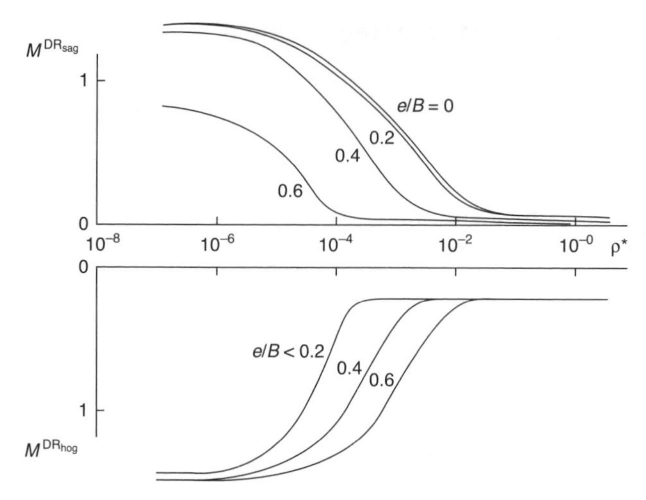

Fig. 3.51 Modification factors for deflection ratio according to relative building stiffness (after Potts and Addenbrooke, 1996; and Mair and Taylor, 1997)

building, see Fig. 3.50). These vary with the relative bending stiffness, ρ^*. In practice, many buildings have ρ^* values exceeding 10^{-2} and for these cases Fig. 3.51 indicates low modification factors in the range 0.1–0.2, that is, the deflection ratio that would be predicted if the building were perfectly flexible is reduced to only 10–20% of that value by the stiffness of the building. Similar modification factors were produced by Potts and Addenbrooke for horizontal strain.

CHAPTER TWENTY-NINE

Case study: effect of the Jubilee Line Extension on the Big Ben Clock Tower

The completed Jubilee Line Extension

Burland *et al.* (2001) describe this case study. The Jubilee Line Extension (JLE) was built between the existing Green Park station, in the West End of London, to Stratford, in east London, a distance of 15.5 km, of which the western length of some 11.5 km is in twin tunnel. The running tunnels are 4.4 m i.d. (internal diameter) and at depths of 20–30 m below ground level.

The full length of the JLE is shown in the simple map of London of Fig. 3.52. In all, there are 11 JLE stations. Four were constructed in open cut, five as enlarged tunnels or a combination of enlarged tunnel and open cut, and there are two surface stations.

The Jubilee Line Extension Project (JLEP) provided a unique opportunity to capture reliable field measurements about the effects of tunnelling

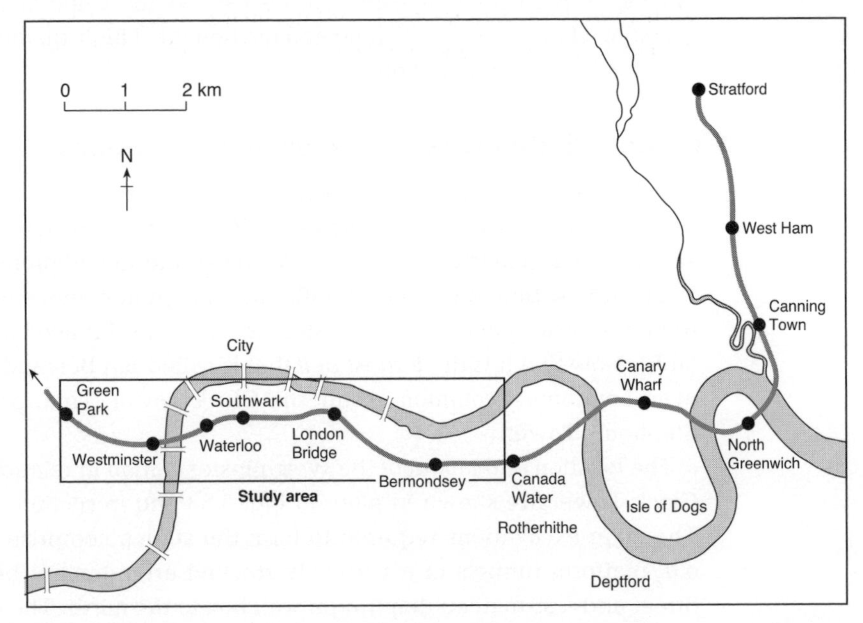

Fig. 3.52 The full route of the Jubilee Line Extension (after Burland et al.*, 2001)*

on a wide range of buildings. Burland *et al.* (2001) present 27 case studies in their geographical sequence from Green Park in the west to London Bridge all on London Clay, and then eastward on the Lambeth Group to Canada Water. Two of the case studies are of instrumented greenfield sites. The case study buildings include the Big Ben Clock Tower, public and commercial buildings of central London and Southwark, and residential buildings, including two tower blocks at Bermondsey and Rotherhithe. Important case studies are those where compensation grouting was used.

The Big Ben Clock Tower and the Palace of Westminster

The construction of the Jubilee Line Extension station at Westminster was predicted to produce significant movements of the Big Ben Clock Tower and the adjoining Palace of Westminster as a result of the excavation of two 7.4 m o.d. (outer diameter) tunnels and the 39 m deep station escalator box. These activities produce two components of movement: an initial, immediate movement directly associated with the progress of excavation and a time-related component due to drainage and consolidation of the London Clay. Protective measures implemented during the construction period, primarily in the form of compensation grouting below the Clock Tower, were extremely effective in controlling settlement and tilt of the structure. The case study not only demonstrates the success of this innovative protective measure, but also shows the value of careful interpretation of numerical modelling and the benefit of high-quality monitoring and reliable instrumentation.

Situation, design criteria and ground conditions

The Houses of Parliament are situated on the north bank of the River Thames adjacent to and upstream of Westminster Bridge. Parts of the structure, particularly Big Ben Clock Tower and the adjoining Ministers' Wing, are within the zone of influence of ground movements arising from the construction of Westminster station on the Jubilee Line Extension (JLE). Note that it is the *largest bell* that is called Big Ben and not, despite even Londoners' common usage, the Clock Tower. That, properly, is St Stephen's Tower.

The location and layout of the Westminster station in relation to Big Ben Clock Tower are shown in plan on Fig. 3.53 and in section on Fig. 3.54. The main excavations required to form the station comprise bored 7.4 m o.d. platform tunnels in a vertically stacked arrangement below Bridge Street and a 39 m deep diaphragm wall box to the north. The lower, westbound tunnel and the upper, eastbound tunnel are at depths of 30 m and 21 m to tunnel axis level below ground level, respectively. The northern

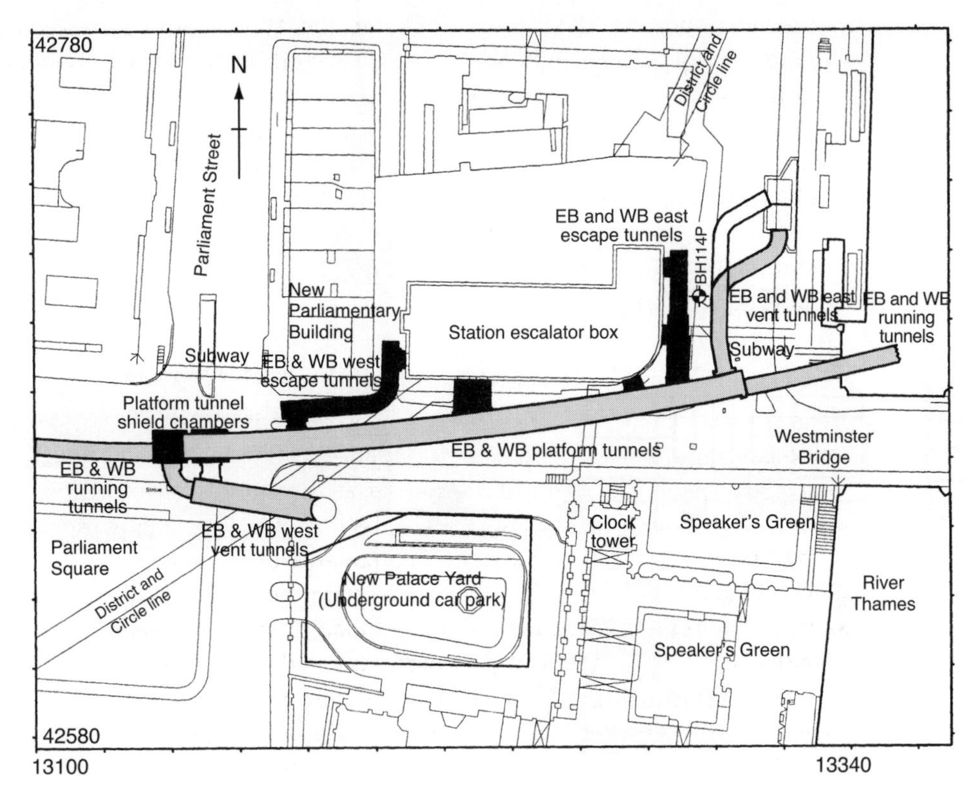

Fig. 3.53 Plan of Jubilee Line Extension works at Westminster (after Burland et al.*, 2001)*

edge of the Clock Tower is 28 m from the centre-line of the tunnels and 34 m from the diaphragm walls of the deep escalator box.

The Big Ben Clock Tower was constructed in 1858, and consists of load-bearing brickwork with a stone cladding to a height of 61 m, supporting a cast-iron-framed spire to the total height of 92 m. The Clock Tower is founded on a mass concrete raft 15 m square and 3 m thick. The raft is founded within the Terrace Gravels about 7 m below the adjacent ground level. The Clock Tower is estimated to have a mass of 8540 t, giving an average bearing pressure of about 400 kPa.

The JLE Westminster station's location arose from sessions with the House of Commons Select Committee, which was considering a proposal for a second phase of the New Parliamentary Building (NPB2, later Portcullis House) at the junction of Bridge Street and Victoria Embankment. It was decided to locate the new station directly below the proposed building and to reconstruct the existing District and Circle Line station, which occupied part of the site. This meant the traditional parallel station tunnel layout could not be accommodated, so a combination of a deep box and tunnels was adopted. The main station structure consists of a shallow

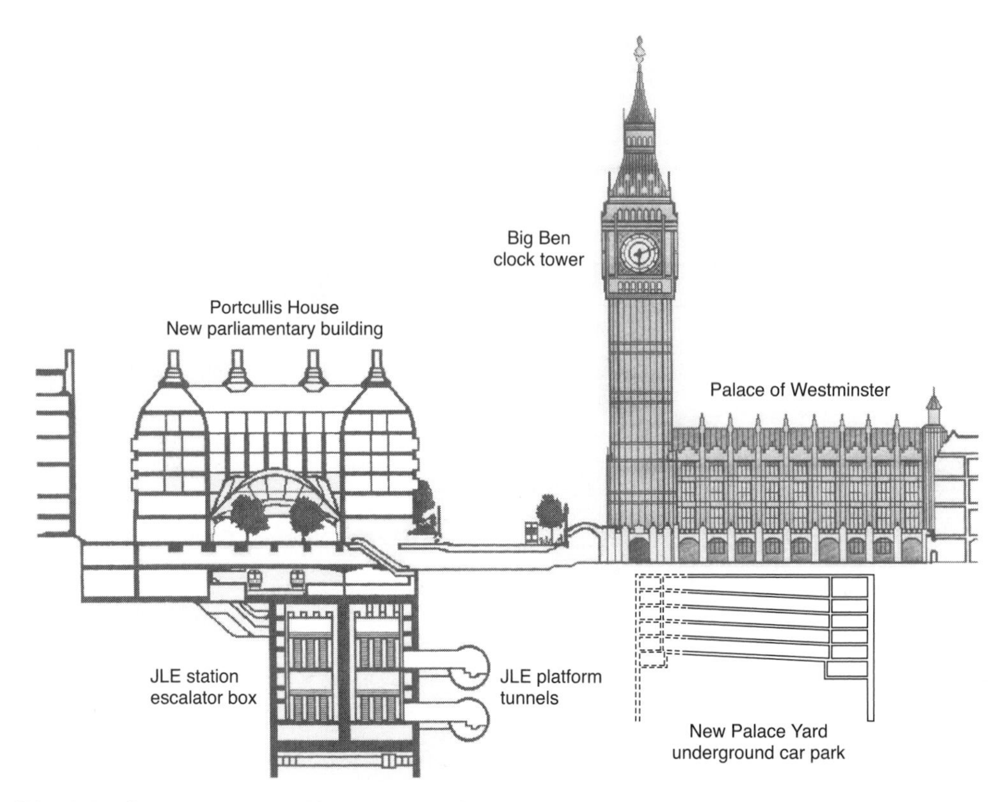

Fig. 3.54 Cross-section at Westminster Station (after Burland et al., *2001)*

basement covering the entire site and underneath a deeper 74 m long by 28 m wide basement structure for the escalators.

The stratigraphy at Westminster is as follows. Made ground of depth varying between 5 m and 8 m overlies Alluvium and Terrace Gravels. Bridge Street is located on one of the former courses of the River Tyburn; consequently, the thickness and nature of the Alluvium varies significantly over short distances. The combined thickness of the Terrace Gravel and Alluvium is about 5 m. The London Clay is 35 m thick between elevations of 94 m and 59 m PD. The Lambeth Group, below the London Clay, lies over Thanet Beds, which are above the Chalk. The Lambeth Group here is 18 m thick. It is of predominantly clayey rather than granular layers, comprising 8 m of Upper Mottled Clay dissected by a thin layer of the Laminated beds, over Lower Mottled Clay about 5 m thick. The lowermost 5 m includes a thin layer of the Pebble Bed over Glauconitic Sand. The Thanet Beds are about 8 m thick on the top of the Chalk, which is at an elevation of 34 m PD, that is, some 73 m below ground level. The log of the closest borehole to the Clock Tower (BH 114P) is summarized in Fig. 3.55.

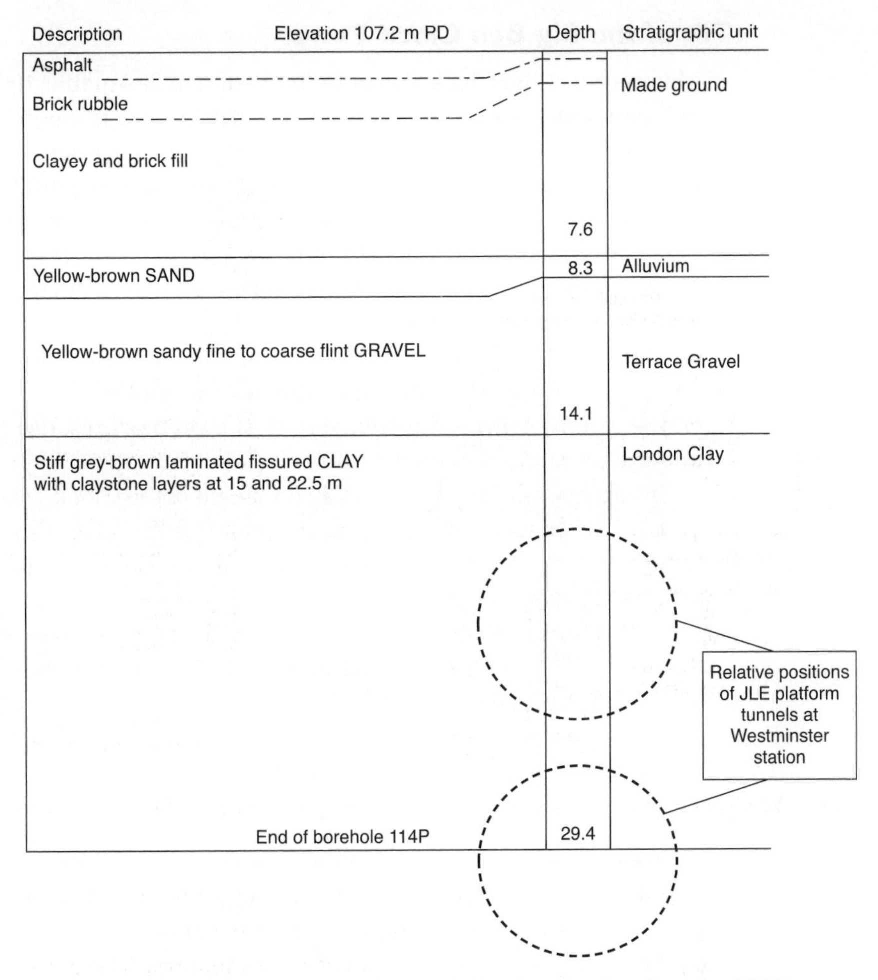

Fig. 3.55 Summary log of Borehole 114P (after Burland et al., *2001)*

The groundwater level in the Terrace Gravel is at about 98 m PD with little tidal variation. The water pressure in the Chalk is substantially reduced (below hydrostatic) and was recorded at an elevation of 58 m PD in 1992. Pore pressures are close to hydrostatic throughout the London Clay and within the Upper Mottled Clay of the Lambeth Group.

Potential damage assessments were undertaken for all structures within the zone of potential ground movement adjacent to the proposed Westminster station. These assessments indicated that protective measures would be required for several of them. Compensation grouting was specified as the way to protect them and they were designated as areas of mandatory ground treatment (MGT).

Tilt of the Big Ben Clock Tower

Monitoring of the Clock Tower for tilt began in November 1994 using five independent systems. Monitoring was carried out throughout construction and was continued for three years after construction ended. The measured tilts of the Clock Tower from the optical plumb throughout the construction period and for six months thereafter are shown in Fig. 3.56, which also gives the timings of the main construction activities. The passage of the four tunnel drives is shown across the top of the figure and the dates of installation of the props at various levels within the escalator box excavation are shown at the bottom. The thick vertical line at December 1995 indicates the start of grouting to control the tilt of the Clock Tower directly and it shows the episodes of grouting undertaken up to September 1997.

Performance control levels (PCL) on the tilt of the Clock Tower were set. Initially, these were conservative at 1 : 6000 for the Amber and 1 : 4000 for the Red. These limits are equivalent to about 9 mm and 14 mm relative movement over the 55 m gauge length monitored.

As predicted, the first activity to affect the tilt of the Clock Tower was the westbound running tunnel drive, which was undertaken in March 1995. Limited compensation grouting was undertaken during the passage of the tunnel past the Clock Tower. Grouting was limited to the arrays within the 45° splay settlement trough and no restriction was placed on

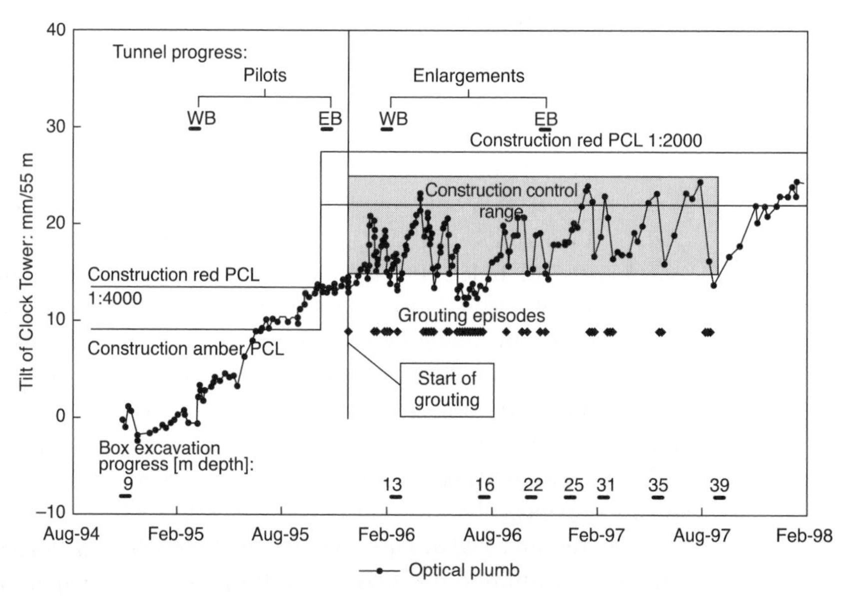

Fig. 3.56 Tilt of the Big Ben Clock Tower measured by optical plumb (after Burland et al., *2001)*

the rate of tunnel advance. As the tunnelling volume losses were greater than expected, the grouting was not effective in mitigating settlements above the tunnel fully or in preventing tilting of the Clock Tower. A tilt to the north of about 4 mm over the 55 m gauge length was observed. No grouting directly below the Clock Tower was carried out during this tunnel drive.

The northwards tilt continued to increase significantly following the tunnel drive and reached the Amber PCL of 9 mm in July 1995. This initiated reappraisal of the potential movements. It was evident that measures to minimize further movements and a strategy for implementing compensation grouting to control tilt of the Clock Tower were required.

A back-analysis of the recorded settlements associated with the westbound running tunnel drive was carried out. This indicated that volume losses in this area were about 3%, that is, significantly greater than had been allowed for in the settlement assessments (2%). The observed increase in tilt after the westbound tunnel drive confirmed that substantial time-dependent movements should also be expected both during the construction period and subsequently.

It was clear at this early stage of construction, with three further tunnel drives and virtually the full depth of the escalator box still to be excavated, that protective measures to control the tilt of the Clock Tower were essential.

Compensation grouting

As noted above, a contingency protective measure was needed to ensure that the tilt of the Clock Tower did not exceed the maximum acceptable value of 1 : 2000. Compensation grouting was adopted as the most appropriate method. The Clock Tower is predominantly outside the volume loss settlement trough associated with the tunnels, so to enable direct control of its movements (in particular the tilt) the grouting arrays were extended below the full footprint of the tower's foundations. The full extent of the grouting arrays installed at Westminster, requiring 10.5 km of drilling, is shown on Fig. 3.57. The arrays installed below the tower initially comprised six *tubes à manchette* (TAMs) from Shaft 4/4 with a maximum spacing of 5 m. Subsequently, as a consequence of changes in other construction activities, these arrays had to be replaced; 16 TAMs were installed from Shaft 4/6, with the maximum spacing reduced to 2.5 m.

As the predicted settlement of the Clock Tower from volume loss of the shallower eastbound running tunnel was minimal (even at greater values of volume loss), this tunnel drive was permitted before the trial grouting which was carried out to demonstrate control of the tilt of the Clock

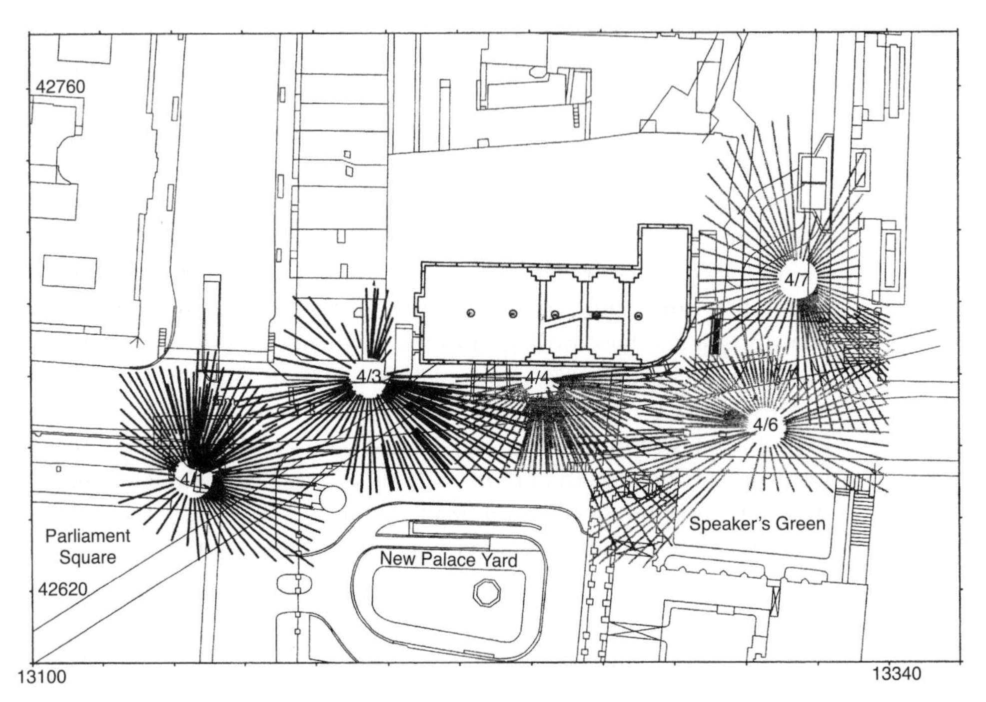

Fig. 3.57 Plan showing shafts and arrays of tubes à manchette *(TAMs) for grouting (after Burland* et al., *2001)*

Tower. Nevertheless, compensation grouting within the settlement trough was undertaken both before and during the progress of the tunnel. The advance rate of the tunnel was controlled such that sufficient grout could be injected to keep the movement to within a few millimetres. The Clock Tower was effectively outside the zone of influence, and the tunnel construction in October 1995 produced no noticeable increase in the tilt of the Tower.

The trial episode of grouting was undertaken in December 1995, at which time the tilt to the north had increased to 14 mm over the 55 m gauge length. The trial was inconclusive and a further trial was delayed until February 1996 by the need to replace the TAMs below the Clock Tower. The tilt to the north was reduced by 5 mm over the 55 m gauge length (about 1 : 11 000) in this second trial, confirming both the suitability of the method and that significant control could be exercised.

The next grouting episode below the Clock Tower was associated with the enlargement of the westbound platform tunnel. Grouting within the settlement trough was again fully coordinated with tunnel advances, but was augmented by additional injections below the Clock Tower. Grout quantities were estimated from experience on the eastbound running tunnel and the first episode of grouting below the Clock Tower.

The aim was to produce full compensation for the tunnelling-induced movements and a small reduction in the tilt. This was successfully achieved, with the tilt reduced by 5 mm. Thereafter, during the box excavation from 96 m down to 71 m PD and the enlargement for the eastbound platform tunnel, grouting was undertaken in response to the observed tilts of the Clock Tower rather than being directly related to construction activities.

The absolute maximum permissible limit on tilt had been set at 1 : 2000, which is equivalent to about 27.5 mm over the 55 m gauge length used to compare the various instrumentation results. A trigger for initiation of grouting was also agreed at 1 : 2500 (or about 22 mm). In practice, a construction control range of 15–25 mm increase in tilt was adopted. Figure 3.56 shows that the upper limit of this control range was not exceeded throughout the construction period. In total, 24 episodes of grouting were undertaken between January 1996 and September 1997, in which a total volume of 122 m^3 of grout was injected.

PART 3

Summary

1 Soil arching is considered and it is shown how easily soil arching can contribute significantly to the support of tunnels, making sprayed concrete linings as used by the New Austrian Tunnelling Method (NATM) quite acceptable.
2 Modern tunnel construction techniques relating to open-face and closed-face tunnels are described briefly.
3 The most important requirements for the successful design and construction of a tunnel are summarized under three headings. These are: stability (both short- and long-term), ground movements and their effects, and the performance of tunnel linings. Positive measures are necessary to prevent water inflow when tunnelling below the water table.
4 Five primary components of ground movement associated with tunnelling are identified. These are: deformation of the ground towards the face, the passage of the shield, the tail void, the deflection of the lining as ground loading develops, and volume change in the ground as a result of changes in effective stress from the short-term to the long-term condition.
5 An important case study, the Heathrow Express trial tunnel, is described in some detail. Data are presented covering both surface and sub-surface deformations together with pore water pressure and earth pressure changes with time, see Table 3.5.
6 A three-dimensional finite-element analysis has been carried out to investigate the effects of anisotropy and the initial effective stress state on deformations due to an idealized open-face tunnel. The higher the ratio of horizontal effective stiffness to vertical effective stiffness, the greater was the settlement. A similar analysis was performed to investigate the multiple interactions between large parallel twin tunnels in stiff clay.
7 The use of soil nails for stabilizing tunnel headings has been studied using finite-element analyses and it is concluded that soil nails induce stabilizing effects on a tunnel face resulting in significant reductions of the face deformations.

8 Finite-element modelling of a tunnel advancing on an existing loaded pile shows that tunnelling near to a pile can lead to an increase in pile deformation and a reduction in pile bearing capacity, the magnitude of the changes being governed by the geometry of the system.
9 A study of the effects of ground movement due to tunnelling on buildings indicates that the stiffness of a building greatly affects the settlement of the building compared with the settlement of a corresponding flexible loaded area.
10 The case study of the Jubilee Line Extension in London is briefly considered. The excavation of two 7.4 m o.d. tunnels and a 34 m deep station escalator box were predicted to produce significant movements of the Big Ben Clock Tower and the adjoining Palace of Westminster. Accordingly, a comprehensive deformation monitoring system was set up and regular readings were taken as the excavations proceeded. It became clear that protective measures to control the tilt of the Clock Tower were essential. Compensation grouting was adopted and proved to be successful in limiting the tilt to an acceptable value.

Recommended list of units, unit abbreviations, quantity symbols and conversion factors for use in soil and rock mechanics

Part 1. SI base units, derived units and multiples.

Quantity and symbol	Units and multiples	Unit abbreviations	Conversion factors for existing units	Remarks
Length (various)	kilometre metre millimetre micrometre	km m mm μm	1 mile = 1.609 km 1 yard = 0.9144 m 1 ft = 0.3048 m 1 in = 25.40 mm	1 micrometre = 1 micron
Area (A)	square kilometre square metre square millimetre	km^2 m^2 mm^2	1 $mile^2$ = 2.590 km^2 1 yd^2 = 0.8361m^2 1 ft^2 = 0.09290 m^2 1 in^2 = 645.2 mm^2	
Volume (V)	cubic metre cubic centimetre cubic millimetre	m^3 cm^3 mm^3	1 yd^3 = 0.7646 m^3 1 ft^3 = 0.02832 m^3 1 in^3 = 16.39 cm^3 1 UK gallon = 4546 cm^3	To be used for solids and liquids
Mass (m)	megagram (or tonne) kilogram gram	Mg (t) kg g	1 ton = 1.016 Mg 1 lb = 0.4536 kg	Megagram is the SI term
Unit weight (γ)	kilonewton per cubic metre	kN/m^3	100 lb/ft^3 = 15.708 kN/m^3 (62.43 lb/ft^3 pure water = 9.807 kN/m^3 = specific gravity 1.0 approx.)	Unit weight is weight per unit volume
Force (various)	Meganewton kilonewton Newton	MN kN N	1 tonf = 9.964 kN 1 lbf = 4.448 N 1 kgf = 9.807 N	
Pressure (p, u)	Meganewton per square metre Megapascal	MN/m^2 MPa	1 tonf/in^2 = 15.44 MN/m^2 (1 MN/m^2 = 1 N/mm^2)	To be used for shear strength, compressive strength, bearing capacity, elastic moduli and laboratory pressures of rock
Stress (σ, τ) and Elastic moduli (E, G, K)	kilonewton per square metre kilopascal	kN/m^2 kPa	1 lbf/in^2 = 6.895 kN/m^2 1 lbf/ft^2 = 0.04788 kPa 1 tonf/ft^2 = 107.3 kPa 1 bar = 100 kPa 1 kgf/cm^2 = 98.07 kPa	Ditto for soils

Quantity and symbol	Units and multiples	Unit abbreviations	Conversion factors for existing units	Remarks
Coefficient of volume compressibility (m_v) or swelling (m_s)	square metre per meganewton square metre per kilonewton	m^2/MN m^2/kN	$1\,ft^2/tonf = 9.324\,m^2/MN = 0.009324\,m^2/kN$	
Coefficient of water permeability (*kw*)	metre per second	m/s	$1\,cm/s = 0.01\,m/s$	This is a velocity, depending on temperature and defined by Darcy's law $V = k_w \frac{\delta h}{\delta s}$ V = velocity of flow $\frac{\delta h}{\delta s}$ = hydraulic gradient
Absolute permeability (*k*)	square micrometre	μm^2	$1\,Darcy = 0.9869\,\mu m^2$	This is an area which quantifies the seepage properties of the ground independently of the fluid concerned or its temperature $V = \frac{kpg}{\eta}\frac{\delta h}{\delta s}$ p = fluid density g = gravitational acceleration η = dynamic viscosity
Dynamic viscosity (η)	millipascal second (centipoise)	mPas (cP)	$1\,cP = 1\,mPas$ $(1\,Pa = 1\,N/m^2)$	Dynamic viscosity is defined by Stokes' Law. A pascal is a kilonewton per square metre
Kinematic viscosity (ν)	square millimetre per second (centistoke)	mm^2/s (cSt)	$1\,cSt = 1\,mm^2/s$	$\nu = \eta/\rho$
Celsius temperature (t)	degree Celsius	°C	$t°F = 5(t-32)/9°C$	The Celsius temperature t is equal to the difference $t = T - T_0$ between two thermodynamic temperatures T and T_0 where $T_0 = 273.15\,K$ (K = Kelvin)

Part 2. Other units

Quantity and symbol	Units and multiples	Unit abbreviations	Conversion factors for existing units	Remarks
Plane angle (various)	Degree Minute second (angle)	° ′ ″		To be used for angle of shearing resistance (ϕ) and for slopes
Time (t)	year	year	1 year = 31.557×10^6 s	'a' is the abbreviation for year
	day	d	1 d = 86.40×10^3 s	
	hour	h	1 h = 3600 s	
	second (time)	s		The second (time) is the SI unit
Coefficient of consolidation (c_v) or swelling (c_s)	square metre per year	m^2/year	1 ft^2/year = 0.09290 m^2/year	

References and bibliography

Agerschou, H. A. (1962). Analysis of the *Engineering News* pile formula. *J. Soil Mech. & Fdn Engng, ASCE*, **88**(5), 1–11.

Aldridge, T. R. and Schnaid, F. (1992). Degradation of skin friction for driven piles in clay. *Proc. Conf. on Recent Large-scale Fully Instrumented Pile Tests in Clay*, ICE, London.

Anagnostou, G. and Kovari, K. (1996). Face stability in slurry and EPB shield tunnelling. *Proc. Int. Symp. on Geotechnical Aspects of Underground Construction in Soft Ground, London* (eds R. J. Mair and R. N. Taylor), Balkema, Rotterdam, pp. 453–458.

Anderson, J. M. (1996). Reducing risk and improving safety with particular reference to NATM. *Proc. North American Tunnelling '96* (ed. L. Ozdemir), Balkema, Rotterdam, pp. 35–42.

Antao, A., Leca, E. and Magnan, J. P. (1995). Finite element determination of upper bound solutions for the stability of shallow tunnels, based on a regularization principle. *Proc. 10th Danube Eur. Conf. on Soil Mech. & Fdn Engng*, Vol. 3, pp. 479–486.

API (1984). *Recommended Practice for Planning, Designing and Constructing Fixed Offshore Platforms*. American Petroleum Institute, API RP2A, 15th edn., Washington, USA.

ASCE (1993). *Design of Pile Foundations*. American Society of Civil Engineers, Reston, Va.

Aschenbrenner, T. B. and Olsen, R. E. (1984). Prediction of settlement of single piles in clay. *Anal. Design Pile Fdns* (ed. J. R. Meyer), *ASCE*, 41–58.

ASTM (1995). *Standard Method of Testing Piles under Static Axial Compressive Load*. American Society for Testing and Materials, West Conshohocken, Pa., D1143–81.

Ata, A. A. (1996). Ground settlements induced by slurry shield tunnelling in stratified soils. *Proc. North American Tunnelling '96* (ed. L. Ozdemir), **1**, pp. 43–50.

Atahan, C., Leca, E. and Guilloux, A. (1996). Performance of a shield driven sewer tunnel in the Val-de-Marne, France. *Proc. Int. Symp. on Geotech. Aspects of Underground Construction in Soft Ground, London* (eds R. J. Mair and R. N. Taylor), Balkema, Rotterdam, pp. 641–646.

Atkinson, J. H. and Potts, D. M. (1977). Subsidence above shallow tunnels in soft ground. *Proc. ASCE Geotech. Engng Div.* **103**(GT 4), pp. 307–325.

Atkinson, J. H. and Sällfors, G. (1991). Experimental determination of stress–strain–time characteristics in laboratory and in-situ tests. *Proc. 10th Eur. Conf. Soil Mech. & Fdn Engng, Florence*, **3**, Balkema, Rotterdam, pp. 915–956.

Atkinson, J. H. and Stallebrass, S. E. (1991). A model for recent history and nonlinearity in the stress–strain behaviour of overconsolidated soil. *Proc. 7th Int. Conf. Comp. Methods & Adv. in Geotechnics*, Balkema, Rotterdam, pp. 555–560.

Atkinson, J. H., Richardson, D. and Stallebrass, S. E. (1990). Effect of recent stress history on the stiffness of overconsolidated soil. *Géotechnique*, **40**(4), 531–540.

Attewell, P. B. (1978). Ground movements caused by tunnelling in soil. *Proc. Int. Conf. on Large Movements and Structures* (ed. J. D. Geddes), Pentech Press, London, pp. 812–948.

Attewell, P. B. and Farmer, I. W. (1974). Ground deformations resulting from shield tunnelling in London Clay. *Can. Geotech. J.*, **2**(3), 380–395.

Attewell, P. B., Glossop, N. H. and Farmer, I. W. (1978). Ground deformations caused by tunnelling in a silty alluvial clay. *Ground Engng*, **2**(8), 32–41.

Attewell, P. B. and Woodman, J. P. (1982). Predicting the dynamics of ground settlement and its derivatives caused by tunnelling in soil. *Ground Engng*, **15**(8), pp. 13–22 and p. 36.

Attewell, P. B., Yeates, J. and Selby, A. R. (1986). *Soil Movements Induced by Tunnelling and their Effects on Pipelines and Structures*. Blackie, Glasgow.

Baguelin, F., Bustamante, M., Frank, R. and Jezequel, J. F. (1975). La capacité portante des pieux. *Annales de l'Institut Technique du Bâtiment et des Travaux Publics*, Suppl. 330, série SF/116, pp. 1–22.

Baker, C. N., Jr, Azam, T. and Joseph, L. S. (1994). Settlement analysis for 450 meter tall KLCC Towers. *Vertical and horizontal deformations of foundations and embankments: Proc. Settlement '94, Geotech. Special Publication No. 40*, **2**, 1650–1671, ASCE, New York.

Balaam, N. P., Poulos, H. G. and Booker, J. R. (1975). Finite element analysis of the effects of installation on pile load–settlement behaviour. *Geotech. Engng*, **6**(1), 33–48.

Balakrishnan, E. G., Balasubramaniam, A. S. and Phien-Wej, N. (1999). Load deformation analysis of bored piles in residual weathered formation. *J. Geotech. & Geoenvir. Engng, ASCE*, **125**(2), 122–131.

Banerjee, P. K. (1970). *A contribution to the study of axially loaded pile foundations*. PhD thesis, University of Southampton.

Banerjee, P. K. (1978). Analysis of axially and laterally loaded pile groups. *Developments in Soil Mechanics* (ed. C. Scott), Ch. 9. Applied Science Publishers, London.

Banerjee, P. K. and Davies, T. G. (1978). The behaviour of axially and laterally loaded single piles embedded in non-homogeneous soils. *Géotechnique*, **28**(3), 309–326.

Barcham, M. C. and Gillespie, B. J. (1988). New bank of China: Hong Kong foundations and substructure design. *Proc. 4th Int. Conf. on Tall Buildings, Hong Kong and Shanghai*, **1**, pp. 128–135.

Barley, A. D. and Graham, M. (1997). Trial soil nails for tunnel face support in London Clay and the detected influence of tendon stiffness and bond length on load transfer. *Ground Improvement Geosystems – Densification and Reinforcement* (eds M. C. R. Davies and F. Schlosser). Thomas Telford, London, pp. 432–444.

Barratt, D. A. and Tyler, R. G. (1976). *Measurements of ground movement and lining behaviour on the London Underground at Regent's Park*. TRRL Report 684.

BD (1997a). *Pile Foundations*. Practice Note for Authorised Persons and Registered Structural Engineers, No. 66, Buildings Ordinance Office, 1997 rev., Buildings Department, Hong Kong Government.

BD (1997b). *Development in the Area Numbers 2 and 4 of Scheduled Areas*. Practice Note for Authorised Persons and Registered Structural Engineers, No. 161, Buildings Department, Hong Kong Government.

Belcotec (1985). Pile foundation problems. *Belgian Geotechnical Volume*, Ch. 1, Pub. for 1985 Golden Jubilee of ISSFME. Belgian Soc. Soil Mech. Fdn Engng.

Berezantzev, V. G., Khristoforov, V. and Golubkov, V. (1961). Load bearing capacity and deformation of piled foundations. *Proc. 5th Int. Conf. Soil Mech. & Fdn Engng, Paris*, **2**, pp. 11–15.

Bjerrum, L. and Eide, O. (1956). Stability of strutted excavations in clay. *Géotechnique*, **6**(1), 32–47.

Boden, J. B. and McCaul, C. (1974). *Measurement of ground movements during a bentonite tunnelling experiment*. TRRL Report 653.

Bolton, M. D. (1979). *A Guide to Soil Mechanics* (eds M. D. Bolton and K. Bolton), Macmillan Press, Cambridge, 439 pp.

Bolton, M. D. (1993). Mechanisms of ground deformation due to excavation in clay. *Proc. Excavation in Urban Areas, KIGForum '93* (ed. T. Adachi). Japanese Soc. Soil Mech. & Fdn Engng, pp. 1–33.

Bolton, M. D. and Sun, H. W. (1991a). Designing foundations on clay to limit immediate movements. *Proc. 4th Int. Conf. on Ground Movements and Structures, Cardiff*.

Bolton, M. D. and Sun, H. W. (1991b). The behaviour of bridge abutments on spread foundations. *Proc. 10th Eur. Conf. on Soil Mech. & Fdn Engng, Florence*, pp. 319–322.

Bolton, M. D., Powrie, W. and Symons, L. F. (1989–90). The design of in-situ walls retaining overconsolidated clay. Part I. *Ground Engineering*, **22**(8), 44–48; **22**(9), 34–40; Part II. *Ground Engineering*, **23**(2), 22–28.

Bolton, M. D., Sun, H. W. and Britto, A. M. (1993). Finite element analyses of bridge abutments on firm clay. *Comp. & Geotech*.

BOO (1990). *Foundation Design. Building (Construction) Regulations 1990 – Part VI*, Practice Note for Authorised Persons and Registered Structural Engineers, No. 141, Buildings Ordinance Office, 1995 rev., Buildings Department, Hong Kong Government.

Boscardin, M. D. and Cording, E. J. (1989). Building response to excavation-induced settlement. *ASCE J. Geotech. Engng*, **115**(1), 1–21.

Bowles, J. E. (1988). *Foundation Analysis and Design*, 4th edn. McGraw-Hill.

Brinch Hansen, J. (1963). Discussion of hyperbolic stress–strain response: cohesive soil, by Robert L. Kondner. *J. Soil Mech., Fnd* Div., *ASCE*, **89**(SM4), Pro. Paper 3429, 241–242.

Broms, B. B. and Bennermark, H. (1967). Stability of clay at vertical openings. *ASCE J. Soil Mech. & Fdn Engng Div.*, **93**(SM1), 71–94.

Broms, B. B. and Lim, P. C. (1988). A simple pile driving formula based on stress wave measurements. *Proc. 3rd Int. Conf. on Application of Stress-wave Theory to Piles*, pp. 591–600.

Broms, B. B. and Shirlaw, J. N. (1989). Settlements caused by earth pressure balance shields in Singapore. *Tunnels et Micro-tunnels en Terrain Meuble – du Chantier à la Théorie*. Presse de l'Ecole Nationale des Ponts et Chaussées, Paris, pp. 209–219.

BSI (1985). *Structural Use of Concrete.* BS 8110: Part I and Part II. British Standards Institution, London.

BSI (1986). *British Standard Code of Practice for Foundations.* BS 8004. British Standards Institution, London.

BSI (1995). *Geotechnical Design.* Eurocode 7, British Standards Institution, London.

Burland, J. B. (1973). Shaft friction of piles in clay – a simple fundamental approach. *Ground Engng*, **6**(3), 30–42.

Burland, J. B. (1995). Assessment of risk of damage to buildings due to tunnelling and excavations. Invited Special Lecture to IS-Tokyo '95. *Proc. 1st Int. Conf. on Earthquake Geotech. Engng.*

Burland, J. B. and Hancock, R. J. R. (1977). Underground car park at the House of Commons, London: geotechnical aspects. *Struct. Engr*, **55**(2), 87–100.

Burland, J. B., Simpson, B. and St. John, H. D. (1979). Movements around excavations in London clay. *Proc. 7th Eur. Conf. Soil Mech.*, Balkema, Rotterdam, **1**, pp. 13–19.

Burland, J. B., Standing, J. R. and Jardine, F. M. (2001). *Building Response to Tunnelling.* Thomas Telford, London.

Bustamante, M. and Gianeselli, L. (1982). Pile bearing capacity prediction by means of static penetrometer CPT. *Proc. ESOPT 2, Amsterdam*, **2**, pp. 493–500.

Butcher, A. P. and Powell, J. J. M. (1995). The effects of geological history on the dynamic stiffness in soils. *Proc. 11th Eur. Conf. Soil Mech. & Fdn Engng, Copenhagen*, **1**, pp. 27–36.

Butler, H. D. and Hoy, H. E. (1977). *Users Manual for the Texas Quick Load Method for Foundation Load Testing.* Office of Development, Federal Highway Administration, Washington, DC.

Butler, R. A. and Hampton, D. (1975). Subsidence over soft ground tunnel. *J. Geotech. Engng, ASCE*, **100**(GT1), 35–49.

Butterfield, R. and Abdrabbo, F. M. (1983). Static working load tests on small scale pile groups in clay. *Developments in Soil Mechanics and Foundation Engineering.* Applied Science Publishers, London, **1**, pp. 111–168.

Butterfield, R. and Banerjee, P. K. (1971). The elastic analysis of compressible piles and pile groups. *Géotechnique*, **21**(1), 43–60.

Butterfield, R. and Douglas, R. A. (1981). *Flexibility coefficients for the design of piles and pile groups*, CIRIA: Technical Note 108, Construction Industry Research and Information Association, London.

Butterfield, R. and Ghosh, N. (1977). The response of single piles in clay to axial load. *Proc. 9th Int. Conf. Soil Mech. & Fdn Engng, Tokyo*, **1**, pp. 451–457.

Buttling, S. (1976). Estimates of shaft and end loads in piles in chalk using strain gauge instrumentation. *Géotechnique*, **26**, 133–147.

Bykovtsev, G. I. (1961). On the velocity field in the problem of indentation of a flat punch into a plastic half-space. *Prikl. Mat. Mekh.*, **25**, 552–553 (in Russian).

Callanan, J. F. and Kulhawy, F. H. (1985). *Evaluation of procedures for predicting foundation uplift movements.* Report to Electric Power Research Institute, No. EPRI EL-4107, Cornell University.

Carrubba, P. (1997). Skin friction of large-diameter piles socketed into rock. *Can. Geotech. J.*, **34**(2), 230–240.

Carter, J. P. and Kulhawy, E. H. (1987). *Analysis and design of foundations socketed into rock*. Research Report 1493-4, Geotech. Engng Group, Cornell University.

Cernak, B. (1976). The time effect of suspension on the behaviour of piers. *Proc. 6th Eur. Conf. Soil Mech. & Fdn Engng*, **1**, pp. 111–114.

CGS (1995). *Canadian Foundation Engineering Manual – Fourth Edition*. Can. Geotech. Society, Technical Committee on Foundations, Bitech Publishers, Richmond, British Columbia.

Chambon, I. F. and Corte, J. F. (1994). Shallow tunnels in cohesionless soil: stability of tunnel face. *J. Geotech. Engng, ASCE*, **120**(7), 1150–1163.

Chambosse, G. (1972). Das verformungsverhalten des Frankfurter Tons beim Tunnelvortrieb mitt versuchanstalt bodenmech. *Grundbau, Tech. Huchschule Darmstadt*, No. 10, p. 103.

Chang, M. F. and Wong, I. H. (1995). Axial load test behaviour of bored piles in weathered granite. *Proc. 10th Asian Reg. Conf. Soil Mech. & Fdn Engng, Beijing*, **1**, pp. 185–188.

Chellis, R. D. (1961). *Pile Foundations*. McGraw Hill, New York.

Cheng, E. and Haberfield, C. M. (1998). Laboratory study of the effect of drilling fluids on piles rocketed into soft rock. *Proc. 2nd Int. Symp. on the Geotechnics of Hard Soils–Soft Rocks* (eds A. Evangelista and L. Picarelli), Balkema, Rotterdam, **I**, pp. 93–101.

Chin, F. K. (1970). Estimation of the ultimate load of piles from tests not carried to failure. *Proc. 2nd SE Asian Conf. Soil Engng, Singapore*, pp. 81–92.

Chiu, H. K. and Johnston, I. W. (1980). The effects of drainage conditions and confining pressures on the strength of Melbourne mudstone. *Foundations on Melbourne Mudstone: Rock Socketed Piles*. Monash University, Melbourne.

Chow, Y. K. (1986). Analysis of vertically loaded pile groups. *Int. J. Numer. Anal. Meth. Geomechs*, **10**, 59–72.

Christian, J. T. and Carrier, W. D. (1978). Janbu, Bjerrum and Kjaernsli's chart reinterpreted. *Can. Geotech. J.*, **15**, 123–128.

Christoulas, S. G. (1988). Dimensionnement de pieux. Quelques experiences et recherches en Grèce. *Bull. Liaison Labs Ponts et Chaussées*, **154**, 5–10.

Christoulas, S. G. and Pachakis, M. (1987). Pile settlement based on SPT results. *Bull. Res. Centre of Public Works Dept of Greece*, **3**, July–Sept., 221–226.

Clayton, C. R. I. and Milititsky, J. (1983). Installation effects and the performance of bored piles in stiff clay. *Ground Engng*, **16**(2), 17–22.

Clear, C. A. and Harrison, T. A. (1985). *Concrete pressure on formwork*. CIRIA Report 108, Construction Industry Research and Information Association, London.

Clough, G. W. and Hansen, L. (1981). Clay anisotropy and braced wall behaviour. *J. Geotech. Engng Div.*, ASCE, **107**(7), 893–914.

Clough, G. W. and Leca, E. (1989). With focus on use of finite element methods for soft ground tunnelling. *Tunnels et Micro-tunnels en Terrain Meuble – du Chantier à la Théorie*. Presse de l'Ecole Nationale des Ponts et Chaussées, Paris, pp. 531–573.

Clough, G. W. and O'Rourke, T. D. (1990). Construction induced movements of insitu walls. *Proc. ASCE Conf. on Design and Performance of Earth-Retaining Structures*, Cornell University, Geoetch. Special Publication 25, pp. 439–470.

Clough, G. W. and Schmidt, B. (1981). Design and performance of excavations and tunnels in soft clay. *Soft Clay Engineering*. Elsevier, Amsterdam, pp. 569–634.

Clough, G. W., Sweeny, B. P. and Finno, R. J. (1983). Measured soil response to EPB shield tunnelling. *J. Geotech. Engng, ASCE*, **109**(2), Feb., 131–149.

Cooke, R. W. (1974). The settlement of friction pile foundations. *Proc. Conf. on Tall Buildings, Kuala Lumpur.*

Cooke, R. W. and Price, G. (1974). Horizontal inclinometers for the measurement of vertical displacement in the soil around experimental foundations. *Field Inst. in Geotechnical Engineering*. Butterworth, London, pp. 112–125.

Cording, E. J. (1991). Control of ground movements around tunnels in soil. General report. *Proc. 9th Pan-American Conf. on Soil Mech. & Fdn Engng, Chile.*

Cording, E. J. and Hansmire, W. H. (1975). Displacements around soft ground tunnels. General report *Proc. 5th Pan American Conf. on Soil Mech. & Fdn Engng, Buenos Aires*, Session IV, pp. 571–632.

Coyle, H. M. and Reese, L. C. (1966). Load transfer for axially loaded piles in clay. *J. Soil Mech. Fdn Engng, ASCE*, **92**(SM2), 1–26.

D'Appolonia, D. J., D'Appolonia, E. and Brissette, R. F. (1970). Closure to settlement of spread footings on sand. *J. Soil Mech. & Fdn Engng, ASCE*, **96**(SM2), 754–761.

Das, B. M. (2000). *Fundamentals of Geotechnical Engineering*. Brookes/Cole, Thomson Learning.

Dasari, G. R. (1996). *Modeling the variation of soil stiffness during sequential construction*. PhD thesis, University of Cambridge, UK.

Datta, M., Gulhati, S. K. and Rao, G. V. (1980). An appraisal of the existing practise of determining the axial load capacity of deep penetration piles in calcareous sands. *Proc. 12th Ann. OTC, Houston*, Paper OTC 3867, pp. 119–130.

Davis, A. G. (1974). Contribution to discussion IV, Rocks. *Proc. Cambridge Conf. Settlement of Struct.*, pp. 757–759.

Davis, E. H., Gunn, M. J., Mair, R. J. and Seneviratne, H. N. (1980). The stability of shallow tunnels and underground openings in cohesive material. *Géotechnique*, **30**(4), 397–419.

Davisson, M. T. (1972). High capacity piles. *Proc. ASCE Lecture Series, Innovations in Fdn Constr.*, ASCE, New York.

Day, P. W., Wates, J. A. and Knight, K. (1981). Skin friction on underslurry piles. *Proc. 10th Int. Conf. on Soil Mech. & Fdn Engng, Stockholm*, **2**, pp. 689–692.

De Beer, E. E. (1967). Proefondervindelijke bijdrage tot de studie vanhet grensdraag vennogen van zand onder fundering op staal. *Tijdshrift der Openbar Werken van Belgie*, 6.

De Moor, E. K. and Taylor, R. N. (1991). Ground response to construction of a sewer tunnel in very soft ground. *Proc. Conf. Tunnelling '91*, Inst. Min. and Metallurgy, London, pp. 43–54.

De Ruiter, J. and Beringen, F. L (1979). Pile foundations for large North Sea structures. *Mar. Geotechnol.*, **3**(3), 267–314.

De Silva, S., Cheung, C. T., Pratt, M. and Walsh, N. (1998). Instrumented bored and barrette pile load tests for western harbour crossing. *Proc. 17th Ann. Sem. Geotech. Div. Hong Kong Instn Engrs, Hong Kong*, pp. 79–94.

Deane, A. P. and Bassett, R. H. (1995). The Heathrow Express Trial Tunnel. *Proc. Instn Civ. Engrs Geotech. Engng*, **113**, July, 144–156.

Decourt, L. (1982). Prediction of the bearing capacity of piles based exclusively on N values of the SPT. *Proc. ESOPT2, Amsterdam*, **1**, pp. 29–34.

Denver, H. (1982). Modulus of elasticity determined by SPT and CPT. *Proc. ESOPT2, Amsterdam*, **1**, pp. 35–40.

Desai, C. S. (1974). Numerical design-analysis for piles in sands. *J. Geotech. Engng, ASCE*, **100**(GT6), 613–635.

DiBiaggio, E. and Roti, J. A. (1972). Earth pressure measurements on a braced slurry-trench wall in soft clay. *Proc. 5th Eur. Conf. Soil Mech. & Fdn Engng, Madrid*, **1**, pp. 473–483.

Dutt, R. N. and Ingram, W. B. (1984). Jackup rig siting in calcareous soils. *Proc. 16th Ann. OTC, Houston*, Paper OTC 4840, pp. 541–548.

Dyer, M. R., Hutchinson, M. T. and Evans, N. (1996). Sudden Valley Sewer: a case history. *Proc. Int. Symp. Geotech. Aspects of Underground Construction in Soft Ground, London* (eds R. J. Mair and R. N. Taylor), Balkema, pp. 671–676.

Eadie, H. S. (1977). Settlements observed above a tunnel in sand. *Tunnels and Tunnelling*, **9**(5), 93–94.

Eden, W. J. and Bozozuk, M. (1968). Earth pressures on Ottawa outfall sewer tunnel. *Can. Geotech. J.*, **6**(1), 17–32.

Edil, T. B. and Mochtar, I. B. (1988). Creep response of model pile in clay. *J. Geotech. Engng, ASCE*, **114**(11), 1245–1260.

Eide, O., Aas, G. and Josang, T. (1972). Special application of cast-in-place walls for tunnels in soft clay in Oslo. *Proc. 5th Eur. Conf. on Soil Mech. & Fdn Engng, Madrid*, **1**, pp. 485–498.

Farmer, L. W. and Attewell, P. B. (1973). Ground movements caused by a bentonite-supported excavation in London clay. *Géotechnique*, **23**(4), 577–581.

FDT (1999). *Florida Department of Transportation 1999 Standard Specifications for Road and Bridge Construction*. FDT, Tallahassee, Florida.

Fellenius, B. H. (1980). The analysis of results from routine pile loading tests. *Ground Engng*, **13**(6), 19–31.

Fellenius, B. H. (1990). *Guidelines for the Interpretation and Analysis of the Static Loading Test*. Deep Foundations Institute, Sparta, NJ.

Findlay, J. D. (1984). *Piling and Ground Treatment. Discussion*. ICE, Thomas Telford, London, pp. 189–190.

Flaate, K. S. (1964). *An investigation of the validity of three pile driving formulae in cohesionless material*. Pub. No. 56. Norwegian Geotechnical Institute, Oslo.

Fleming, W. G. K. and Sliwinski, Z. J. (1977). *The use and influence of bentonite in bored pile construction*. CIRIA Report (PG3), London.

Fleming, W. G. K. and Thorburn, S. (1984). Recent piling advances. *Piling and Ground Treatment*. ICE, Thomas Telford, London, pp. 1–16.

Fleming, W. G. K., Weltman, A. J., Randolph, M. F. and Elson, W. K. (1985). *Piling Engineering*. Surrey University Press, Glasgow and London; and Halstead Press, New York.

Fletcher, M. S. and Mizon, D. H. (1984). Piles in chalk for Orwell bridge. *Piling and Ground Treatment*. ICE, Thomas Telford, London, pp. 203–209.

Focht, J. A. (1967). Discussion to paper by Coyle and Reese. *J. Soil Mech. & Fdn Engng, ASCE*, **93**(SM1), 133–138.

Frank, R. A. (1974). *Etude théorique du comportement des pieux sous charge verticale, introduction de la dilatance.* Dr-Eng. thesis, University Paris VI (Pierre et Marie Curie University).

Frank, R. A. (1985). Recent developments in the prediction of pile behaviour from pressuremeter results. *Proc. Symp. from Theory to Practice on Deep Foundations, Porto Alegre, Brazil,* **1**, pp. 69–99.

Fraser, R. A. (1985). Design of large diameter piles. *Hong Kong Engr,* **13**(1), 37–49.

Fraser, R. A. and Kwok, D. (1986). Mobilisation of skin friction in bored piles in Hong Kong. *Proc. Int. Conf. Deep Found., Beijing,* **2**, pp. 2.65–2.74.

Fraser, R. A. and Wardle, L. J. (1976). Numerical analysis of rectangular rafts on layered foundations. *Géotechnique,* **26**(4), 613–630.

Fujita, K. (1981). On the surface settlements caused by various methods of shield tunnelling. *Proc. 11th Int. Conf. on Soil Mech. & Fdn Engng,* **4**, pp. 609–610.

Fujita, K. (1989). Special lecture B: Underground construction, tunnel, underground transportation. *Proc. 12th Int. Conf. on Soil Mech. & Fdn Engng, Rio de Janeiro,* **4**, pp. 2159–2176.

Fujita, K. (1994). Soft ground tunnelling and buried structures. State-of-the-art report. *Proc. 13th Int. Conf. on Soil Mech. & Fdn Engng, New Delhi,* **5**, pp. 89–108.

Fuller, E. M. and Hoy, H. E. (1970). *Pile load tests including quick load test method, conventional methods and interpretations.* Highway Research Board, No. 333, Washington, DC, pp. 74–86.

GCO (1984). *Geotechnical Manual for Slopes (2nd edn).* Geotechnical Control Office, Hong Kong.

GCO (1988). *Guide to Rock and Soil Descriptions. Geoguide* 3, 1994 edn. Geotechnical Control Office, Hong Kong.

GEO (1991). *Foundation design of caissons on granitic and volcanic rocks.* GEO Report No. 8, Geotechnical Engineering Office, Civil Engineering Department, Hong Kong.

GEO (1993). *Behaviour of four instrumented machine-bored piles for stage 3 of the West Kowloon Corridor.* Special Project Report 5/93, Geotechnical Engineering Office, Civil Engineering Department, Hong Kong.

GEO (1996). *Pile design and construction.* GEO Publication 1/96, Geotechnical Engineering Office, Civil Engineering Department, Hong Kong.

Giroud, J. P., Tran-Vo-Nhiem, and Obin, J. P. (1973). *Tables pour le Calcul Desfondations.* Dunod, Paris, **3**.

Glossop, H. H. (1978). *Ground movements caused by tunnelling in soft soils.* PhD Thesis, University of Durham.

Glossop, N. H. and O'Reilly, M. P. (1982). Settlement caused by tunnelling through soft marine silty clay. *Tunnels and Tunnelling,* **14**(9), 13–16.

Glossop, N. H., Saville, D. R., Moore, J. S., Benson, A. P. and Farmer, I. W. (1979). Geotechnical aspects of shallow tunnel construction in Belfast estuarine deposits. *Proc. Tunnelling '79*, London, pp. 45–56.

Goble, G. G. and Rausche, F. (1976). *Wave equation analysis of pile driving – WEAP Program.* US Department of Transport Report, FHWA-IP-4.2.

Goble, G. G., Garland, L. and Rausche, F. (1970). *Dynamic studies on the bearing capacity of piles, Phase III.* Report No. 48, Division of Solid Mechanics, Structures and Mechanical Design, Case Western Reserve University, Cleveland, USA.

Goble, G. G., Likins, G. E. and Rausche, F. (1975). Bearing capacity of piles from dynamic measurements. Final report No. OHIO-DOT-05–75, Case Western Reserve University, Cleveland, USA.

Goble, G. G., Scanlan, R. H. and Tomko, J. J. (1967). Dynamic studies on the bearing capacity of piles. *Highway Research Record* No. 67, *Bridges and structures*, Highway Research Board, Washington, DC.

Grant, R. J. and Taylor, R. N. (1996). Centrifuge modelling of ground movements due to tunnelling in layered ground. *Proc. Int. Symp. on Geotech. Aspects of Underground Construction in Soft Ground, London* (eds R. J. Mair and R. N. Taylor), Balkema, Rotterdam, pp. 507–512.

Grasso, P., Mahtab, A. and Pelizza, S. (1989). Reinforcing a rock zone for stabilizing a tunnel in complex formations. *Proc. Int. Congr. on Progress and Innovation in Tunnelling. Toronto*, **2**, pp. 663–670.

Grose, W. J. and Eddie, C. M. (1996). Geotechnical aspects of the construction of the Heathrow Transfer Baggage System tunnel. *Proc. Int. Symp. on Geotech. Aspects of Underground Construction in Soft Ground, London* (eds R. J. Mair and R. N. Taylor), Balkema, Rotterdam, pp. 269–276.

Gunn, M. J. (1993). The prediction of surface settlement profiles due to tunnelling. *Predictive Soil Mechanics, Proc. Wroth Memorial Symp., Oxford*, 1992, Thomas Telford, London, pp. 304–316.

Gunn, M. J. and Clayton, C. R. I. (1992). Installation effects and their importance in the design of earth-retaining structures. *Géotechnique*, **42**(1), 137–141.

Gunn, M. J., Satkunananthan, A. and Clayton, C. R. I. (1993). Finite element modelling of installation effects. *Retaining Structures*. Thomas Telford, London, pp. 46–55.

HA (1998). *Hong Kong Housing Authority Specification Library for Building Works*, 1998 edn, PIL 1 piling. Hong Kong Housing Authority, Hong Kong Government.

Hamza, M. and Ibrahim, M. H. (2000). Base and shaft grouted large diameter pile and barrettes load tests. *Proc. Geotech-Year 2000: Developments in Geotech. Engng* (eds A. S. Balasubramaniam *et al.*), **2**, pp. 219–228.

Hanya, T. (1977). Ground movements due to construction of shield-driven tunnel. *Proc. 9th Int. Conf. Soil Mech. & Fdn Engng, Tokyo*, Case History Vol., pp. 759–790.

Harrison, T. A. (1983). *Pressure on vertical formwork when concrete is placed in wide sections.* Research Report 22, Cement and Concrete Association.

Heerema, E. P. (1979). Relationships between wall friction, displacement velocity, and horizontal; stresses in clay and sand for pile driveability analysis. *Ground Engng*, **12**(1), 55–65.

Henry, K. (1974). Grangemouth tunnel sewer. *Tunnels and Tunnelling*, **6**(1), 25–29.

Hibbitt, Karlsson and Sorensen Inc. (1998). *ABAQUS User's Manual, Version 5.8.*

Hiley, A. (1922). The efficiency of the hammer blow. *Engineering*, 673–674, 711–714, 745.

Hirany, A. and Kulhawy, E. H. (1989). Interpretation of load tests on drilled shafts. *Proc. Fnd Engng: Current Principles and Practices*, **2**, 1132–1149.

HKG. (1992). *General specification for civil engineering works*, Vol. 1. Civil Engineering Department, Hong Kong Government, Hong Kong.

Ho, C. E. (1992). *The assessment of shaft friction of bored piles in Hong Kong.* MSc dissertation, University of London.

Ho, C. E. (1993). Deep barrette foundation in Singapore weathered granite. *Proc. 11th Southeast Asian Geotech. Conf.* pp. 529–534.

Ho, C. E. and Lim, C. H. (1998). Barrettes designed as friction foundations: a case history. *Proc. 4th Int. Conf. on Case Histories in Geotech. Engng*, pp. 236–241.

Ho, K. K. S. (1993a). *Behaviour of the instrumented trial barrette for the Rumsey Street Flyover Project*. Special Project Report No. SPR4/93, Geotechnical Engineering Office, Civil Engineering Department, Hong Kong Government, Hong Kong.

Ho, K. K. S. (1993b). *Behaviour of the instrumented trial barrette for the Hang Seng Bank Headquarters Building*. Special Project Report No. SPRI2/93, Geotechnical Engineering Office, Civil Engineering Department, Hong Kong Government, Hong Kong.

Ho, K. K. S. (1993c). *Behaviour of the instrumented trial barrette for the Development at N.K.I.L. 6056, Carpenter Road*. Special Project Report No. SPR13/93, Geotechnical Engineering Office, Civil Engineering Department, Hong Kong Government, Hong Kong.

Ho, K. K. S. (1994). *Behaviour of the instrumented trial barrette for the Trademart Development at N.K.I.L. 6032, Kowloon Bay*. Special Project Report No. SPR8/94, Geotechnical Engineering Office, Civil Engineering Department, Hong Kong Government, Hong Kong.

Hobbs, N. B. (1977). Behaviour and design of piles in chalk – an introduction to the discussion of the papers on chalk. *Piles in Weak Rock*. ICE, London, pp. 149–175.

Holeyman, A. (1985). Dynamic nonlinear skin friction of piles. *Proc. Symp. on Penetrability, Drivability of Piles, 11th Int. Conf. Soil Mech. & Fdn Engng, San Francisco*.

Holt, D. N., Lumb, P. and Wong, P. K. K. (1982). Site control and testing of bored piles at Telford Gardens, an elevated township at Kowloon Bay, Hong Kong. *Proc. 7th Southeast Asian Geotech. Conf., Hong Kong*, **1**, pp. 349–361.

Hong, S. W. and Bae, G. J. (1995). Ground movements associated with subway tunnelling in Korea. *Underground Construction in Soft Ground* (eds K. Fujita and O. Kusakabe). Balkema, Rotterdam, pp. 229–232.

Horvath, R. G. (1978). *Field load test data on concrete-to-rock bond strength for drilled pier foundations*. Publn 78/07, Dept of Civil Engineering, University of Toronto, Canada.

Horvath, R. G. (1982). *Drilled piers socketed into weak shale – methods of improving performance*. PhD thesis, University of Toronto, Canada.

Horvath, R. G. and Kenney, T. C. (1979). Shaft resistance of rock socketed drilled piers. *Proc. Symp. on Deep Fdn, ASCE, New York*, pp. 182–214.

Horvath, R. G., Kenney, T. C. and Kozicki, P. (1983). Methods of improving the performance of drilled piers in weak rock. *Can. Geotech. J.*, **20**, 758–772.

Horvath, R. G., Trow, W. A. and Kenney, T. C. (1980). Results of tests to determine shaft resistance of rock-socketed drilled piers. *Proc. Int. Conf. on Struct. Fdn on Rock*. pp. 349–361.

Housel, W. S. (1965). Michigan study of pile driving hammers. *J. Soil Mech. & Fdn Div., ASCE*, **91**(SM5), 37–64.

Housel, W. S. (1966). Pile load capacity – estimates and test results. *J. Soil Mech. & Fdn Engng, ASCE*, **92**(SM4), 1–29.

HSE (1996). *Safety of New Austrian Tunnelling Method (NATM) Tunnels – a review of sprayed concrete lined tunnels with particular reference to London Clay*. Health and Safety Executive, HSE Books, Sudbury, 87 pp.

Hull, T. S. (1987). *The static behaviour of laterally loaded piles*. PhD thesis, University of Sydney, Australia.

Humpheson, C., Fitzpatrick, A. J. and Anderson, J. M. D. (1986). The basements and substructure for the new headquarters of the Hong Kong and Shanghai Banking Corporation, Hong Kong. *Proc. Instn Civ. Engrs*, Part I, **80**, 851–883. (Discussion, **82**, 1987, 831–858).

Humpheson, C., Fitzpatrick, A. J. and Anderson, J. M. D. (1987). Discussions on the basements and substructure for the new headquarters of the Hong Kong and Shanghai Banking Corporation, Hong Kong. *Proc. Instn Civ. Engrs*, Part I, **82**, 831–858.

Imai, T. and Tonouchi, K. (1982). Correlation of N value with S-wave velocity and shear modulus. *Proc. Eur. Symp. on Penetration Testing 2, Amsterdam*, **1**, pp. 67–72.

Institution of Civil Engineers (1996a). *Specification for piling and embedded retaining walls*. Thomas Telford, London.

Institution of Civil Engineers (1996b). *Sprayed concrete linings (NATM) for tunnels in soft ground*. Thomas Telford, London.

ISRM (1978). Suggested method for determining the uniaxial compressive strength and deformability of rock materials. *Int. J. Rock Mech. Mining Sci. and Geomech. Abstracts*, **16**(2), 135–140.

ISRM (1985). Suggested method for determining point load strength. *Int. J. Rock Mech. & Mining Sci.*, **22**, 51–60.

ISSMFE (1985). Axial pile loading test 1: static loading. *Geotech. J.*, **8**(2), 79–89.

Itasca (1995). *Fast Lagrangian Analysis of Continua (FLAC)*, Version 3.3, user's manuals. Itasca Consulting Group, Inc., Minnesota, USA.

Itasca (1996). *Fast Lagrangian Analysis of Continua (FLAC-3D)*, Version 1.1, User's manuals. Itasca Consulting Group, Inc., Minnesota, USA.

Jacobsz, S. W., Standing, J. R, Mair, R. J., Hagiwara, T. and Sugiyama, T. (2002). Centrifuge modeling of tunneling near driven piles. *Proc. 3rd Int. Symp. on Geotech. Aspects of Underground Construction in Soft Ground, IS-Toulouse*, Pre-print volume, Session 6, pp. 89–94.

Jaeger, J. C. and Cook, N. G. W. (1969). *Fundamentals of Rock Mechanics*. Methuen, London.

Janbu, N., Bjerrum, L. and Kjaernsli, B. (1956). Veiledning ved Løsning av Fundamenteringsoppgaver. *NGI Publication No. 16*, Norwegian Geotechnical Institute, Oslo.

Jancsecz, S. and Steiner, W. (1994). Face support for a large mix-shield in heterogeneous ground conditions. *Proc. Tunnelling '94, IMM, London*, pp. 531–550,.

Jardine, R. J., Potts, D., Fourie, A. B. and Burland, J. B. (1986). Studies of the influence of non-linear stress–strain characteristics in soil–structure interaction. *Géotechnique*, **36**(3), 377–396.

JGJ (1995). *Technical Code for Building Pile Foundations*. People's Republic of China.

Johnson, E. G., Johnson, K. E. and Erikson, C. M. (1992). Deep foundation elements installed by slurry wall techniques. *Slurry walls: design, construction, and quality control, ASTM STP 1129* (eds D. B. Paul, R. R. Davidson and N. J. Cavalli). Philadelphia, pp. 207–224.

Johnston, I. W. and Donald, I. B. (1979). *Final report on rock socket pile test.* Contract No. 703 Flinders St.–Spencer St. Overpass. Melbourne underground rail loop project. Dept of Civil Engng, Report No. 7816/G, Monash University, Melbourne.

Johnston, I. W. and Lam, T. S. K. (1984). Frictional characteristics of planar concrete–rock interfaces under constant normal stiffness conditions. *Proc. 4th Australian–New Zealand Conf. on Geomechs, Perth,* **2**, pp. 397–401.

Johnston, I. W., Carter, J. P., Novello, E. A. and Ooi, L. H. (1988). Constant normal stiffness direct shear testing of calcarenite. *Engineering for Calcareous Sediments.* Balkema, Rotterdam, **2**, pp. 541–553.

Kimura, T. and Mair, R. J. (1981). Centrifugal testing of model tunnels in soft clay. *Proc. 10th Int. Conf. on Soil Mech. & Fdn Engng, Stockholm,* **1**, pp. 319–322.

Komornik, A. (1974). Penetration testing in Israel. State-of-the-art report. *Proc. ESOPT, Stockholm,* **1**, pp. 183–192.

Kraft, L. M. and Lyons, C. G. (1974). State-of-the art: ultimate axial capacity of grouted piles. *Proc. 6th Annual OTC, Houston,* Paper OTC 2081, pp. 487–503.

Kraft, L. M., Ray, R. P. and Kagawa, T. (1981). Theoretical *t–z* curves. *J. Geotech. Engng, ASCE,* **107**(GTI1), 1543–1561.

Kulhawy, F. H. (1984). Limiting tip and side resistance: fact or fallacy? *Proc. Symp. Analysis and Design of Pile Foundations, ASCE, New York,* pp. 80–98.

Kulhawy, E. H. and Phoon, K. K. (1993). Drilled shaft side resistance in clay soil to rock. *Proc. Conf. on Design and Performance of Deep Fdns: Piles and Piers in Soil and Soft Rock, ASCE, New York,* Geotech. Spec. Publn No. 38, pp. 172–183.

Kuwamura, N. (1997). *Shield tunnelling in Chicago clay.* Research Proposal, University of Illinois.

Ladanyi, B. (1977). Discussion on friction and end bearing tests. *Can. Geotech. J.,* **14**, 153.

Lake, L. M., Rankin, W. J. and Hawley, J. (1992). *Prediction and effects of ground movements caused by tunnelling in soft ground beneath urban areas.* CIRIA Project Report 30, Construction Industry Research and Information Association, London.

Lam, T. S. K., Yau, J. H. W. and Premchitt, J. (1991). Side resistance of a rock-socketed caisson. *Hong Kong Engr,* Feb., 17–28.

Lambe, T. W. (1967). Stress path method. *J. Geotech Engng Div., ASCE,* **93**(6), 309–331.

Leblais, Y. and Bochon, A. (1991). Villejust Tunnel: slurry shield effects on soils and lining behaviour and comments on monitoring equipment. *Proc. Tunnelling '91, London,* pp. 65–77, IMM.

Leca, E. (1989). *Analysis of NATM and shield tunnelling in soft ground.* PhD thesis, Virginia Polytechnic Institute and State University, Blacksburg.

Leca, E. and Dormieux, L. (1990). Upper and lower bound solutions for the face stability of shallow circular tunnels in frictional material. *Géotechnique,* **40**(4), 581–605.

Leca, E. and Dormieux, L. (1992). Contribution à l'étude de la stabilité du front de taille d'un tunnel en milieu cohérent. (Contribution to the analysis of the face stability of a tunnel excavated in cohesive ground.) *Revue Française de Géotechnique,* **61**, 5–16 (in French).

Lee, G. T. K. and Ng, C. W. W. (2002). Three-dimensional analysis of ground settlements due to tunnelling: role of K_0 and stiffness anisotropy. *Proc. 4th Int. Symp. Geotech. Aspects of Underground Consruction in Soft Ground, Toulouse*, SPECIFIQUE, Lyon, pp. 617–622.

Lee, G. T. and Ng, C. W. W. (2004). The effects of advancing open face tunnelling on an existing loaded pile. *J. Geotech. Geoenvironmental Engng, ASCE*. Accepted.

Lehane, B. M. and De Cock, F. (1999). Standard European practice for estimating ultimate pile capacities from laboratory test data. *Ground Engng (Eur. Fdns)*, **4**(1), 24–27.

Lei, G. H., Ng, C. W. W. and Rigby, D. (2001). Stress and displacement around an elastic artificial rectangular hole. *J. Engng Mech., ASCE*, **127**(9), 880-890.

Leung, C. E. (1996). Case studies of rock-socketed piles. *Geotech. Engng*, **27**, 51–67.

Li, J. H. M. (2000). *Side shear resistance of large diameter bored piles in weathered geomaterials.* MPhil thesis, Hong Kong University of Science and Technology.

Li, K. S., Chan, S. T. and Lam, J. (2003). An alternative pile driving formula – a preliminary study. Case histories in geotechnical engineering in Hong Kong. *Proc. 23rd Ann. Sem., Geotech. Div.*, The Hong Kong Institution of Engineers, pp. 179–189.

Liang, R. Y. (1991). In situ determination of Smith soil model parameters for wave equation analysis. *Proc. Geotech. Engng Congr., Colorado*, pp. 64–75.

Lings, M. L., Nash, D. F. T., Ng, C. W. W. and Boyce, M. D. (1991). Observed behaviour of a deep excavation in Gault Clay: a preliminary appraisal. *Proc. 10th Eur. Conf. Soil Mech. & Fdn Engng, Florence*, Balkema, Rotterdam, **2**, pp. 467–470.

Lings, M. L., Nash, D. F. T. and Ng, C. W. W. (1993). Reliability of earth pressure measurements adjacent to a multi-propped diaphragm wall. *Proc. Conf. Ret. Struct., Cambridge*, Thomas Telford, London, pp. 258–269, and discussion pp. 301–304 and pp. 312–313,.

Lings, M. L, Ng, C. W. W. and Nash, D. F. T. (1994). The lateral pressure of wet concrete in diaphragm wall panels cast under bentonite. *Proc. Instn Civ. Engrs Geotech. Engng*, **107**, pp. 163–172.

Litkouhi, S. and Poskitt, T. J. (1980). Damping constants for pile driveability calculations. *Géotechnique*, **30**(1), 77–86.

Littlechild, B. D. and Plumbridge, G. D. (1998). Effects of construction technique on the behaviour of plain bored cast in-situ piles constructed under drilling slurry. *Proc. 7th Int. Conf. and Exhibition on Piling and Deep Foundations, Deep Foundation Institute, Vienna*, pp. 1.6.1–1.6.8.

Liu, J. L., Zhu, Z. C. and Zhang, Y. (1997). The technology and application of post-grouting for slurry bored piles. *Proc. 14th Int. Conf. Soil Mech. & Fdn Engng, Hamburg*, **2**, pp. 831–834.

Lo, D. O. K. (1997). *Behaviour of the instrumented test barrette for the West Kowloon Corridor Stage IV Project.* Special Project Report No. SPR2I97, Geotechnical Engineering Office, Civil Engineering Dept, Hong Kong Government, Hong Kong.

Loganathan, N. and Poulos, H. G. (1998). Analytical prediction for tunnelling-induced ground movements in clays. *J. Geotech. & Geoenvir. Engng*, **124**(9), 846–856.

Loganathan, N., Poulos, H. G. and Stewart, D. P. (2000). Centrifuge model testing of tunnelling-induced ground and pile deformations. *Géotechnique*, **50**(3), 283–294.

Lui, J. Y. H. and Yau, P. K. F. (1995). The performance of the deep basement for Dragon Centre. Instrumentation in Geotechnical Engineering. *Proc. 1994 Ann. Sem. of the Geotech. Div. of the Hong Kong Inst. of Engrs*, Hong Kong Institution of Engineers, pp. 183–201.

Lundardi, P., Focaracci, A., Giorgi, P. and Papacella, A. (1992). Tunnel face reinforcement in soft ground design and controls during excavation. *Towards New World in Tunnelling*. Balkema, Rotterdam, pp. 897–908,.

MacPherson, H. H. (1978). *Settlements around tunnels in soil: three case histories*. Final report by University of Illinois to Dept. of Transportation, Washington DC, Report No. UMTA-IL-06-0043-78-1.

Mair, R. J. (1979). *Centrifugal modelling of tunnelling construction in soft clay*. PhD thesis, University of Cambridge.

Mair, R. J. (1987). Tunnel face pressure in soft clay. *Proc. 8th Asian Reg. Conf. on Soil Mech. & Fdn Engng, Kyoto*, **2**, p. 290.

Mair, R. J. (1989). Discussion Leader's report on Session 9: Selection of design parameters for underground construction. *Proc. 12th Int. Conf. on Soil Mech. & Fdn Engng, Rio de Janeiro*, **5**, pp. 2891–2893.

Mair, R. J. (1993). Developments in geotechnical engineering research: application to tunnels and deep excavations. *Proc. Instn Civ. Engrs, Civ. Engng*, **97**(1), 27–41.

Mair, R. J. (1996). Settlement effects of bored tunnels. Session Report. *Proc. Int. Symp. Geotech. Aspects of Underground Construction in Soft Ground, London* (eds R. J. Mair and R. N. Taylor). Balkema, Rotterdam, pp. 43–53.

Mair, R. J. and Taylor, R. N. (1993). Prediction of clay behaviour around tunnels using plasticity solutions. *Predictive Soil Mech. Proc. Wroth Memorial Symp., Oxford*, Thomas Telford, London, pp. 449–463.

Mair, R. J. and Taylor, R. N. (1997). Theme lecture: Bored tunnelling in the urban environment. *Proc. 14th Int. Conf. Soil Mech. & Fdn Engng, Hamburg*, **4**, pp. 2353–2385.

Mair, R. J., Gunn, M. J. and O'Reilly, M. P. (1981). Ground movements around shallow tunnels in soft clay, *Proc. 10th Int. Conf. Soil Mech. & Fdn Engng, Stockholm*, **1**, pp. 323–328.

Mair, R. J., Taylor, R. N. and Bracegirdle, A. (1993). Subsurface settlement profiles above tunnels in Clay. *Géotechnique*, **43**(2), 315–320.

Mair, R. J., Taylor, R. N. and Burland, J. B. (1996). Prediction of ground movements and assessment of risk of building damage due to bored tunnelling. *Proc. Int. Symp. on Geotech. Aspects of Underground Construction in Soft Ground, London* (eds Mair, R. J. and Taylor, R. N.), Balkema, Rotterdam, pp. 713–718.

Malone, A., Ng, C. W. W. and Pappin, J. (1997). Invited paper: Displacements around deep excavations in completely decomposed granite. *Proc. 14th Int. Conf. Soil Mech. & Fdn Engng, Hamburg*, **4**, pp. 2325–2328.

Mana, A. I. and Clough, G. W. (1981). Prediction of movements for braced cut in clay. *J. Geotech. Engng Div., ASCE*, **107**(GT8), 759–777.

Martin, R. E., Seli, J. J., Powell, G. W. and Bertoulin, M. (1987). Concrete pile design in Tidewater, Virginia. *J. Geotech. Engng, ASCE*, **113**(6), 568–585.

Martos, F. (1958). Concerning an approximate equation of the subsidence trough and its time factors. *Proc. Int. Strata Control Cong.*, Leipzig, (Berlin: Deutsche Akadamie der Wissenshaften zu Berlin, Sektion fûr Bergbau), pp. 191–205.

Mason, R. C. (1960). Transmission of high loads to primary foundations by large diameter shafts. *Proc. ASCE Convention, New York.*

Matich, M. A. J. and Kozicki, P. (1967). Some load tests on drilled cast-in-place concrete caissons. *Can. Geotech. J.*, **4**, 367–375.

Mattes, N. S. and Poulos, H. G. (1971). Model tests on piles in clay. *Proc. 1st Australian–New Zealand Conf. Geomechs, Melbourne*, pp. 254–259.

McCaul, C. (1978). *Settlements caused by tunnelling in weak ground at Stockton-on-Tees.* TRRL Supplementary Report 383.

McClelland, B. (1974). Design of deep penetration piles for ocean structures. *J. Geotech. Engng, ASCE*, **100**(GT7), 705–747.

McVay, M. C., Birgisson, B., Zhang, L. M., Perez, A. and Putcha, S. (2000). Load and resistance factor design (LRFD) for driven piles using dynamic methods – a Florida perspective. *Geotech. Test. J., ASTM*, **23**(1), 55–66.

Meigh, A. C. and Wolski, W. (1979). Design parameters for weak rocks. *Proc. 7th Eur. Conf. on Soil Mech. & Fdn Engng, Brighton*, **5**, pp. 59–79.

Melis, M., Arnaiz, M., Oteo, C. S. and Mendana, F. (1997). Ground displacements in Madrid soils due to tunnel excavation with earth pressure T.B.M. *Proc. 14th Int. Conf. on Soil Mech. & Fdn Engng, Hamburg*, **3**, pp. 1433–1436.

Meyerhof, G. G. (1956). Penetration tests and bearing capacity of cohesionless soils. *J. Soil Mech. & Fdn Engng, ASCE*, **82**(SM1), 1–19.

Meyerhof, G. G. (1959). Compaction of sands and bearing capacity of piles. *J. Soil Mech. & Fdn Engng, ASCE*, **85**(SM6), 1–29.

Meyerhof, G. G. (1963). Some recent research on the bearing capacity of foundations. *Can. Geotech. J.*, **1**(1), 16–26.

Meyerhof, G. G. (1976). Bearing capacity and settlement of pile foundations. *J. Geotech. Engng, ASCE*, **102**(GT3), 195–228.

Michigan State Highway Commission (1965). *A Performance Investigation of Pile Driving Hammers and Piles.* Lansing, Michigan.

Milovic, D. and Stevanovic, S. (1982). Some soil parameters determined by cone penetration tests. *Proc. ESOPT2, Amsterdam*, **2**, pp. 709–714.

Moh, Z.-C., Ju, D. H. and Hwang, R. N. (1996). Ground movements around tunnels in soft ground. *Proc. Int. Symp. Geotech. Aspects of Underground Construction in Soft Ground, London* (eds R. J. Mair and R. N. Taylor), Balkema, Rotterdam, pp. 725–730.

Moretto, O. (1969). Deep excavations and tunnelling in soft ground. *Proc. 7th Int. Conf. Soil Mech. & Fdn. Engng, Mexico City*, **3**, Discussion, pp. 311–315.

Muir Wood, A. M. and Gibb, F. R. (1971). Design and construction of the cargo tunnel at Heathrow Airport, London. *Proc. Instn Civ. Engrs*, January, **48**, Paper 7357, pp. 11–34.

Nauroy, J.-F., Brucy, F., Le Tirant, P. and Kervadec, J.-P. (1986). Design and installation of piles in calcareous formations. *Proc. 3rd Int. Conf. Numerical Methods in Offshore Piling, Nantes*, pp. 461–480.

Negro, A. and Eisenstein, Z. (1991). Shallow tunnels in soft ground. General Report: *Proc. 9th Int. Conf. SM and FE*, **4**, pp. 2245–2275.

Negro, A., Sozio, L. E. and Ferreira, A. A. (1996). Tunnelling in São Paulo, Brazil. *Proc. Int. Symp. Geotech. Aspects of Underground Construction in Soft Ground, London* (eds R. J. Mair and R. N. Taylor), Balkema, Rotterdam, pp. 295–300.

New, B. M. and Bowers, K. H. (1994). Ground movement model validation at the Heathrow Express trial tunnel. *Tunnelling '94, Proc. 7th Int. Symp. Inst. Mining and Metallurgy and British Tunnelling Society, London*, Chapman and Hall, pp. 310–329.

New, B. M. and O'Reilly, M. P. (1991). Tunnelling induced ground movements; predicting their magnitude and effects. *Proc. 4th Int. Conf. on Ground Movements and Structures, Cardiff*, invited review paper, Pentech Press, pp. 671–697.

Ng, C. W. W. (1992). *An evaluation of soil–structure interaction associated with a multi-propped excavation.* PhD thesis, University of Bristol.

Ng, C. W. W. (1993). Nonlinear modelling of wall installation effects. *Retaining Structures*. Thomas Telford, London, pp. 160–163.

Ng, C. W. W. (1998). Observed performance of multi-propped excavation in stiff clay. *J. Geotech. & Geoenvir. Engng, ASCE,* **124**(9), 889–905.

Ng, C. W. W. (1999). Stress paths in relation to deep excavations. *J. Geotech. & Geoenvir. Engng, ASCE,* **125**(5), 357–363.

Ng, C. W. W. and Lee, G. T. K. (2002). A three-dimensional parametric study of the use of soil nails for stabilising tunnel faces. *Comp. & Geotech.*, **29**(8), 673–697.

Ng, C. W. W. and Lee, G. T. K. (2004). Three-dimensional ground settlements and stress transfer mechanisms due to tunnelling. Tentatively accepted, *Can. Geotech. J.*

Ng, C. W. W. and Lei, G. H. (2003a). Performance of long rectangular barrettes in granitic saprolites. *J. Geotech. & Geoenvir. Engng, ASCE,* **129**(8), 685–696.

Ng, C. W. W. and Lei, G. H. (2003b). An explicit analytical solution for calculating horizontal stress changes and displacements around an excavated diaphragm wall panel. *Can. Geotech. J.*, **40**(4), 780–792.

Ng, C. W. W. and Lings, M. L. (1995). Effects of modelling soil nonlinearity and wall installation on back-analysis of a deep excavation in stiff clay. *J. Geotech. & Geoenvir. Engng, ASCE,* **121**(10), 687–695.

Ng, C. W. W. and Wang, Y. (2001). Field and laboratory measurements of small strain stiffness of decomposed granites. *Soils & Fdns*, **41**(3), 57–71.

Ng, C. W. W. and Yan, W. M. (1998a). Three-dimensional modelling of a diaphragm wall construction sequence. *Géotechnique*, **49**(6), 825–834.

Ng, C. W. W. and Yan, W. M. (1998b). Prediction of ground deformations during a diaphragm wall construction. *Proc. 13th Southeast Asian Geotech. Conf., Taiwan*, **1**, pp. 631–636.

Ng, C. W. W. and Yan, W. M. (1998c). Stress transfer and deformation mechanisms around a diaphragm wall panel. *J. Geotech. & Geoenvir. Engng, ASCE,* **124**(7), 638–648.

Ng, C. W. W. and Zhang, L. M. (2001). Three-dimensional analysis of performance of laterally loaded and sleeved piles in sloping ground. *J. Geotech. & Geoenvir. Engng, ASCE,* **127**(6), 499–509.

Ng, C. W. W., Bolton, M. D. and Dasari, G. R. (1995a). The small strain stiffness of a carbonate stiff clay. *Soils and Fdns*, **35**(4), 109–114.

Ng, C. W. W., Lee, K. M. and Tang, D. K. W. (2004). 3D numerical investigations of NATM twin tunnel interactions. *Can. Geotech. J.*, **41**(3), 523–539.

Ng, C. W. W., Lings, M. L., Simpson, B. and Nash, D. F. T. (1995b). An approximate analysis of the three-dimensional effects of diaphragm wall installation. *Géotechnique*, **45**(3), 497–507.

Ng, C. W. W., Lings, M. L. and Nash, D. F. T. (1992). Back-analysing the bending moment in a concrete diaphragm wall. *Struct. Engr*, **70** (23 and 24), 421–426.

Ng, C. W. W., Nash, D. F. T. and Lings, M. L. (1990). *Underground car park at Lion Yard, Cambridge. Factual report on field monitoring*. Report No. UBCE-SM.90-1, Department of Civil Engineering, University of Bristol.

Ng, C. W. W., Pun, W. K. and Pang, R. P. L. (2000a). Small strain stiffness of natural granitic saprolites in Hong Kong. *J. Geotech. & Geoenvir. Engng, ASCE*, **126**(9), 819–833.

Ng, C. W. W., Rigby, D., Ng, S. W. L. and Lei, G. (2000b). Field studies of a well-instrumented barrette in Hong Kong. *J. Geotech. & Geoenvir. Engng, ASCE*, **126**(1), 60–73.

Ng, C. W. W., Rigby, D., Lei, G. and Ng, S. W. L. (1999). Observed performance of a short diaphragm wall panel. *Géotechnique*, **49**(5), 681–694.

Ng, C. W. W., Simpson, B., Lings, M. L. and Nash, D. F. T. (1998a). Analysis of a multi-propped excavation in stiff clay. *Can. Geotech. J.*, **35**(1), 115–130.

Ng, C. W. W., Sun, Y. F. and Lee, K. M. (1998b). Laboratory measurements of small strain stiffness of granitic saprolites. *Geotech. Engng*, Southeast Asian Geotechnical Society.

Ng, C. W. W., Yau, T. L. Y., Li, J. H. M. and Tang, W. H. (2001a). Side resistance of large diameter bored piles socketed into decomposed rocks. *J. Geotech. & Geoenvir. Engng, ASCE*, **127**(8), 642–657.

Ng, C. W. W., Li, J. H. M. and Yau, T. L. Y. (2001b). Behaviour of large diameter floating bored piles in saprolitic soils. *Soils and Fdns*, **41**(6), 37–52.

Ng, C. W. W., Yau, T. L. Y., Li, J. H. M. and Tang, W. H. (2001c). New failure load criterion for large diameter bored piles in weathered geomaterials. *J. Geotech. & Geoenvir. Engng, ASCE*, **127**(6), 488–498.

Ng, C. W. W., Zhang, L. M. and Ho, K. K. S. (2001d). Influence of laterally loaded sleeved piles and pile groups on slope stability. *Can. Geotech. J.*, **38**(3), 553–566.

Ng, C. W. W., Zhang, L. M. and Nip, D. C. N. (2001e). Response of laterally loaded large-diameter bored pile groups. *J. Geotech. & Geoenvir. Engng, ASCE*, **127**(8), 658–669.

Ng, T. P., Yeung, K. H. and O'Kelly, F. J. (1987). Silica hazard of caisson construction in Hong Kong. *J. Soc. Occ. Med.*, **37**, 62–65.

Nyren, R. J. (1998). *Field measurements above twin tunnels in London Clay*. PhD thesis, Imperial College of Science Technology and Medicine, London.

Nystrom, G. A. (1984). Finite-strain axial analysis of piles in clay. *Anal. Design Pile Fdns, ASCE*, 1–20.

Ohio Department of Transportation (1975). *Bearing capacity of piles from dynamic measurements*. Research Report OHIO-DOT-05-75, Final Report.

Olsen, R. E. and Flaate, K. S. (1967). Pile driving formulas for piles in sands. *J. Soil Mech. & Fdn Engng, ASCE*, **93**(SM5), 59–73.

O'Neill, M. W. (1983). Group action in offshore piles. *Proc. ASCE Conf., Geotech. Practice in Offshore Engng, Austin*. pp. 25–64.

O'Neill, M. W. and Reese, L. C. (1970). *Behaviour of axially loaded drilled shafts in Beaumont clay*. Research Report No. 89–8, Center for Highway Research, University of Texas, Austin.

O'Neill, M. W. and Reese, R. C. (1999). *Drilled shaft: construction procedures and design methods*. Federal Highway Administration, Washington, DC.

O'Neill, M. W., Hawkins, R. A. and Mahar, L. J. (1982). Load transfer mechanisms in piles and pile groups. *J. Geotech. Engng, ASCE,* **108**(GT12), 1605–1623.

Ooi, L. H. and Carter, J. P. (1987). A constant normal stiffness direct shear device for static and cyclic loading. *Geotech. Testing, J., ASTM*, **10**, 3–12.

O'Reilly, M. P. (1988). Evaluating and predicting ground settlements caused by tunnelling in London Clay. *Proc. Tunnelling 88, IMM, London,* pp. 231–241.

O'Reilly, M. P. and New, B. M. (1982). Settlements above tunnels in the United Kingdom – their magnitude and prediction. *Proc. Tunnelling 82, IMM, London*, pp. 173–181.

O'Reilly, M. P., Ryley, M. D., Barratt, D. A. and Johnson, P. E. (1981). Comparison of settlements resulting from three methods of tunnelling in loose cohesionless soil. *Proc. 2nd Int. Conf. on Ground Movements and Structures* (ed. J. D. Geddes), Pentech Press, Plymouth, pp. 359–376.

O'Rourke, T. D. (1992). Base stability and ground movement prediction for excavations in soft clay. *Proc. Int. Conf. on Retaining Structures, Cambridge*, ICE, London, pp. 657–686.

O'Rourke, T. D., Cording, E. L. and Boscardin, M. (1976). *The ground movements related to braced excavation and their influence on adjacent buildings*. Final report prepared for US Dept of Transport, Washington, DC.

Osterberg, J. O. (1989). New device for load testing driven piles and drilled shafts separates friction and end bearing. *Proc. 3rd Int. Conf. on Piling and Deep Fdn*, London, **1**, pp. 421–428.

Osterberg, J. O. and Gill, S. A. (1973). Load transfer mechanism for piers socketed in hard soils or rock. *Proc. 9th Can. Rock Mech. Symp.*, pp. 235–262.

Ottaviani, M. (1975). Three-dimensional finite element analysis of vertically loaded pile groups. *Géotechnique*, **25**(2), 159–174.

Paikowsky, S. G. (1982). *Use of dynamic measurements to predict pile capacity under local conditions*. MSc thesis, Dept of Civil Engineering Technion, Israel Institute of Technology.

Paikowsky, S. G. and Chernauskas, L. R. (1992). Energy approach for capacity evaluation of driven piles. *Proc. 4th Conf. Application of Stress-wave Theory to Piles* (ed. F. B. I. Barends), Balkema, Rotterdam, pp. 595–601.

Peck, R. B. (1969). Deep excavations and tunnelling in soft ground. *Proc. 7th Int. Conf. on Soil Mech. & Fdn Engng, Mexico City*, State-of-the-art Volume, pp. 225–290.

Pelia, D. (1994). A theoretical study of reinforcement influence on the stability of a tunnel face. *Geotech. & Geol. Engng*, **12**, 145–168.

Peila, D., Oreste, P. P., Pelizza, S. and Poma, A. (1996). Study of the influence of sub-horizontal fiber-glass pipes on the stability of a tunnel face. *Proc. North American Tunnelling '96* (ed. L. Ozdemir), Balkema, Rotterdam, **I**, pp. 425–432.

Pells, P. J. N. and Turner, R. M. (1979). Elastic solution for the design and analysis of rock-socketed piles. *Can. Geotech. J.*, **16**, 481–487.

Pells, P. J. N., Douglas, D. J., Rodway, B., Thorne, C. and McManhon, B. K. (1978). *Design loadings for foundations on shale and sandstone in the Sydney region*. Research Report No. R315, University of Sydney, Australia.

Pells, P. J. N., Rowe, R. K. and Turner, R. M. (1980). An experimental investigation into side shear for socketed piles in sandstone. *Proc. Int. Conf. on Struct. Fdn on Rock*, **1**, pp. 291–302.

Pile Dynamics Inc. (1990). *Model GCPC Pile Driving Analyzer*. Manual. Cleveland, OH.

Potts, D. M. and Addenbrooke, T. I. (1996). The influence of an existing surface structure on the ground movements due to tunnelling. *Proc. Int. Symp. on Geotech. Aspects of Underground Construction in Soft Ground, London* (eds R. J. Mair and R. N. Taylor), Balkema, Rotterdam, pp. 573–578.

Potts, D. M. and Addenbrooke, T. I. (1997). A structure's influence on tunnelling-induced ground movements. *Proc. Instn Civ. Engrs, Geotech. Engng*, **125**, Apr., 109–125.

Potts, D. M. and Bond, A. J. (1994). Calculation of structural forces for propped retaining walls. *Proc. 13th Int. Conf. Soil Mech. & Fdn Engng, New Delhi*, **2**, pp. 823–826.

Potts, D. M. and Zdravković, L. (2000). *Finite Element Analysis in Geotechnical Engineering. Vol. 1: Theory. Vol. 2: Application*. Thomas Telford, London.

Potts, D., Axelsson, K., Grande, L., Schweiger, H. and Long, M. (2002). *Guidelines for the Use of Advanced Numerical Analysis*. Thomas Telford, London.

Poulos, H. G. (1968). Analysis of the settlement of pile groups. *Géotechnique*, **18**(4), 449–471.

Poulos, H. G. (1979). Development of an analysis for cyclic axial loading of piles. *Proc. 3rd Int. Conf. Numer. Methods* in *Geomechs, Aachen*, **4**, pp. 1513–1530.

Poulos, H. G. (1987). Analysis of residual stress effects in piles. *J. Geotech. Engng, ASCE*, **113**(3), 216–229.

Poulos, H. G. (1988a). Modified calculation of pile-group settlement interaction. *J. Geotech. Engng Div., ASCE*, **114**(6), 697–706.

Poulos, H. G. (1988b). Cyclic stability diagram for axially loaded piles. *J. Geotech. Engng, ASCE*, **114**(8), 877–895.

Poulos, H. G. (1988c). *Marine Geotechnics*. Unwin Hyman, London.

Poulos, H. G. (1988d). The mechanics of calcareous sediments. Jaeger Memorial Lecture. *Proc. 5th Australia–New Zealand Geomechs Conf., Aust. Geomechs*, pp. 8–41.

Poulos, H. G. (1989). Twenty-ninth Rankine Lecture: Pile behaviour – theory and application. *Géotechnique*, **39**(3), 363–415.

Poulos, H. G. and Davis, E. H. (1980). *Pile Foundation Analysis and Design*. John Wiley and Sons, New York.

Poulos, H. G. and Hull, T. S. (1989). The role of analytical geomechanics in foundation engineering. *Proc. ASCE Fdn Engng Cong., Chicago*.

Poulos, H. G. and Mattes, N. S. (1969). The behaviour of axially loaded end-bearing piles. *Géotechnique*, **19**(2), 285–300.

Poulos, H. G. and Mattes, N. S. (1971). Displacements in a soil mass due to pile groups. *Aust. Geomech. J. G1*, No. 1, 18–28.

Powell, J. J. M. (1990). A comparison of four different pressuremeters and their methods of interpretation in a stiff heavily overconsolidated clay. *Proc. Int. Symp. on Pressuremeter Testing*, Balkema, Rotterdam, pp. 287–298.

Powell, J. J. M. and Butcher, A. P. (1991). Assessment of ground stiffness from field and laboratory tests. *Proc. 10th Eur. Conf. Soil Mech. & Fdn Engng, Florence*, Balkema, Rotterdam, **I**, pp. 153–156.

Powrie, W. (1997). *Soil Mechanics: Concepts and Applications.* E & F N Spon, London, 420 pp.

Powrie, W. and Kantartzi, C. (1996). Ground response during diaphragm wall installation in clay: centrifuge model tests. *Géotechnique*, **46**(4), 725–739.

Powrie, W. and Li, E. S. E (1991). Finite element analyses of an in situ wall propped at formation level. *Géotechnique*, **41**(4), 499–514.

Prandtl, L. (1920). Uber die Harte plastischer Korper. Gotinger Nachrichten, matphys., p. 74.

Pratt, M. (1989). The bored pile Olympics. *Hong Kong Engr*, **17**(12), 12–16.

Pratt, M. and Sims, M. J. (1990). The application of new techniques to solve deep basement and foundation problems. *Proc. Int. Conf. on Deep Fdn Practice*, pp. 189–195.

Pun, W. K. and Ho, K. K. S. (1996). *Analysis of Triaxial Tests on Granitic Saprolite Performed at Public Works Central Laboratory*, Discussion Note DN 4/96. Geotechnical Engineering Office, the Hong Kong Government of the Special Administrative Region.

Ramaswamy, S. D. and Pertusier, E. M. (1986). Construction of barrettes for high-rise foundations. *J. Constr. Engng., ASCE*, **112**(4), 455–462.

Randolph, M. F. (1977). *A theoretical study of the performance of piles.* PhD thesis, University of Cambridge, UK.

Randolph, M. F. (1983). Design considerations for offshore piles. *Proc. ASCE Spec. Conf. Geotech. Practice in Offshore Engng*, Austin, pp. 422–439.

Randolph, M. F. and Wroth, C. P. (1978a). Analyses of deformation of vertically loaded piles. *J. Geotech. Engng, ASCE*, **104**(GT12), 1465–1488.

Randolph, M. F. and Wroth, C. P. (1978b). A simple approach to pile design and the analysis of pile tests. *Proc. Am. Soc. Test. & Mat. Symp. Behaviour of Deep Foundations, Boston.*

Randolph, M. F. and Wroth, C. P. (1979). An analysis of the vertical deformation of pile groups. *Géotechnique*, **29**(4), 423–439.

Randolph, M. F., Carter, J. P. and Wroth, C. P. (1979). Driven piles in clay – the effects of installation and subsequent consolidation. *Géotechnique*, **29**(4), 361–393.

Rankin, W. J. (1988). Ground movements resulting from urban tunnelling; prediction and effects. *Proc. Conf. on Engng Geol. of Underground Movements, Nottingham*, BGS, pp. 79–92.

Reese, L. C. and O'Neill, M. W. (1988). *Drilled shafts: construction procedures and design methods.* Publn No. FHWA-HI-88–042, Federal Highway Administration, Washington, DC, Publn No. ADSC-TL-4, International Association of Foundation Drilling.

Richardson, D. (1988). *Investigations of threshold effects in soil deformation.* PhD thesis, The City University, London.

Robertson, P. K., Campanella, R. G. and Brown, P. T. (1985). Design of axially and laterally loaded piles using in-situ tests: a case history. *Can. Geotech. J.*, **22**(4), 518–527.

Rosenberg, P. and Journeaux, N. L. (1976). Friction and end bearing tests on bedrock for high capacity socket design. *Can. Geotech. J.*, **13**, 324–333.

Rowe, R. K. and Armitage, H. H. (1987). A design method for drilled piers in soft rock. *Can. Geotech. J.*, **24**(1), 126–142.

Rowe, R. K. and Pells, P. J. N. (1980). A theoretical study of pile–rock socket behaviour. *Proc. Int. Conf. on Struct. Fdn on Rock*, pp. 253–264.

SAA (1995). *Piling design and installation.* AS-2159, Sydney, Australia.

Schlosser, F. and Guilloux, A. (1995). Exemples Français récents de tunnels dans des conditions difficiles (Recent French cases of tunnelling under difficult conditions). *Proc. 11th Afr. Reg. Conf. on Soil Mech. & Fdn Engng, Cairo*, **I**, pp. 56–84 (in French).

Schmertmann, J. H. (1975). Measurement of in situ shear strength. *Spec. Conf. on In Situ Measurement of Soil Properties, ASCE*, Raleigh, NC, **2**, pp. 57–138.

Schmertmann, J. H. (1978). *Guidelines for cone penetration test – in performance and design.* US Dept of Transportation, Federal Highways Administration, Washington, DC.

Schmidt B. (1966). Discussion of Earth pressures at rest related to stress history, by Brooker and Ireland. *Can. Geotech. J.*, **3**(4), 239–242.

Schmidt, B. (1969). *Settlements and ground movements associated with tunnelling in soils.* PhD thesis, University of Illinois, Urbana.

Selby, A. R. (1988). Surface movements caused by tunnelling in two-layer soil. *Engineering Geology of Underground Movements* (eds F. G. Bell, *et al.*), Geological Society Engineering Geology Special Publication No. 5, pp. 71–77.

Semple, R. M. and Rigden, W. J. (1984). Shaft capacity of driven piles in clay. *Anal. Design Pile Fdns* (ed. J. R. Meyer), *ASCE*, 59–79.

Shioi, Y. and Fukui, J. (1982). Application of N-value to design of foundations in Japan. *Proc. ESOPT2, Amsterdam*, **1**, pp. 159–164.

Shirlaw, J. N. (1995). Observed and calculated pore pressures and deformations induced by an earth pressure balance shield: Discussion. *Can. Geotech. J.*, **32**, 181–189.

Shirlaw, J. N., Doran, S. and Benjamin, B. (1988). A case study of two tunnels driven in the Singapore 'Boulder Bed' and in grouted coral sands. *Engineering Geology of Underground Movements* (eds F. G. Bell *et al.*), Geological Society Engineering Geology Special Publication No. 5, pp. 93–103.

Simons, N. E. and Menzies, B. K. (2000). *A Short Course in Foundation Engineering.* Thomas Telford, London, 244 pp.

Simons, N. E., Menzies, B. K. and Matthews, M. C. (2001). *A Short Course in Soil and Rock Slope Engineering.* Thomas Telford, London, 432 pp.

Simons, N. E., Menzies, B. K. and Matthews, M. C. (2002). *A Short Course in Geotechnical Site Investigation.* Thomas Telford, London, 353 pp.

Simpson, B. (1992). The Thirty-second Rankine lecture. Retaining structures: displacement and design. *Géotechnique*, **42**(4), 541–576.

Simpson, B., O'Riordan, N. J. and Croft, D. D. (1979). A computer model for the analysis of ground movements in London clay. *Géotechnique*, **29**(2), 149–175.

Skempton, A. W. (1953). Discussion: Piles and pile foundations, settlement of pile foundations. *Proc. 3rd Int. Conf. Soil Mech. & Fdn Engng*, **3**, p. 172.

Skempton, A. W. (1959). Cast-in-situ bored piles in London clay. *Géotechnique*, **9**(4),153–173.

Sloan, S. W. and Assadi, A. (1993). Stability of shallow tunnels in soft ground. *Predictive Soil Mechanics, Proc. Wroth Memorial Symp., Oxford*, 1992, Thomas Telford, London, pp. 644–663.

Smith, E. A. L. (1960). Pile driving analysis by the wave equation. *J. Soil Mech. & Fdn Engng, ASCE*, **86**(SM4), 35–61.

Sorensen, T. and Hansen, B. (1957). Pile driving formulae: an investigation based on dimensional considerations and a statistical analysis. *Proc. 4th Int. Conf. Soil Mech. & Fdn Engng*, **2**, pp. 61–65.

Stas, C. V. and Kulhawy, F. H. (1984). *Critical evaluation of design methods for foundations under axial uplift and compression loading.* Report for the Electrical Power Research Institute, No. EL 3771, Cornell University.

Stewart, J. F. and Kulhawy, F. H. (1981). Interpretation of uplift load distribution data. *Proc. 10th Int. Conf. Soil Mech. & Fdn Engng, Stockholm*, **2**, pp. 277–280.

Strange, P. J. (1990). The classification of granitic rocks in Hong Kong and their sequence of emplacement in Sha Tin, Kowloon and Hong Kong Island. *Geological Society of Hong Kong Newsletter*, **8**, 18–27.

Symons, I. E. and Carder, D. R. (1993). Stress changes in stiff clay caused by the installation of embedded retaining walls. *Retaining Structures*. Thomas Telford, London, pp. 227–236.

Szczepinski, W. (1979). *Introduction to the Mechanics of Plastic Forming of Metals.* Sijthoff and Noordhoff International Publishers, the Netherlands (first published in Polish by PWN, Warsaw, in 1967).

Tan, Y. C., Chen, C. S. and Liew, S. S. (1998). Load transfer behaviour of cast-in-place bored piles in tropical residual soils in Malaysia. *Proc. 13th South-east Asian Geotech. Conf., Taipei, Taiwan*, pp. 563–571.

Tatsuoka, F. and Kohata, Y. (1994). Stiffness of hard soils and soft rocks in engineering application. *Proc. Int. Symp. on Prefailure Deformation Characteristics of Geomaterials, Hokkaido, Japan*, Balkema, Rotterdam, No. (2), pp. 947–1063.

Taylor, R. N. (1995). Tunnelling in soft ground in the UK. *Underground Construction in Soft Ground* (eds K. Fujita and O. Kusakabe). Balkema, Rotterdam, pp. 123–126.

Tedd, P., Chard, B. M., Charles, J. A. and Symons, I. F. (1984). Behaviour of a propped embedded retaining wall in stiff clay at Bell Common Tunnel. *Géotechnique*, **34**(4), 513–532.

Terzaghi, K. (1939). Soil mechanics – a new chapter in engineering science. *J. Instn Civ. Engrs*, **12**, 106–141.

Terzaghi, K. (1943). *Theoretical Soil Mechanics.* John Wiley and Sons, New York.

Terzaghi, K. and Peck, R. B. (1967). *Soil Mechanics in Engineering Practice*, 2nd edn. John Wiley and Sons, New York.

Texas Highway Department (1973). *Bearing capacity for axially loaded piles.* Research Report 125-8-F (covered period Sept. 1967 to Aug. 1973), p. 134.

Thompson, P. (1991). *A review of retaining wall behaviour in overconsolidated clay during early stages of construction.* MPhil thesis, University of London.

Thompson, C. D. and Devata, M. (1980). Evaluation of ultimate bearing capacity of different piles by wave equation analysis. *Proc. Semin. on Applic. Stress Wave Theory of Piles, Stockholm.*

Thompson, C. D. and Goble, G. G. (1988). High case damping constants in sand. *Proc. 3rd Int. Conf. of Stress Wave Theory in Piles, Ottawa*, pp. 464–555.

Thorburn, S. (1966). Large diameter piles founded on bedrock. *Proc. Symp. on Large Bored Piles*, ICE, London, pp. 120–129.

Thorburn, S. and McVicar, S. L. (1971). Pile load tests to failure in the Clyde alluvium. *Behaviour of Piles.* 1–7, 53–54, ICE, London.

Thorne, C. P. (1980). The capacity of piers drilled into rock. *Proc. Int. Conf. on Struct. Fdn on Rock*, **1**, pp. 223–233.

Timoshenko, S. P. and Goodier, J. N. (1970). *Theory of Elasticity*, 3rd edn. McGraw-Hill, New York.

Tomlinson, M. J. (1957). The adhesion of piles driven into clay soils. *Proc. 4th Int. Conf. Soil Mech. & Fdn Engng, London*, **2**, pp. 66–71.

Tomlinson, M. J. (1977). *Pile Design and Construction Practice*. Cement and Concrete Association, London.

Tomlinson, M. J. (1994). *Pile Design and Construction Practice*. E & F N Spon, London.

Tomlinson, M. J. (2001). *Foundation Design and Construction*, 7th edn. Prentice Hall, London.

Toombs, A. F. (1980). *Settlement caused by tunnelling beneath a motorway embankment*. TRRL Supplementary Report.

Touma, F. T. and Reese, L. C. (1974). Behaviour of bored piles in sand. *J. Geotech. Div., ASCE*, **100**, 749–761.

Triantafyllidis, T. (2001). On the application of the Hiley's formula in driving long piles. *Géotechnique*, **51**(10), 891–895.

Uriel, A. O. and Sagaseta, C. (1989). General Report: Discussion Session 9; Selection of design parameters for underground construction. *Proc. 12th Int. Conf. on Soil Mech. & Fdn Engng, Rio de Janeiro*, Balkema, Rotterdam, **4**, pp. 2521–2551.

Uriel, S. and Oteo C. S. (1977). Stress and strain beside a circular trench wall. *Proc. 9th Int. Conf. Soil Mech. & Fdn Engng, Tokyo*, **1**, pp. 781–788.

Valliappan, S., Lee, I. K. and Boonlualohr, P. (1974). Settlement of piles in layered soils. *Proc. 7th Bien. Conf. Aust. Rd Res. Bd, Adelaide*, **7**, pt 7, pp. 144–153.

Van Impe, W. F. (1986). Evaluation of deformation and bearing capacity parameters of foundations from static CPT results. *4th Int. Geotech. Seminar on Field Instrumentation and In Situ Measurements, Nanyang Tech. Univ., Singapore*.

Van Impe, W. F. (1991). Developments in pile design. *Proc. 4th Int. Conf. on Piling and Deep Fdns, Stresa, Italy*.

Verbrugge, J. C. (1982). The evaluation of pile cap load–settlement diagram from the CPT results. *Proc. ESOPT2, Amsterdam*, **2**, pp. 923–926.

Vesic, A. S. (1972). Expansion of cavities in infinite soil mass. *J. Soil Mech. & Fdn Engng, ASCE*, **98**(8M3), 265–290.

Vesic, A. S. (1977). *Design of Pile Foundations*. National Cooperative Highway Research Program. Synthesis of Highway Practice, Publication No. 42.

Vinnel, C. and Herman, A. (1969). Tunnel dans le sable de Bruxelles par la méthod du bouclier. *Proc. 7th Int. Conf. Soil Mech. & Fdn Engng, Mexico City*, **2**, pp. 487–494; **3**, Discussion, pp. 369–370.

Vogan, R. W. (1977). Friction and end bearing tests on bedrock for high capacity socket design: Discussion. *Can. Geotech. J.*, **14**, 156–158.

Walter, D. J., Burwash, W. J. and Montgomery, R. A. (1997). Design of large-diameter drilled shafts for Northumberland Strait bridge project. *Can. Geotech. J.*, **34**, 580–587.

Ward, W. H. (1969). Discussion on Peck, R. B. Deep excavations and tunnelling in soft ground. *Proc. 7th Int. Conf. Soil Mech. & Fdn Engng, Mexico City*, **3**, pp. 320–325.

Ward, W. H. and Pender, M. J. (1981). Tunnelling in soft ground. General Report. *Proc. 10th Int. Conf. on Soil Mech. & Fdn Engng, Stockholm*, **4**, pp. 261–275.

Wates, J. A. and Knight, K. (1975). The effect of bentonite on the skin friction in cast-in-place piles and diaphragm walls. *Proc. 6th Reg. Conf. for Afr. on Soil Mech. & Fdn Engng*, **I**, pp. 183–188.

Webb, D. L. (1976). The behaviour of bored piles in weathered diabase. *Géotechnique*, **26**, 63–72.

Weltman, A. J. (1980). *Pile load testing procedures.* DOE and CIRIA Piling Development Group Report PG7, Construction Industry Research and Information Association, London.

West, G., Heath, W. and McCaul, C. (1981). Measurements of the effects of tunnelling at York Way, London. *Ground Engng*, **14**(5), 45–53.

Whitaker, T. (1970). *The Design of Piled Foundations.* Pergamon, Oxford.

Whitaker, T. and Cooke, R. W. (1966). An investigation of the shaft and base resistances of large bored piles in London Clay. *Proc. Symp. on Large Bored Piles*, pp. 7–49.

Williams, A. E. (1980). *The design and performance of piles socketed into weak rock.* PhD thesis, Monash University, Melbourne.

Williams, A. E. and Pells, P. J. N. (1981). Side resistance rock sockets in sandstone, mudstone, and shale. *Can. Geotech. J.*, **18**, 502–513.

Williams, A. E., Donald, I. B. and Chiu, H. K (1980a). Stress distributions in rock socketed piles. *Proc. Int. Conf. on Struct. Fdn on Rock*, pp. 317–326.

Williams, A. E., Johnston, I. W. and Donald, I. B. (1980b). The design of socketed piles in weak rock. *Proc. Int. Conf. on Struct. Fdn on Rock*, pp. 327–347.

Wilson, L. C. (1976). Tests of bored and driven piles in cretaceous mudstone at Port Elizabeth, South Africa. *Géotechnique*, **26**, 5–12.

Withiam, J. L. and Kulhawy, F. H. (1979). Analytical model for drilled shaft foundations. *Proc. 3rd Int. Conf. Numer. Meth. in Geomechs, Aachen*, **3**, pp. 1115–1122.

Wittke, K. (1995). German national report on tunnelling in soft ground. *Proc. Int. Symp. on Underground Constr. in Soft Ground, New Delhi* (eds K. Fujita and O. Kusakabe), Balkema, Rotterdam, pp. 101–106.

Worssam, B. C. and Taylor, J. H. (1975). *Geology of the Country around Cambridge*, 2nd edn. HM Stationery Office, London.

Wright, S. J. and Reese, L. C. (1979). Design of large diameter bored piles. *Ground Engng*, 17–50.

Yamashita, K., Tomono, M. and Kakurai, M. (1987). A method for estimating immediate settlement of piles and pile groups. *Soils and Fdns*, **27**(1), 61–76.

Yau, T. L. Y. (2000). *Capacity and failure criteria of bored piles in soils and rocks.* MPhil thesis, Hong Kong University of Science and Technology.

Yiu, T. M. and Lam, S. C. (1990). Ultimate load testing of driven piles in metasedimentary decomposed rocks. *Proc. Conf. Deep Fdn Practice, Singapore*, pp. 293–300.

Yoo, C. S. and Shin, H. K. (1999). Behavior of tunnel face pre-reinforced with sub-horizontal pipes. *Proc. Geotech. Aspect of Underground Constr. in Soft Ground*, Balkema, Rotterdam, pp. 441–446.

Yoshikoshi, W., Watanabe, O. and Takagi, N. (1978). Prediction of ground settlements associated with shield tunnelling. *Soils and Fdns, J. of Jap. Soc. Soil Mech. & Fdn Engng*, **18**(4), 47–59.

Zhang, L. and Einstein, H. H. (1998). End bearing capacity of drilled shafts in rock. *J. Geotech. & Geoenvir. Engng, ASCE*, **124**(7), 574–584.
Zhang, L. M., McVay, M. C. and Ng, C. W. W. (2001a). A possible physical meaning of Case damping in pile dynamics. *Can. Geotech. J.*, **38**(1), 83–94.
Zhang, L. M., Tang, W. H. and Ng, C. W. W. (2001b). Reliability of axially loaded driven pile groups. *J. Geotech. & Geoenvir. Engng, ASCE*, **127**(12), 1051–1060.

Index